T0293770

Particle Accelerators and Colliders: Volume III (Particle Physics Essentials)

Particle Accelerators and Colliders: Volume III (Particle Physics Essentials)

Josh Hudson

Published by Willford Press,
118-35 Queens Blvd., Suite 400,
Forest Hills, NY 11375, USA

ISBN: 978-1-64728-463-3

Cataloging-in-Publication Data

Particle accelerators and colliders : Volume III (particle physics essentials) / Josh Hudson.
p. cm.
Includes bibliographical references and index.
ISBN 978-1-64728-463-3
1. Particle accelerators. 2. Colliders (Nuclear physics). 3. Particles (Nuclear physics).
4. Accelerator mass spectrometry. I. Hudson, Josh.
QC787.P3 P37 2023
539.73--dc23

For information on all Willford Press publications
visit our website at www.willfordpress.com

Contents

Preface

This book has been a concerted effort by a group of academicians, researchers and scientists, who have contributed their research works for the realization of the book. This book has materialized in the wake of emerging advancements and innovations in this field. Therefore, the need of the hour was to compile all the required researches and disseminate the knowledge to a broad spectrum of people comprising of students, researchers and specialists of the field.

A particle accelerator refers to a machine that produces a beam of fast-moving, electrically charged atomic or subatomic particles. Large accelerators are employed for fundamental research in particle physics. The largest accelerator in the world is the Large Hadron Collider. The two fundamental types of accelerators are electrostatic and electrodynamic (or electromagnetic) accelerators. Electrostatic particle accelerators make use of static electric fields to accelerate particles. The most common types of electrostatic particle accelerators are the Cockcroft–Walton generator and the Van de Graaff generator. Electrodynamic or electromagnetic accelerators involve the use changing electromagnetic fields to accelerate particles. These accelerators can be linear or circular. The particles are accelerated in a straight line in linear accelerators. In the circular accelerator, particles move in a circle until they attain sufficient energy. Some applications of accelerators are particle therapy for oncological purposes, radioisotope production for medical diagnostics, and ion implanters for the manufacturing of semiconductors. This book brings forth some of the most innovative concepts and elucidates the unexplored aspects of particle accelerators and colliders. It is meant for students who are looking for an elaborate reference text on this topic.

At the end of the preface, I would like to thank the authors for their brilliant chapters and the publisher for guiding us all-through the making of the book till its final stage. Also, I would like to thank my family for providing the support and encouragement throughout my academic career and research projects.

Josh Hudson

10

Major Colliders and Accelerators

K. Hübner, S. Ivanov, R. Steerenberg, T. Roser, J. Seeman, K. Oide, Karl Hubert Mess, Peter Schmüser, R. Bailey, and J. Wenninger

10.1 Proton Accelerators and Colliders

K. Hübner · S. Ivanov · R. Steerenberg

10.1.1 CERN Proton Synchrotron (CPS)

The Study Group for a GeV-scale Proton Synchrotron was launched in 1952 at CERN. Initially, an up-scaled version of the 3 GeV Cosmotron was considered but soon a new design based on the newly discovered alternating-gradient principle and promising a proton energy of 30 GeV was adopted by the CERN Council in the same year. In order to limit cost the energy was subsequently limited to 25 GeV

K. Hübner · R. Steerenberg · K. Oide · K. H. Mess · R. Bailey (✉) · J. Wenninger
CERN (European Organization for Nuclear Research) Meyrin, Genève, Switzerland
e-mail: Rende.Steerenberg@cern.ch; roger.bailey@cern.ch; jorg.wenninger@cern.ch

S. Ivanov
Institute of High Energy Physics, Moscow, Russia
e-mail: sergey.ivanov@ihep.ru

T. Roser
Brookhaven National Laboratory, New York, NY, USA
e-mail: roser@bnl.gov

J. Seeman
SLAC National Accelerator Laboratory, Stanford University, Menlo Park, CA, USA
e-mail: seeman@slac.stanford.edu

P. Schmüser
DESY, Hamburg, Germany
e-mail: Peter.Schmueser@desy.de

Table 10.1 Basic parameters of the CPS [5]

Accelerated particles	Protons, lead ions
Momentum protons/lead ions	26 GeV/c, 5.9 GeV/c nucleon
Circumference [m]	200 π
Magnetic lattice	Alternating-gradient focusing, combined-function
Focusing order	FOFDOD
Magnetic field index	$n = 288$
Number of main magnets	100
Bending magnetic field	0.1013 T (inj. at 1 GeV), 1.25 T (extr. at 26 GeV/c)
Betatron oscillations/turn	6.24 (h), 6.26 (v)
Rise time/flat top time	0.7 s/0.3 s (26 GeV/c)
Long straight sections	Number = 20, length = 3.0 m
RF system (tunable)	11 cavities, 2.8–9.55 MHz, 220 keV/turn total maximum
Auxiliary RF systems	13, 20, 40, 80, 200 MHz
Vacuum chamber	Stainless steel, 146 × 70 mm^2 in the bending magnets

and the project led by J.B. Adams was approved in 1953. The final parameters were fixed in 1954 and construction started in 1955. The CPS [1] became operational towards the end of 1959 reaching an energy of 28 GeV [2, 3]. It has turned out to be an extremely versatile facility [4] (Table 10.1).

Initially, the proton injector was a 50 MeV Alvarez-type linear accelerator (linac L1) operating at 200 MHz. In order to increase the intensity a four-ring synchrotron booster (PSB) was inserted in 1972 between L1 and CPS raising the kinetic injection energy to 0.8 GeV. Over the years, it had its top kinetic energy raised in steps to 1 GeV in 1985 and 1.4 GeV in 1999 to allow for the production of the LHC beams. As part of the LHC Injectors Upgrade (LIU) project [6] a further increase to 2 GeV is being prepared at present for first beam in 2020. In 1979, L1 was replaced by linac 2 (L2) of a more modern and robust design. The 50 MeV proton linac 2 is presently being replaced by the new linac 4 (L4) [7] that will provide H^- ions at 160 MeV to the charge exchange injection equipped PS Booster.

The CPS provided initially secondary beams by means of internal targets. Since 1967 a fast extraction system over one turn became available for fixed-target physics complemented in 1969 by a system providing slow-extracted beams spilling out particles over a large number of turns. The fast extraction was used to produce neutrino beams with protons and, between 1970 and 1983, 26 GeV/c protons for the Intersecting Storage Rings (ISR). At present, the fast extraction provides a 26 GeV/c proton beam (1.5×10^{13} per pulse) for the Antiproton Decelerator (AD) after compression of the four equidistant bunches that occupy half the CPS circumference. This manipulation by the radio-frequency accelerating system makes this proton bunch train so short that the secondary antiproton bunch train fits into the AD circumference being one quarter of that of the CPS. Further, a 20 GeV/c single high-intensity bunch containing up to 9×10^{12} protons is extracted after a non-adiabatic bunch shortening to produce neutrons from a lead-target. Since 2010, a large variety of LHC beams have been produced that over time have become much

Table 10.2 Pre- and post-LIU main LHC beam parameters at 26 GeV/c PS extraction

	Beam type	N [$\times 10^{11}$ p]	ε [mm mrad]	N_b/extr
2018	Standard	1.4	2.3	72
	BCMS	1.3	1.2	48
Post-LIU	Standard	2.6	1.85	72
	BCMS	2.6	1.45	48

brighter. As foreseen in the LHC design report [8] the PS produces single bunch beams varying in intensity from 0.05×10^{11} to 1.2×10^{11} protons. For the initial multi-bunch beams 12–72 bunches per extraction at 26 GeV/c with respective bunch spacings of 25, 50, 75 and 150 ns were routinely produced. The highest LHC bunch intensity of 1.7×10^{11} protons per bunch in a transverse emittance of 1.6 mm mrad and a bunch train length of 36 bunches was reached with the 50 ns bunch spacing that was initially used to limit electron cloud effects in the LHC. By mid-July 2015 the LHC requested the 25 ns bunch spacing with 1.15×10^{11} protons per bunch in a transverse emittance of 2.5 mm mrad and 72 bunches per extraction. Thanks to the versatile PS RF system an even brighter LHC beam, based on Bunch Merging, Compression and Splitting (BCMS), was established. Up to 1.3×10^{11} protons per bunch in a transverse emittance of 1.1 mm mrad and 48 bunches per extraction are delivered to the SPS since mid-2016. The LIU project aims at increasing the bunch intensity to 2.6×10^{11} protons per bunch for both beam types as given in Table 10.2.

For fixed-target experiments at the Super Proton Synchrotron (SPS), a 14 GeV/c proton beam (3×10^{13} per pulse) was spilled out over five turns by cutting it with an electro-static septum in horizontal phase space in order to fill by box-car stacking 5/11 parts of the SPS circumference being 11 times longer than the one of the CPS. A second CPS beam pulse fills further 5/11 parts, thus leaving 1/11 for the kicker rise-times. Since 2010 a novel scheme, using a fourth order betatron resonance for capturing the beam in five stable islands in the horizontal phase space, avoids the losses at the electrostatic extraction septum [9].

The slow-extraction based on a third-order betatron resonance is still in use to produce primary 24 GeV/c proton beams. Up to 5×10^{11} protons can be spilled out in up to 450 ms for the production of secondary beams for fixed-target experiments at the CPS, but also for primary protons to the IRRAD and CHARM irradiation facilities [10].

As soon as L2 was available, L1 was modified to provide Deuterium and α-particle beams for collisions in the ISR and, later, Sulphur beams for fixed target experiments in the SPS. In 1994, L1 was replaced by linac 3 (L3) providing 4.2 MeV/amu Pb^{+53} ions which, fully stripped after CPS extraction, are used for fixed-target experiments in the SPS. Since 2010 the lead ions are also fast-ejected to the SPS for lead-lead collisions in LHC. They no longer pass through the PSB but a small storage ring, the Low Energy Ion Ring (LEIR), acts as accumulator between the fast-cycling L3 and the slow-cycling CPS and is equipped with stochastic and electron cooling to decrease the transverse emittance of the accumulated beam. The CPS provides a total of 8×10^{10} ions per pulse in four bunches.

The acceleration cycles for the different users are grouped in a supercycle depending on the user requirements allowing for a quick, reproducible switching from one to another mode of operation [11].

10.1.2 Brookhaven Alternating Gradient Synchrotron (AGS)

After successful completion of the 3 GeV Cosmotron in 1952 the design study for a more powerful accelerator was launched coincident with the invention of the alternating-gradient principle [12, 13]. Construction led by G.K. Green and J.P. Blewett started in 1953 and commissioning was completed with the first proton beam accelerated to 31 GeV in July 1960 [14]. In order to test experimentally whether the beam would pass transition energy, an electron analogue had been built in 1954 and operated until 1957. This model had a circumference of 43.1 m accelerating electrons from 1 to 10 MeV with transition energy at 3.5 MeV. In order to reduce cost, the alternating-gradient, strong-focusing was provided by electrostatic lenses and bending by electrostatic fields [15]. The test showed that transition can be crossed without problems but the price was a delay in the AGS construction relative to the CERN PS, which however was at the end compensated by better preparation of the experimental programme compared to CERN.

The first injector was a 50 MeV Alvarez-type proton linear accelerator. The present 200 MeV linear accelerator began operation in 1970. In 1982 H^- charge exchange injection into the AGS was introduced [16] and in 1991 a 1.5 GeV booster synchrotron was commissioned [17]. The Booster can provide 1.5×10^{13} protons per pulse at 1.9 GeV at the design repetition frequency of 7.5 Hz. The acceleration harmonic schemes (Booster harmonic, AGS harmonic, transfers) evolved from (3, 4, 12) to (2, 4, 8) and finally to (1, 6) in pursuit of higher intensity [18] (Table 10.3).

Secondary beams from the AGS were initially provided from internal targets. This also creates high beam loss and activation in the accelerator not compatible with high-intensity operation. The first fast-extraction was installed in the mid-60s followed by slow-extraction in 1967 which served up to six target stations and spilling out protons with repetition periods from 1.8 s to 5.8 s. To cope with the intensity increases, the AGS underwent a series of upgrades including a new main magnet power supply, addition of transverse feed-back, special magnets to provide fast crossing of the transition energy, and a high power RF system [20]. In the early 2000s the AGS provided a slow-extracted beam of 7×10^{13} protons per pulse at 24 GeV [21].

With the appropriate source added to the 200 MeV linear accelerator, polarized protons have been produced by the injector chains from 1985 onward for fixed target experiments. To meet injector requirements for RHIC—intensity and polarization—the polarized source underwent a major upgrade [22]. The polarization transmission efficiency in the AGS has been substantially improved with the installation of two

Table 10.3 Basic parameters of the AGS [19]

Accelerated particles	Protons, polarized protons, heavy ions (up to Au)
Particle energy	30 GeV, 25 GeV, 14.5 GeV/n
Circumference [m]	256.9 π
Magnetic lattice	Alternating-gradient focusing, combined-function
Focusing order	(F/2)O(F/2)(D/2)O(D/2)
Magnetic field index	$n = 365$
Number of main magnets	240
Bending magnetic field	0.105 T at injection, 1.31 T at maximum particle momentum
Betatron oscillations/turn	8.75 (h), 8.75 (v)
Rise time/flat top time	0.6 s/0.5 to 2.5 s
Long straight sections	Number = 24, length = 3.15 m
RF system (tunable)	10 cavities, 1.8 to 4.5 MHz, 200 KeV/turn total maximum
Auxiliary RF system (fixed RF)	92 MHz
Vacuum chamber	Inconel, 173×78 mm^2 in the bending magnets

"partial Siberian" snakes [23, 24] and a system to rapidly cross weak resonances. Polarization at transfer to RHIC (24 GeV) is 70% (with 82% at 200 MeV) and with intensity 2×10^{11} protons per bunch [25, 26].

Since 1986 the Booster has also accelerated ions (d to Au) using a Tandem Van de Graaff as injector. For RHIC operation about 5×10^9 Au^{31+} ions at 41.6 MeV/n are injected over 60 turns into the Booster, accelerated to 101 MeV/n, stripped to Au^{77+} and injected into the AGS. The ions are fully stripped before injection into RHIC [27]. A new pre-injector is being commissioned based on an EBIS source followed by a new linear accelerator [28]. The new system increases the available ions for RHIC to include Uranium [29].

10.1.3 The 70 GeV Proton Synchrotron (U-70) of NRC "Kurchatov Institute": IHEP (Protvino)

The study of a powerful synchrotron started in the mid-60s in the then Soviet Union and focused onto the Protvino site since 1958 [30]. The project was led by V.V. Vladimirski from 1958 and A.A. Logunov from 1963 after the foundation of the Institute for High-Energy Physics. The construction started in 1961 and the commissioning [31] took place in 1967 culminating in a test run at 76 GeV in the same year, which was the world record at that time (Table 10.4).

The first proton injector has been the 100 MeV Alvarez-type DTL proton linear accelerator (I-100) providing a pulse current of 100 mA over five turns in U-70. It served until 1985 as injector [34] and is still in operation as light ion injector [33]. In order to increase the intensity, the 1.5 GeV booster synchrotron (U-1.5) cycling at $16^2/_3$ Hz came on line in 1985 [35]. Its injector is a 30 MeV RFQ-type DTL linear

Table 10.4 Basic parameters of the U-70 [32, 33]

Accelerated particles	Protons, carbon nuclei
Particle energy	70 GeV (34 GeV/n carbon nuclei)
Circumference	1483.7 m
Magnetic lattice	Alternating-gradient focusing, combined-function
Focusing order	FODO
Magnetic field index	$n = 443$
Number of main magnets	120
Bending magnetic field	0.035 T at injection, 1.2 T at maximum energy
Betatron oscillations/turn	9.9 (h), 9.8 (v)
Rise time/flat top time	2.8 s/2 s
Long straight sections	Number = 24, length = 4.87 m
RF system (tunable)	38 (+2 spare) cavities, 2.6–6.1 MHz, 150 keV/turn total maximum
Auxiliary RF system (fixed RF)	2 cavities at 200 MHz, 500 kV peak total voltage
Vacuum chamber	stainless steel, 200×100 mm^2 in the bending magnets

accelerator (URAL-30) injecting 1–4 turns into the booster with a pulse current up to 80 mA. Twenty nine single-bunch pulses of the booster with up to 8×10^{11} protons per pulse build up the beam in U-70 within 1.8 s using bunch-to-bucket transfer. The U-1.5 has a circumference 1/15 times the one of U-70. It operates now at 1.3 GeV limited by the power supply.

The U-70 accelerator has been equipped with a new vacuum chamber in 1997 and the control system has been modernized from 1998 onwards. Thanks to all these upgrades and, in particular, to the addition of the booster and the new linear accelerator, the beam intensity has reached 1.5×10^{13} protons per pulse with a repetition time of 9.8 s, which has to be compared with the initially planned 1×10^{12} protons per pulse [32]. In recent years, U-70 operates with 1.1×10^{13} protons per pulse at 50 GeV to save energy [36].

Initially, only internal targets have provided a large variety of secondary beams and some internal targets are still in use providing spills up to 1.8 s. A fast-extraction system has been added soon. A slow-extraction system based on a third order resonance came into operation in 1979 and it has been upgraded for higher intensity in 1989. Its spill-length could be extended up to 1.3 s. Now, a stochastic slow extraction system provides smooth spills over up to 3 s at top energy. Extraction, beam splitting and collimation using bent crystals have been achieved being under study since 1990 [37].

Deuterons have been accelerated from 2008 onwards in I-100 and U-1.5 to 16.7 MeV/n and 455 MeV/n, respectively. In 2009 deuterons were further accelerated in U-70 to 23.6 GeV/n corresponding to 50 GeV protons. The latter has the potential to accelerate them to 34 GeV/n. An intensity of 5×10^{10} deuterons per pulse has been achieved.

In 2011, carbon ions have been first accelerated in I-100, U-1.5 and U-70 to 34 GeV/n [38]. In 2013, validation tests of all the top-energy extractions available with carbon beam have been accomplished successfully. Carbon beam intensity is 3–5 × 10^9 ions per pulse (8.2 s), in a single bunch. At the DC flat-bottom, U-70 now also operates in a beam storage- and stretcher-ring mode for 455 MeV/n carbon ions enabling their square-wave slow stochastic extraction (0.5–1 s long) via a Piccioni-Wright technique for an applied fixed-target research [39, 40].

10.1.4 The CERN Intersecting Storage Rings (ISR)

The first ideas for a realistic proton-proton collider were publicly discussed in 1956 [41, 42]. The Accelerator Research Group set up by the CERN Council in 1956 formulated in 1960 the first proposal for a proton-proton collider attached to the CERN PS. In 1960 construction began on a small-proof-of-principle 1.9 MeV electron storage ring, the CERN Electron Storage and Accumulation Ring (CESAR) which experimentally proved the accumulation of particles by RF stacking in 1964, an essential technique [43] to build up intense beams, a prerequisite for getting the required luminosity. CESAR also was an important test bed for the for Ultra-High Vacuum (UHV) technology which had to developed to achieve a very low vacuum pressure, indispensable for a long lifetime of stored beams. The Design report [44] was issued in 1964, the project was approved in 1965, and construction lead by K. Johnsen started without delay. The first proton-proton collisions took place in 1971 with a beam momentum up to 26.5 GeV/c, the maximum momentum available from the CPS. The ISR consisted of two independent storage rings intersecting at eight points at an angle of 14.8°. To create space for long straight sections in the interaction regions, the circumference of the ring was 1.5 times of that of the CPS which supplied particles to the ISR through two long transfer lines. The ISR operated for physics as collider from 1971 to 1983. It was decommissioned after 1984 (Table 10.5).

The CPS was the injector for the ISR supplying mainly protons up to 26.5 GeV/c. Typical intensities were 3 × 10^{12} protons per pulse in 20 bunches every 2.4 s during the filling process. The filling of one ISR ring took less than 10 min. Later, also deuterons, alpha particles and antiprotons were accelerated for the ISR.

The ISR team had to tackle a number of technological challenges but the most important was to assure UHV imperative for a long beam lifetime. The stainless steel vacuum chamber was in situ bakeable eventually to 300 °C. Pumping was provided by sputter ion-pumps and, at critical places, Ti-sublimation pumps. All vacuum chambers had to be glow-discharge cleaned. The continuous effort eventually resulted in average pressure below 10^{-11} Torr (N_2 equivalent) reducing beam loss rates to typically around one part per million per minute during physics runs. Beams of physics quality could last 40–50 h. Beam currents of 10 A were achieved already after start-up. Later up to 57 A were stored per ring with 30–40 A as typical value. The proton-proton initial luminosity (design 4 × 10^{30} cm^{-2} s^{-1}) was increased from 1.6 × 10^{30} cm^{-2} s^{-1} in 1971 to 1.4 × 10^{32} cm^{-2} s^{-1} in the superconducting low-beta section installed in one of the interaction points in 1982, which stayed the

Table 10.5 Basic parameters of the ISR [45]

Colliding particles	pp, dd, pd, αα, αp, p$\bar{p}$
Particle momentum	3.5–31.4 GeV/c, typically 26 GeV/c
Circumference [m]	300 π
Magnetic lattice	Alternating-gradient focusing, combined-function
Focusing order	FODO
Number of main magnets	132 per ring
Magnetic dipole field	1.33 T at maximum momentum
Length of main magnets	4.88/2.44 m
Magnetic field index	$n = 248$
Betatron oscillations/turn	8.90 (h), 8.88 (v)
Long straight sections	Number = 8, length = 16.8 m
β_* (h/v)	21 m/12 m
β_* (h/v)	2.5 m/0.28 m in superconducting low-beta section
RF system per ring	7 cavities, 9.5 MHz, 16 kV RF peak voltage
Auxiliary RF system	3rd harmonic
Vacuum chamber	Stainless steel, 160 mm/52 mm full width (h/v)

world record luminosity until 1991. The superconducting low beta-section had been preceded by an insertion based on conventional magnets, operational from 1974. From 1973 onwards, the beams could be accelerated in the ISR to 31.4 GeV/c by phase displacement acceleration [46].

From 1976 onward deuterons were stored in the ISR so that dd and pd collisions became available. Alpha particles were stored in 1980 for αα and αp collisions. Initial dd luminosities reached 1.6×10^{30} cm^{-2} s^{-1} and were 4×10^{28} cm^{-2} s^{-1} in the αα case. Antiprotons were stored as soon as the antiproton injector complex had become operational in 1981 (see Sect. 10.1.6). This required a new transfer line from the CPS.

The ISR will be remembered for a number of breakthroughs in accelerator physic and technology: UHV-technology for a large scale facility, control of intense coasting beams, discovery of Schottky scans, experimental demonstration of stochastic cooling, and absolute luminosity measurement by van der Meer scans [46].

10.1.5 The CERN Super Proton Synchrotron (SPS)

Design studies of a powerful proton synchrotron started at CERN in 1961 when the Accelerator Research Division had been created, not long after the CPS had become operational in 1960. The design energy was 300 GeV and the specified flux 10^{13} protons/s. A study group lead by K. Johnsen presented a first design report in 1964 [47]. However, the idea that the facility should be constructed on a green field, i.e. not in Switzerland near CERN, made the choice of the site difficult and funding

Table 10.6 Basic parameter of the SPS in proton fixed-target mode [51]

Accelerated particles	Protons
Momentum protons	450 GeV/c
Circumference [m]	2200 π
Magnetic lattice	Alternating-gradient focusing, separated-function
Focusing order	FODO
Number of dipole magnets	744
Dipole magnetic field	0.056 T at injection, 1.8 T at maximum particle momentum
Betatron oscillations/turn	26.6 (h), 26.6 (v)
Rise time/flat top time	0.75 s/2.5s
Long straight sections	Number = 6, length = 128 m
RF system	4 traveling-wave structures at 200 MHz, 4 MeV/turn total
Vacuum chamber	Stainless steel, 150 $\times$ 50 mm^2

problems delayed the decision over many years. In 1969, J.B. Adams was appointed leader of the 300 GeV Programme and the project gathered new momentum with the suggestion to construct the facility close to the existing laboratory and using the CPS as injector, which had been advocated already in 1961 by C. Ramm [48]. This and a new design based on the alternating-gradient principle but with the function of bending and focusing separated instead of combined-function allowed a considerable cost reduction [49]. The project was approved in 1971 and a beam energy of 400 GeV, exceeding the design energy, was reached in 1976 after almost a decade of planning and decision making [50] (Table 10.6).

The CPS acts as proton injector having 1/11th circumference of the SPS. Initially, the SPS beam was created by peeling off the required beam from the CPS beam by an electrostatic septum over 10 turns of the CPS at 10 GeV/c, leaving 1/11th of the circumference for the SPS injection kicker fall-time and ejection kicker rise-time, in a process called continuous transfer. From 1978 onwards, after the CPS intensity had been increased by the new linac 2 and the booster synchrotron, two CPS pulses were consecutively sent to the SPS, each CPS pulse was peeled five times. The CPS acted also as positron-electron injector during lepton operation for LEP and as injector during the fixed-target ion runs. It provides also a 26 GeV/c proton beam and 5.9 GeV/u ion beams to the SPS when the latter is used to fill the CERN Large Hadron Collider (LHC).

The intensity of the proton beam for fixed-target experiments had been raised gradually. In 1984, the injection momentum was raised to 14 GeV/c providing a more stable, reproducible injection and a beam of lower emittance. After the kinetic beam energy of the PS-Booster (PSB) had been raised to 1.4 GeV in 1998 resulting in a further emittance reduction, the SPS delivered more than 4×10^{13} protons per pulse with a record value of 5.3×10^{13} ppp.

Since the start-up in 1976 three types of extraction modes have been available towards the two experimental areas: (i) fast extraction of part or the entire beam (spill 3–23 μs); (ii) slow-resonant extraction (0.5–2 s); (iii) fast-resonant extraction (<3 ms). A typical pattern was extraction to the West-Area at an energy of 200 GeV

(250 GeV maximum) during a short pause in the acceleration followed by extraction at top energy to the North Hall. After the upgrading to 450 GeV of the transfer line to the West Hall in 1983, simultaneous resonant extraction to both areas was implemented by an appropriate adjustment of the horizontal betatron phase advance in the ring. The sharing ratio between the two clients was fully adjustable. Extraction towards the West Hall was terminated in 2003 to free resources for LHC.

In 1986 and 1987 two exploratory fixed-target runs with ions took place after linac 2 had become operational freeing linac 1 which had in the meantime been equipped with the appropriate front-end for ion operation. In 1986, fully stripped oxygen ions were accelerated for the first time to 200 GeV/n after the SPS had been set up with a deuteron beam of more convenient higher intensity. In the second run, in 1987, sulphur ions were used. The sulphur runs were resumed from 1990 to 1992 providing 9×10^9 charges per pulse with four batches injected from the CPS. After the construction of the dedicated heavy-ion linac (linac 3) lead ions at 177 GeV/u were available for the experiments from 1994 to 2002 and indium at 158 GeV/u in 2003. No ion runs took place in 1997 and 2001. The SPS delivered up to 6×10^{10} fully-stripped ions in terms of charges per pulse [52].

It is worthwhile to mention that the SPS also accelerated electrons and positrons for injection into LEP in the years 1989 to 2000. This had required substantial modifications. Since the traveling-wave cavities could accelerate the beam only to 14 GeV, 32 copper-based standing-wave cavities operating at 200 MHz and providing a peak-voltage of 30 MV were added. Ejection and injection channels were equipped appropriately and a campaign of meticulously shielding the magnet coils against synchrotron radiation was conducted [53]. The LEP injection energy was first set to 20 GeV and, later, when the standing-wave cavities were replaced by two four-cell superconducting RF structures, the LEP injection energy could be raised to 22 GeV.

The SPS acts as injector into the LHC providing protons at 450 GeV since 2008 and ions at 176 GeV/u since 2010 [54]. Lead is the preferred ion species for the first years of operation. This new role required a number of hardware modifications and, in particular, the addition of a new extraction system for the anti-clockwise LHC ring which serves also the target for the new long-base line neutrino beam towards Gran Sasso in Italy. Two new beam lines towards the LHC had to be built.

10.1.6 The CERN Super Proton Synchrotron (SPS) as Proton-Antiproton Collider

The proposal to use the SPS as proton-antiproton collider was made in 1976 [55]. The required increase of phase space density of the secondary antiprotons was to be produced by stochastic cooling invented by S. van der Meer [56] which had been experimentally demonstrated in the ISR [57]. Simultaneous stochastic cooling in transverse and longitudinal phase space at cooling rates several orders higher than

Table 10.7 Basic parameter of the SPS in proton-antiproton collision mode [60]

Accelerated particles	Protons and antiprotons
Maximum particle energy	315 GeV
Circumference [m]	2200 π
Magnetic lattice	Alternating-gradient focusing, separated-function
Focusing order	FODO
Number of dipole magnets	744
Dipole magnetic field	0.12 T at injection, 1.4 T at maximum particle energy
Long straight sections	Number = 6, length = 128 m (including 2 low-β insertions)
β_* (h/v)	1.0 m/0.5 m
Filling time	29 s
RF system	4 traveling-wave structures at 200 MHz, 3.6 MeV/turn total
Auxiliary RF system	100 MHz, 2 MV/turn
Vacuum chamber	Stainless steel, 150 × 50 mm^2

those achieved in the ISR was proven to work in the Initial Cooling Experiment (ICE) in 1978. ICE was a small storage ring of 74 m circumference operating at 1.73 and 2.1 GeV/c and fed with protons from the CPS [58]. With the project decision taken in 1978, the construction of the Antiproton Accumulator Ring jointly led by R. Billinge and S. van der Meer started as well as the modifications of CPS and SPS. The first collision of protons and antiprotons occurred in 1981 and the data taking of the experiments took place from 1982 to 1991 except 1986 [59]. All modifications of the SPS for collider operation were removed after 1991 (Table 10.7).

The antiproton injector chain [59] consisted of the CPS with its proton injectors and of the Antiproton Accumulator storage ring (AA). The CPS produced an intense pulse of 1.2 × 10^{13} protons at 26 GeV every 4.8 s consisting of five bunches and having a pulse length matched to the circumference of the AA (157 m). The pulse impinged on a tungsten target followed by a magnetic horn or a lithium lens focusing the emerging antiprotons. The latter were collected in the AA operating at 3.5 GeV/c, the momentum where the production was close to the peak. Subsequently, the stochastic cooling systems increased the phase space density by a factor of more than 5 × 10^8. The best daily production was close to 2 × 10^{11} per day. For the SPS fill, the collected and cooled antiprotons were transferred to the CPS, accelerated to 26 GeV/c and injected into the SPS through a new transfer line built for anti-clockwise injection. The fill was terminated by acceleration of the protons in the CPS and their injection through the existing transfer line upgraded from 14 GeV/c to 26 GeV/c. Subsequently, both beams were simultaneously accelerated to collision energy and, after the β_* had been lowered, brought to collision by adjusting the separators. The duration of a coast was between 10 and 20 h.

In order to increase the transverse and longitudinal acceptance after the target, an additional storage ring of 187 m circumference was constructed around the AA in 1986, the Antiproton Collector (AC). This new ring featured an RF system providing 1.5 MV at 9.5 MHz rotating the antiproton bunches in longitudinal phase space to reduce their momentum spread. It took over from AA the stochastic pre-cooling so that AA could be simplified but had its cooling systems upgraded. With these measures the antiproton production rate could be raised to more than 1×10^{12} per day.

The SPS required a number of modifications: the vacuum ion pumps were doubled and Ti-sublimation pumps added resulting in a reduction of pressure from 2×10^{-7} to 6×10^{-9} Torr; an electrostatic deflector separated the beams horizontally at injection; the magnet lattice included two low-beta sections for focusing the beam in the two interaction points; the RF travelling wave structures at 200 MHz had to accelerate particles travelling in both directions, and two underground experimental areas had to be constructed [61]. The initial beam energy was raised from 273 to 315 GeV after an upgrade of the magnet cooling. Towards the end of the operation, the SPS was cycled between 100 and 450 GeV, thus providing collisions at 900 GeV c.m. The performance of the collider was steadily increased during its lifetime by the upgrade of the antiproton injectors, in particular after the commissioning of AC; by adding electrostatic deflectors allowing to raise the number of bunches per beam from three to six avoiding collisions except in the two experiments and in the mid-arc between them; by the installation of a 100 MHz RF system to increase the acceptance at injection. Hence, the average initial luminosity and the centre-of-mass energy rose from 0.05×10^{30} at 546 GeV in 1982 to $3 \times 10^{30} \text{cm}^{-2}\ \text{s}^{-1}$ at 630 GeV with β_* (h/v) at 0.6 m/0.15 m in 1991, the last year of collider operation. The collider collected an integrated luminosity of 17 pb^{-1} during its lifetime [59].

10.1.6.1 Acknowledgement

Thanks to Karel Cornelis (CERN) for his critical reading and useful suggestions.

10.1.7 Tevatron of Fermi National Laboratory (FNAL)

The study of superconducting magnets started in 1972 in view of doubling the proton energy available at FNAL by adding another accelerator in the tunnel of the Main Ring (MR). In the same year, MR equipped with conventional magnets had reached its design energy producing 200 GeV protons. Since a magnet study had shown the feasibility of ramped superconducting magnets of 4–5 T, the official design study of this new accelerator, the Tevatron, with the MR as injector was launched in 1974 under the leadership of R.R. Wilson. Project authorization was granted in 1979 and the accelerator reached 512 GeV during commissioning in 1983 with the MR operating at 150 GeV as injector [62]. The accelerator was then routinely used between 1983 and 2000 in eight runs for fixed-target physics in the three experimental areas dedicated to physics with mesons, neutrinos, and protons respectively. The beam energy was 800 GeV except in the first run (400 GeV) and

Table 10.8 Basic parameters of the Tevatron [63]

Accelerated/colliding particles	Protons/protons—antiprotons
Particle energy	800 GeV/980 GeV
Circumference [m]	2000 π
Magnetic lattice	Alternating-gradient focusing, separated-function
Focusing order	FODO
Number of main bending magnets	774
Bending magnetic field	0.67 T at injection, 4.35 T at maximum energy
Rise time	15 s
Bending magnet	Nb-Ti conductor at 4.3 K, cold-bore, iron at ambient temperature
Quadrupole field gradient	11.4 T/m at injection, 76 T/m at maximum energy
Betatron oscillations/turn	20.59 (h), 20.59 (v)
Long straight sections	Number = 6, length = 50 m (incl. 2 low-β insertions)
β_* (h/v)	0.28 m in low-β sections
RF system (tunable)	8 cavities, 53.1 MHz, 1.2 MeV/turn total maximum
Vacuum chamber	Stainless steel, rounded square 63 mm full aperture (h,v) (in dipoles)

the repetition rate of the order of one per minute. Slow-spills (20 s) and fast beam extraction (2 ms) were available (Table 10.8).

The potential of the accelerator for proton-antiproton collisions with a centre-of-mass energy close to 2 TeV had been realized very early [64]. The bunched proton beam and anti-proton beam counter-rotating in the same vacuum chamber would be simultaneously accelerated in the Tevatron and brought to collision at top energy. Initiated in 1977, a study of the anti-proton beam cooling methods revealed that an anti-proton beam of sufficient intensity and density in phase space could be produced. This led to the construction of the anti-proton source in the period 1982–1985. The MR would also accelerate antiprotons and protons for injection into the Tevatron. It was equipped with two overpasses to provide space for the large detectors located in the two crossing points of the Tevatron. In 1985, the conversion of the Tevatron itself to a collider was started by the installation of the first low-β section in the straight section housing later the CDF detector. First collisions were recorded at a centre-of-mass energy of 1.6 TeV in 1985 and this energy could be raised to 1.8 TeV in 1986. It was increased again to 1.96 TeV for Run II [65] which started in 2001 [66].

In order to increase the average proton beam power on the anti-proton target and to eliminate the perturbation of the experiments by the MR overpasses, a new injector synchrotron, the Main Injector (MI), was constructed between 1993 and 1999 and MR was decommissioned. MI is a synchrotron of 3.32 km circumference and a minimum repetition time of 1.4 s. It is located in a new tunnel and provides the protons at 120 GeV for anti-proton production. It also accelerates the antiprotons and the protons from 8 GeV to 150 GeV for injection into the collider.

For the antiproton production, the full complement of accelerators and storage rings in the injector chain is used. The first accelerator in the injector chain is a 400 MeV H^- linear accelerator (linac) operating at 200 MHz. Its initial energy was 200 MeV but half of the original drift-tube linac was replaced in 1992–1993 by side-coupled accelerating structures providing 300 MeV. The protons are transferred to the booster synchrotron operating with 15 Hz repetition rate and are accelerated to a kinetic energy of 8 GeV. The intensity is up to 5×10^{12} protons per booster pulse. Next the particles are accelerated to 120 GeV in the Main Injector (MI), and hit a solid metal target followed by a lithium lens.

From the emerging secondary particles anti-protons of 8 GeV kinetic energy are selected and their phase space density is stepwise increased in two rings in series, the Debuncher Ring and the Anti-Accumulator Ring; the latter also accumulates the particles. The anti-protons pass then to the latest addition, the Recycler Ring (RR), were they are cooled further and a second step of accumulation takes place. The RR could be fitted into the MI tunnel as its bending and focusing magnets have a small cross-section the magnetic fields being generated by permanent magnets (1.45 T field strength). This new storage ring has therefore the same circumference as MI; it is in operation since 2004. Due to this elaborate anti-proton source the antiproton production rate has reached nearly 3×10^{11} anti-protons/h. Still the production rate of the antiprotons is rather low compared to proton production. Hence, the injector chain produces and accumulates anti-protons all the time except during the period of filling the Tevatron with protons and antiprotons which takes about 1 h.

When the Tevatron needs a new fill, which happens about every 10–20 h, the linac, the booster and MI produce the 36 proton bunches for the Tevatron, then anti-protons collected in RR are transferred to MI for acceleration to 150 GeV and injection into the Tevatron. The 36 bunches are arranged in 3 trains of 12 bunches. Protons and antiprotons circulate on helical orbits produced by high-voltage electrostatic separators to prevent collisions during injection and acceleration. At top energy, after increasing the focusing in the two collision points equipped with detectors, the separator configuration is modified to bring the beams in collision in these two points [67].

A peak luminosity of 4×10^{32} cm^{-2} s^{-1} has been reached by continuously upgrading all the systems, introducing advanced accelerator technology and stream-lining the operational procedures. The Tevatron has delivered an integrated luminosity of more than 12 fb^{-1} to each of the experiments (CDF and D0) before it ceased operation at the end of September 2011. It was one of the most complex research instruments ever built and will be known for its advances in accelerator physics and technological breakthroughs [68].

10.1.7.1 Acknowledgement

Vladimir Shiltsev (FNAL) has contributed with useful suggestions and pertinent comments to this chapter. Sincere thanks are due to him.

10.2 RHIC[1]

T. Roser

10.2.1 The RHIC Facility

With its two independent superconducting rings RHIC is a highly flexible collider of hadron beams ranging from intense beams of polarized protons to fully stripped gold ions [69, 70]. The layout of the RHIC accelerator complex is shown in Fig. 10.1. The collision of 100 GeV/nucleon gold ions probes the conditions of the early universe by producing extreme conditions where quarks and gluons are forming a new state of matter, the strongly interacting quark-gluon plasma. Several runs of high luminosity gold-gold collisions as well as comparison runs using proton, deuteron and copper beams have demonstrated that indeed a new state of matter with extreme density is formed in the RHIC gold-gold collisions. (See also Sect. 11.5).

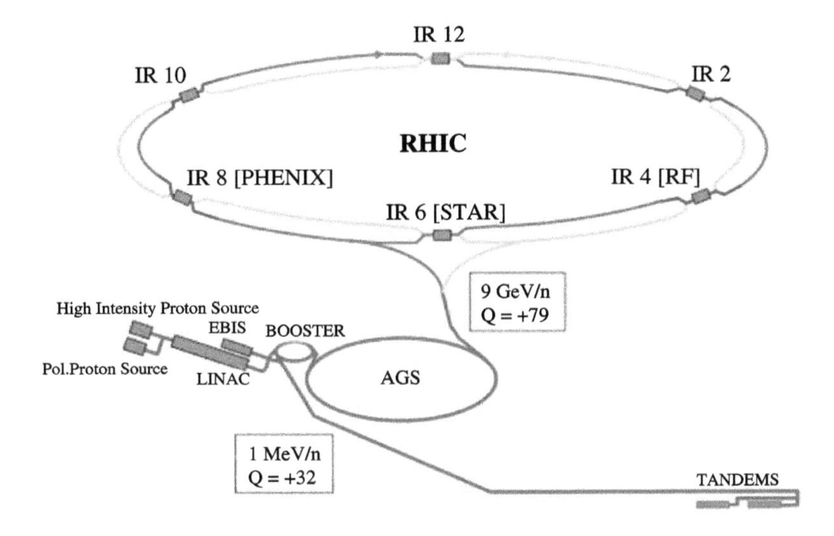

Fig. 10.1 Layout of RHIC and the injector accelerators. The gold ions are stepwise ionized as they are accelerated to RHIC injection energy

[1]This manuscript has been authored by Brookhaven Science Associates, LLC under Contract No. DE-SC0012704 with the U.S. Department of Energy. The United States Government retains and the publisher, by accepting the article for publication, acknowledges that the United States Government retains a non-exclusive, paid-up, irrevocable, world-wide license to publish or reproduce the published form of this manuscript, or allow others to do so, for United States Government purposes.

The RHIC polarized proton collider has opened up the completely unique physics opportunities of studying spin effects in hadronic reactions at high-luminosity high-energy proton-proton collisions. It allows the study of the spin structure of the proton, in particular the degree of polarization of the gluons and anti-quarks, and also verification of the many well-documented expectations of spin effects in perturbative QCD and parity violation in W and Z production. The RHIC center-of-mass energy range of 200–500 GeV is ideal in the sense that it is high enough for perturbative QCD to be applicable and low enough so that the typical momentum fraction of the valence quarks is about 0.1 or larger. This guarantees significant levels of parton polarization.

During its 10 years of operation RHIC has greatly exceeded the design parameters for gold-gold collisions, has successfully operated in an asymmetric mode of colliding deuteron on gold with both beams at the same energy per nucleon, and thereby at different rigidities, and successfully completed a comparison run of colliding copper beams with record luminosities. Operation at unequal rigidities of the two colliding beams is a unique feature of RHIC with its two independent rings. The interaction regions are designed to separate the two beams first before they go through their separate final focus triplets. This is shown in Fig. 10.2 for equal species and for the most unequal species of protons and gold beams with a rigidity ratio of 2.47 for equal energy per nucleon. The necessary increase in distance between the interaction point and the final focus triplet limits the achievable luminosity in RHIC.

In addition to heavy ions, RHIC successfully demonstrated its capabilities as a high luminosity polarized proton collider both at 100 and 250 GeV proton beam energy. For most of the heavy ion runs RHIC was operating with beam energies of 100 GeV/nucleon—the gold beam design energy. Additional running at lower beam energy was also accomplished again demonstrating the high level of flexibility of RHIC. Gold collisions at energies much below the RHIC injection energy of 10 GeV/nucleon is allowing the study of the critical point in the quark-gluon phase diagram. Figure 10.3 shows the achieved integrated nucleon-pair luminosities for the many modes of operation of RHIC since its start of operation in 2000. Using nucleon-pair luminosity, which is calculated as the ion-ion luminosity times the

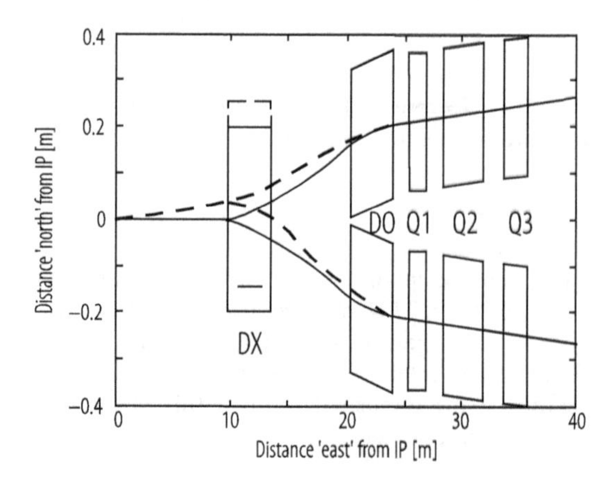

Fig. 10.2 Interaction region geometry in RHIC for equal species (solid line) and the most extreme example of dissimilar species—protons and gold (dashed line)

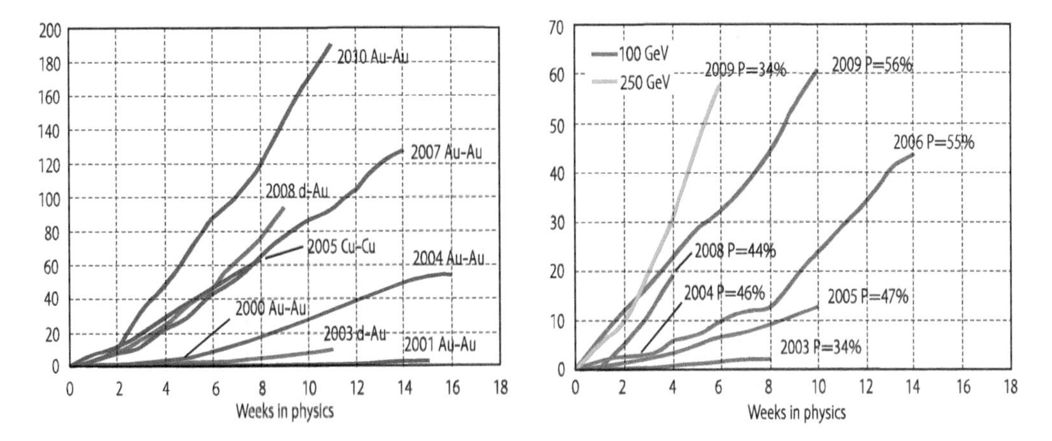

Fig. 10.3 Integrated nucleon-pair luminosity for the heavy ion (left) and the polarized proton (right) running modes since the start of RHIC operation

Table 10.9 Major parameters of RHIC

Parameter	Value
Circumference	3833.845 m
Number of interaction points	6
Harmonic number, acceleration	360
Harmonic number, storage	2520
Typical betatron tunes for Au	28.23/29.22
Transition energy γ_T	26.7 (Au), 22.8 (p)
Maximum magnetic rigidity	839.5 Tm
Total number of dipoles, both rings	396
Total number of quadrupoles, both rings	492
Dipole field at 100 GeV/nucleon, Au	3.458 T
Arc dipole effective length	9.45 m
Arc quadrupole gradient	71.2 T/m

number of nucleons in each of the two beam ions, allows the comparison of the different modes of operation properly reflecting the relative statistical relevance of the data samples and also the degree of difficulty in achieving high luminosity.

For RHIC's major parameters and achieved performance parameters see Tables 10.9 and 10.10, respectively.

10.2.2 *Collider Operation*

Gold-Gold Operation

Starting with Au^{1-} from a sputter source the gold ions are stepwise ionized as they are accelerated in the Tandem Van de Graaff, the AGS Booster and the AGS to RHIC injection energy. The electrostatic acceleration in the Tandem Van de Graff

Table 10.10 Achieved performance parameters of RHIC

Parameter	Unit	Au–Au operation	d–Au operation	Pol. proton–proton operation	
Total energy at store	GeV/nucleon	100	100	100	250
Total energy at injection	GeV/nucleon	9.8	9.9	23.8	23.8
Number of bunches		111	95	109	109
Bunch intensity	10^9	1.1	100d/1.0Au	135	110
IP beta-function	m	0.75	0.85	0.7	0.7
Normalized rms emittance	μm	2.8	2.6	2.5	3.0
Rms bunch length	m	0.3	0.3	0.85	0.6
Hourglass factor		0.93	0.95	0.70	0.80
Beam-beam parameter/IP		0.0015	0.0014	0.0045	0.0054
Peak luminosity	$cm^{-2}\ s^{-1}$	40×10^{26}	27×10^{28}	50×10^{30}	85×10^{30}
Average luminosity	$cm^{-2}\ s^{-1}$	20×10^{26}	14×10^{28}	28×10^{30}	55×10^{30}
Average polarization	%	–	–	55	34
Calendar time in store	%	53	58	54	53
Integrated luminosity per week	nb^{-1}	0.65	40	8300	1840

provides an extremely bright gold beam that can be captured and bunch-merged to provide the necessary bright bunches of 1×10^9 Au ions with a normalized transverse rms emittance of less than 2.5 μm and a total longitudinal emittance of less than 0.3 eVs/nucleon. The final stripping to bare Au^{79+} occurs on the way to RHIC. Recently the Tandem van de Graff pre-injector has been replaced by an Electron Beam Ion Source (EBIS) followed by an RFQ and IH-Linac. The EBIS can produce very bright high-charge state heavy ion beams by accumulating and stripping heavy ions with a 10 A space-charge neutralizing electron beam. The new source can produce beams of many species including uranium.

The two RHIC rings, labeled blue and yellow, are intersecting at six interaction regions (IRs). All IRs can operate at a betastar between 2 and 10 m. In two interaction regions, occupied by the two main detectors STAR and PHENIX, the quality of the triplet quadrupoles allows further reduction of betastar to less than 1 m. Typically betastar is about 10 m at injection energy for all IRs and is then squeezed during the acceleration ramp first to 5 m at the transition energy ($\gamma_T = 26.7$), which minimizes the momentum dependence of the transition energy crossing, and then to less than 1 m for PHENIX and STAR. A typical acceleration cycle consists of filling the blue ring with 111 bunches in groups of 4 bunches, filling the yellow ring in the same way and then simultaneous acceleration of both beams to storage energy. During acceleration the beam bunches are longitudinally aligned but are separated vertically by 10 mm in the interaction regions to avoid beam losses from beam-beam interaction.

The collision rate is measured using identical Zero Degree Calorimeters (ZDC) at all detectors. The ZDC counters detect at least one neutron on each side from mutual Coulomb and nuclear dissociation with a total cross section of about 10 barns. Typical stores in RHIC last about 4–5 h. Due to intra-beam scattering, which is particularly important for the fully stripped, highly charged gold beams, the initial normalized rms emittance of about 2.5 μm grows to about 5 μm at the end of the store and a significant amount of beam is lost from the RF bucket. The resulting luminosity lifetime is about 2.5 h.

The reduced number of gold ions compared to typical proton bunches makes it possible to contemplate stochastic cooling of the 100 GeV/nucleon bunched beam. A 6–9 GHz longitudinal and a 5–8 GHz vertical stochastic cooling system was installed using novel high power multi-cavity kickers and the high energy bunched beam was successfully cooled [71]. Figure 10.4 shows the successful cooling of the transverse emittance in the yellow ring compared to the emittance growth in the blue ring without cooling. Both the vertical and horizontal emittance are reduced due to x-y coupling. Figure 10.5 shows the vertex distribution at the end of the store with and without stochastic cooling. It clearly shows a significant increase in luminosity in the central region. The modulation of the vertex distribution is a result of the 200 MHz storage RF system that provides about 4 MV/turn to compress the bunch length and consequently reduce the length of the vertex distribution to better match the acceptance of the collider detectors. Again due to intra-beam scattering the initially narrow distribution widens during the store. A new 2 MV, 56 MHz superconducting cavity will provide improved RF focusing that, together

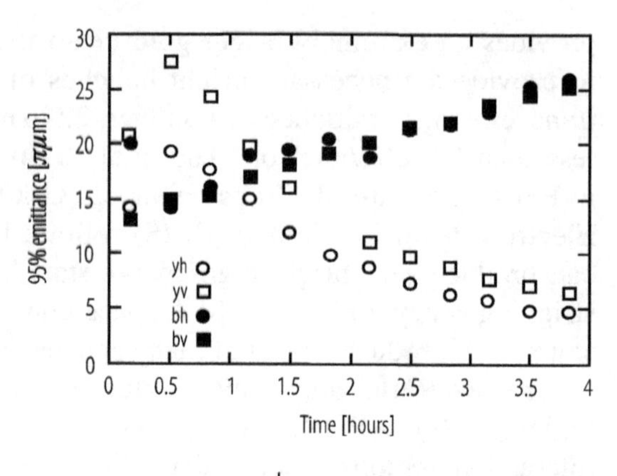

Fig. 10.4 Transverse emittance evolution in the blue (filled) and yellow (open) RHIC rings with vertical stochastic cooling in the yellow ring [71]

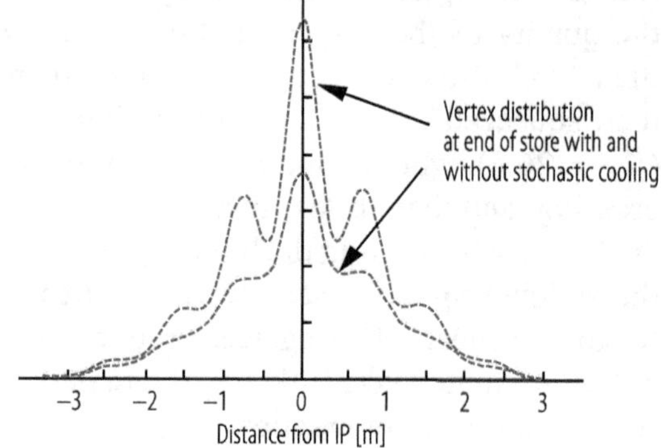

Fig. 10.5 Vertex distribution at the end of a 100 GeV/n Au-Au store with and without longitudinal stochastic cooling

with the longitudinal stochastic cooling, will maintain the short vertex distribution throughout the store.

With the two planes of stochastic cooling operational in both rings a peak luminosity at PHENIX and STAR of up to 40×10^{26} cm^{-2} s^{-1} (160×10^{30} cm^{-2} s^{-1} nucleon-pair luminosity) with an average store luminosity of 20×10^{26} cm^{-2} s^{-1}: ten times the original RHIC design average store luminosity. The full stochastic cooling system will include also the horizontal plane and should result in another doubling of the luminosity.

The total gold beam intensity in RHIC was initially limited by vacuum breakdowns in the room temperature sections of the RHIC rings [72]. This pressure rise is associated with the formation of electron clouds, which in turn appear when the bunch peak intensity is high around transition and after bunch compression, and when the bunch spacing is below about 200 ns. This situation was greatly improved by installing vacuum pipes with an internal coating of non-evaporative getter (NEG) that is properly activated. The resulting residual static pressure is about 10^{-11} Torr. The NEG coating acts as a very effective distributed pump and also suppresses electron cloud formation due to its low secondary electron yield.

Very fast single bunch transverse instabilities that develop near transition, where the chromaticity needs to cross zero, originally limited bunch intensity and is still responsible for occasional emittance growth, especially towards the end of bunch trains. The instability can be stabilized using octupoles [73] and the instability threshold can be increased by lowering the peak current during transition crossing. This instability has a growth rate faster than the synchrotron period and is similar to a beam break-up instability. The instability is also enhanced by the presence of electron clouds [74].

Deuteron-Gold Operation

Colliding 100 GeV/nucleon deuteron beam with 100 GeV/nucleon gold beam will not produce the required temperature to create a new state of matter and therefore serves as an important comparison measurement to the gold-gold collisions. The rigidity of the two beams is different by about 20%, which results in different deflection angles in the beam-combining dipoles on either side of the interaction region. This requires a non-zero angle at the collision point, which slightly reduces the available aperture.

The injection energy into RHIC was also the same for both beams requiring the injector to produce beams with different rigidity. With same energy beams throughout the acceleration cycle in RHIC the beams can remain cogged. Without this the effect of the time-modulated long-range beam-beam interaction in the IRs would lead to rapid beam loss. The typical bunch intensity of the deuteron beam was about 1×10^{11} with a transverse rms emittances of about 2 μm and a total longitudinal emittance of 0.3 eVs/nucleon. The gold beam parameters were similar to the gold-gold operation described above. Recently the ring with the gold beam was operated with increased transverse focusing to reduce the effect of intra-beam scattering on the transverse emittance. As a beneficial side effect the two beams crossed transition at different times, which reduced the development of the fast transverse instability. A peak luminosity of 27×10^{28} cm^{-2} s^{-1} (100×10^{30} cm^{-2} s^{-1} nucleon-pair luminosity) and store-averaged luminosity of 14×10^{28} cm^{-2} s^{-1} was reached at the IRs with a 0.85 m betastar.

Polarized Proton Operation

Figure 10.6 shows the layout of the RHIC accelerator complex highlighting the components required for polarized beam acceleration. The 'Optically Pumped Polarized Ion Source' [75] is producing about 10^{12} polarized protons per pulse. A single source pulse is captured into a single bunch, which is sufficient beam intensity to reach in RHIC the nominal bunch intensity of 2×10^{11} polarized protons.

In the AGS two partial Siberian snakes are installed, an iron-based helical dipole that rotates the spin around the longitudinal direction by 11° and a superconducting helical dipole that can reach a 3 T field and a spin rotation of up to 45°. A view down the magnet gap is shown in Fig. 10.7. With the two partial snakes placed with one third of the AGS ring between them all vertical spin resonances are avoided up to the required RHIC transfer energy of 23.8 GeV as long as the vertical betatron tune is placed at 8.98, very close to an integer [76]. With an 80% polarization from the source 65% polarization was reached at AGS extraction. The remaining polarization loss in the AGS comes from weak spin resonances driven by the horizontal motion of the beam [77]. It is planned to overcome them by quickly shifting the betatron tune during resonance crossing.

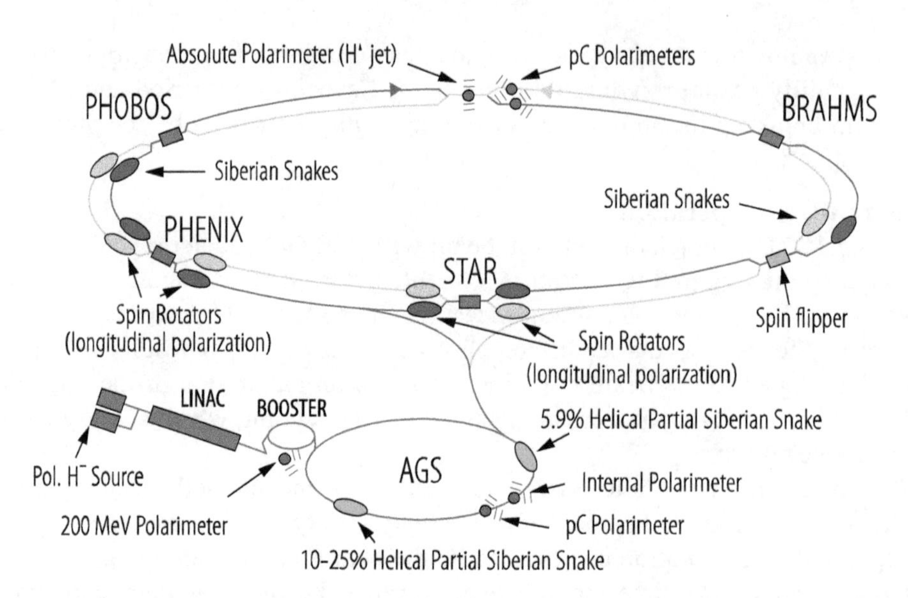

Fig. 10.6 The RHIC accelerator complex with the elements required for the acceleration and collision of polarized protons highlighted

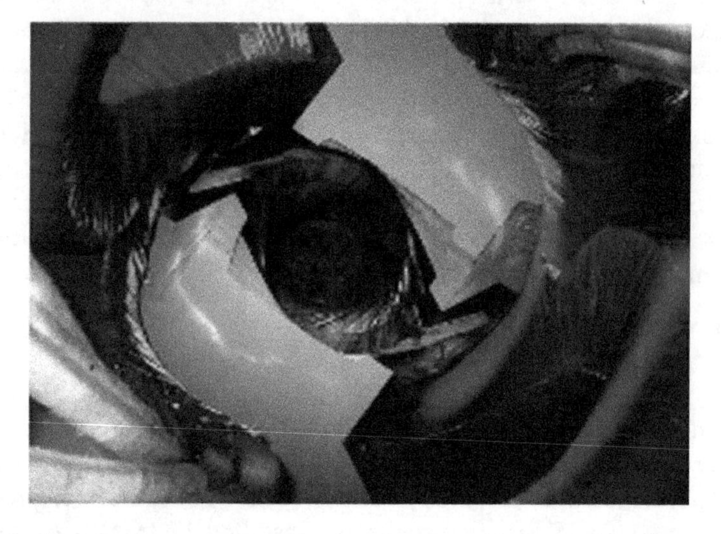

Fig. 10.7 View down the magnet gap of the warm, iron-based helical partial Siberian snake of the AGS

The full Siberian snakes [78], two for each ring, and the spin rotators, four for each collider experiment, in RHIC each consist of four 2.4 m long, 4 T superconducting helical dipole magnet modules each having a full 360° helical twist. Figure 10.8 shows the orbit and spin trajectory through a RHIC snake. The two Siberian snakes are installed in the location of the missing dipole of the dispersion suppression section and the orbit angle is exactly 180° in between the two snakes.

The spin rotation axis of the two snakes is pointing 45° in and out, respectively, relative to the direction of the beam resulting in the required 90° angle between the

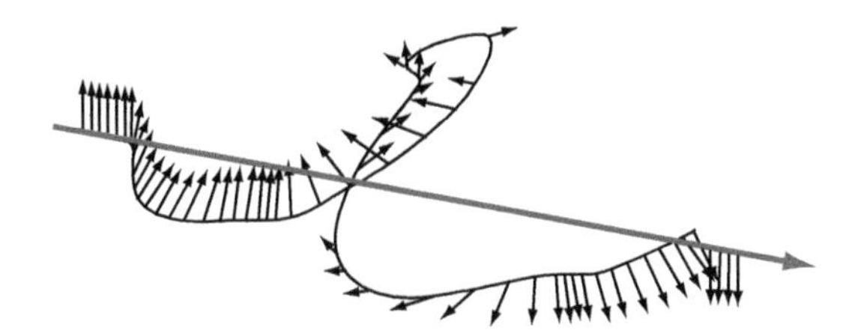

Fig. 10.8 Orbit and spin tracking through the four helical magnets of a Siberian Snake. The spin tracking shows the reversal of the vertical polarization. The spin rotation axis is in the horizontal plane and is pointing 45° away from the beam direction

spin rotation axes. This configuration yields a spin tune $Q_s = 0.5$ and a vertical stable spin direction around the ring. Here, spin tune is defined as the number of spin precessions in one orbital revolution. Since the betatron tunes in circular accelerators are kept away from half integer to keep the beam stable, both intrinsic and imperfection depolarizing spin resonances at $G\gamma = kP \pm Q_y$ and $G\gamma = k$ are avoided [79]. Here, G is the anomalous g factor, γ is the Lorentz factor, k is an integer, P is the super-periodicity of the accelerator, and Q_y is the vertical betatron tune.

The accurate measurement of the beam polarization is required for set-up and operation of the polarized proton collider. Very small angle elastic scattering in the Coulomb-Nuclear interference region offers the possibility for an analyzing reaction with a high figure-of-merit, which is not expected to be strongly energy dependent [80]. For polarized beam commissioning in RHIC an ultra-thin carbon ribbon is used as an internal target, and the recoil carbon nuclei are detected to measure both vertical and radial polarization components. The detection of the recoil carbon with silicon detectors using both energy and time-of-flight information shows excellent particle identification. It was demonstrated that this polarimeter can be used to monitor polarization of the high energy proton beams in an almost non-destructive manner and that the carbon fiber target could be scanned through the circulating beam to measure beam and polarization profiles. A polarized atomic hydrogen jet was also installed as an internal target for small angle proton-proton scattering which allows the absolute calibration of the beam polarization to better than 5%.

Figure 10.9 shows circulating beam current, luminosity and measured circulating beam polarization of a typical store with a beam energy of 100 GeV. The maximum peak luminosity achieved with 100 GeV proton beams is about 50×10^{30} cm^{-2} s^{-1} and store beam polarization of about 55% calibrated at 100 GeV with the absolute polarimeter mentioned above. To preserve beam polarization in RHIC during acceleration and storage the vertical betatron tune has to be controlled to better than 0.005 and the orbit has to be corrected to better than 1 mm rms to avoid depolarizing "snake" resonances.

During acceleration from 100 GeV to the maximum RHIC energy of 250 GeV, several very strong depolarizing spin resonances need to be crossed [81]. In a first running period an average store polarization of 34% was measured at 250 GeV again calibrated with the absolute polarimeter. The peak luminosity

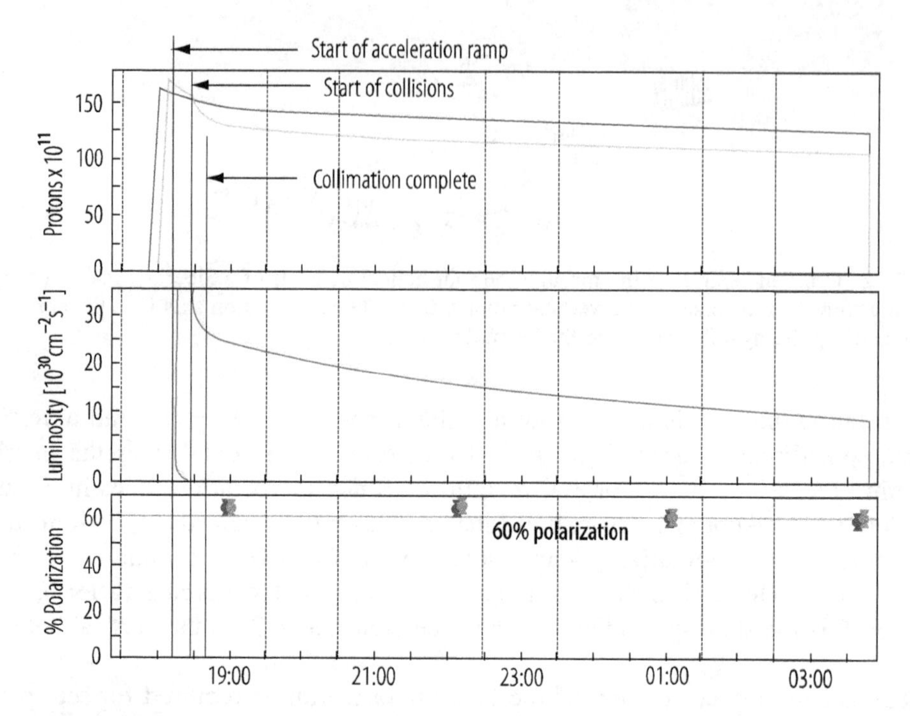

Fig. 10.9 Circulating beam in the blue and yellow ring, luminosity at STAR (red), as well as the measured circulating beam polarization in the blue and yellow RHIC ring (blue (dark) and yellow (light) lines and symbols, respectively) for a typical store with 100 GeV beam energy

reached was about 85×10^{30} cm^{-2} s^{-1} and the average store luminosity was about 55×10^{30} cm^{-2} s^{-1}. The goal for operation of RHIC with polarized 250 GeV proton beams is a peak luminosity of 200×10^{30} cm^{-2} s^{-1} with a beam polarization of 70%.

Head-on beam-beam effects are the main limitation for increasing the luminosity in proton-proton operation. The maximum beam-beam parameter reached was 0.0065 per IP with 100 GeV beams when operating with beams in collision at two detector IPs. This value is somewhat smaller than what was reached at the Tevatron proton-antiproton collider probably due to the smaller momentum aperture available at RHIC as a result of the dipole first IR design. Additional sextupole circuits are available to correct non-linear chromaticity and the beta-function momentum dependence. In the future it is planned to mitigate the beam-beam effects with the installation of two electron lenses in RHIC. Space is available to partially compensate the beam-beam effect with an opposite charge electron beam separated from the proton-proton collisions by the correct phase advance. With the electron lenses and also a more intense polarized proton source it is expected that the proton-proton luminosity could be increased by about a factor of two.

Recent Progress

The performance of RHIC has been continuously improved with store-averaged luminosity reaching 87×10^{26} cm^{-2} s^{-1}, exceeding the design luminosity by more than a factor of 40, and polarized proton operation at 510 GeV center-of-mass energy with peak luminosity of 250×10^{30} cm^{-2} s^{-1} with store-averaged beam polarization of up to 60%. Major upgrades to RHIC included electron lenses to compensate for head-on beam-beam interactions [82] and a superconducting storage RF system. In preparation for operation at very low center-of-mass energies with beam energies below the RHIC injection energy bunched beam electron cooling of both RHIC ion beams is being installed [83].

10.3 Electron Accelerators and Electron–Positron Colliders

J. Seeman

10.3.1 Cyclotrons

Cyclotrons are one of the first accelerators invented and use a fixed magnetic field and a radiofrequency (RF) cavity to accelerate electrons in ever increasing orbits [84]. The electron beam is injected into the center of a circular magnetic field region between magnet poles and then made to circulate. These electrons are accelerated twice on each turn and are extracted at high energy near the outer edge of the accelerator. The cyclotron can produce continuous beams leading to high power applications. The first cyclotron was built at Berkeley in 1931 with a diameter of 11 cm. Many more cyclotrons were to follow over the years. The TRIUMF cyclotron in use now is located in Vancouver, Canada, and has a diameter of 18 m and has a 4000 ton magnet. The beam energy is limited by the field strength and the diameter of the magnetic pole. Cyclotrons are mainly used to produce high beam power for science, for materials analysis, and industrial processing applications. Two cyclotrons are shown in Fig. 10.10.

10.3.2 Synchrotrons

A synchrotron is a fixed circumference accelerator that increases the electron beam energy using RF accelerating cavities but keeps the radius constant by increasing the magnetic fields. The field used to bend the beam in a circle is ramped in proportion to the beam energy [85]. The beam is injected at low energy and magnetic fields, accelerated, and then extracted at high energy and magnetic fields. The highest beam energy is limited by the strength of the magnetic field and the diameter of the accelerator which can be over 1 km and is usually limited by the site or its building. Synchrotrons are pulsed machines due to the cyclic ramping of the magnetic fields. The RF system accelerates the beam during the ramp and keeps the beam bunched.

Fig. 10.10 On the left is the first electron cyclotron constructed at the Berkeley in 1931 (courtesy LBNL Berkeley, CA). The right photograph is a modern cyclotron at PSI (courtesy PSI Lausanne, Switzerland)

Fig. 10.11 The left photograph is the 12 GeV electron synchrotron at Cornell (courtesy Wilson Laboratory Ithaca, NY) used as an injector into the CESR storage ring collider, also shown. The right photograph is the Advanced Light Source injector synchrotron (courtesy LBNL Berkeley, CA)

The magnets are often divided into separated units to allow simplified construction including dipoles for bending, quadrupoles for beam focusing, sextupoles for chromatic corrections, and correction dipoles for trajectory tuning. The quadrupoles often form an alternating focus and defocusing magnetic lattice around the ring to produce "strong focusing" which reduces the beam's excursions. As a result, the cross sections of the magnet gaps can be made significantly smaller and less expensive. The radiofrequency RF system can have a large range of frequencies and cavity designs. A vacuum system in the milli-Torr to nano-Torr level is also needed. Synchrotrons are often used as injectors for higher energy accelerators and storage rings. In Fig. 10.11 are shown the electron synchrotrons at Cornell (12 GeV, ~756 m) and the injector at the Berkeley Advanced Light Source.

10.3.3 *Electron Positron Circular Colliders*

In the late 1950s the accelerator community realized that colliding beams head-on would increase the center of mass energy significantly over fixed target collisions and could significantly expand the reach for new particles and physics. The goal was to design an accelerator that could provide enough centre of mass energy to make new physics, high enough beam currents and small enough collision spot sizes to provide a sufficient data rate, and to provide a volume for the particle physics detector surrounding the collision point with backgrounds sufficiently low to allow clean data collection. A world wide effort was initiated to design these accelerators and has continued to today [86–88]. At least 25 electron-electron or electron-positron colliders have been built and operated for particle physics over the past 50 years with ever increasing beam currents, luminosity, energy, and sophistication. Several colliders are shown in Fig. 10.12. In Table 10.11 are listed in chronological order these 24 colliders along with several of their technical parameters. Following the table are descriptions of each of the colliders discussing several of their

Fig. 10.12 The ADA collider on the top left (courtesy INFN Frascati, Italy). Top right is the PEP-II B-Factory collider (courtesy SLAC Menlo Park, CA). The LEP collider is shown on the bottom left (courtesy CERN Geneva, Switzerland). SuperKEKB is shown on the bottom right (courtesy KEK Tsukuba, Japan)

Table 10.11 Historical listing of electron-electron and electron-positron colliders

Collider	Laboratory	Date (start–end)	Circumf. (type) [m]	Beams	E [GeV]	Luminosity [10^{30} cm^{-2} s^{-1}]
ADA	Frascati, Italy	1961–1964	3 (SR)	e^-/e^+	0.25	Measured
VEP-1	BINP, Russia	1962–1967	2.7 (DR)	e^-/e^-	0.13	0.003
CBX	Stanford, USA	1963–1967	12 (DR)	e^-/e^-	0.5	0.0017
VEPP-2	BINP, Russia	1967–1970	11.5 (SR)	e^-/e^-	0.13	0.02
ACO	Orsay, France	1967–1972	22 (SR)	e^-/e^+	0.5	0.1
VEPP-2M	BINP, Russia	1974–2000	17.8 (SR)	e^-/e^+	0.7	100
ADONE	Frascati, Italy	1969–1993	105 (SR)	e^-/e^+	1.5	0.6
CEA Bypass	Cambridge, USA	1971–1974	225 (SR)	e^-/e^+	3.0	10
SPEAR	SLAC, USA	1972–1988	234 (SR)	e^-/e^+	2.5	12
DORIS	DESY, Germany	1973–1993	288 (DR,SR)	e^-/e^+	6.0	33
DCI	Orsay, France	1977–1984	95 (DR)	e^-/e^+	1.8	1.4
PETRA	DESY, Germany	1978–1986	2304 (SR)	e^-/e^+	19	23
CESR	Cornell, USA	1979–2008	768 (SR)	e^-/e^+	6.0	1100
VEPP-4	BINP, Russia	1979–present	366 (SR)	e^-/e^+	7.0	50
PEP	SLAC, USA	1980–1988	2200 (SR)	e^-/e^+	15	59
Tristan	KEK, Japan	1986–1995	3016 (SR)	e^-/e^+	32	140
SLC	SLAC, USA	1989–1998	4000 (linear)	e^-/e^+	49	2.8
BEPC	IHEP, China	1989–2004	240 (SR)	e^-/e^+	2.8	8
LEP	CERN, Switzerland	1989–2000	26,659 (SR)	e^-/e^+	104	100
DAFNE	Frascati, Italy	1998–present	98 (DR)	e^-/e^+	0.7	453
PEP-II	SLAC, USA	1998–2008	2200 (DR)	e^-/e^+	9.0 × 3.1	12,069
KEKB	KEK, Japan	1999–2009	3016 (DR)	e^-/e^+	8.0 × 3.5	21,083
BEPC-II	IHEP, China	2008–present	240 (DR)	e^-/e^+	2.1	1000
VEPP-2000	BINP, Russia	2006–present	24.4 (SR)	e^-/e^+	1.0	120
SuperKEKB	KEK, Japan	2016–present	2200 (DR)	e^-/e^+	7.0 × 4.0	NA

The energy E shown is for each beam and is doubled for the center of mass energy
SR single ring, *DR* double ring

unique features and results. The intent is to illustrate the advancement of collider technology over the years. Several of the technical advances are high power RF systems, superconducting RF systems, high field magnets, superconducting magnets, high field permanent magnets, ultra-high vacuum systems, low emittance beam lattices, complex interaction region designs, and sophisticated diagnostics and controls. Many of the colliders in later life became synchrotron radiation sources and several serve in this capacity today.

10.3.3.1 ADA

ADA at Frascati was a pioneering electron-positron collider [89]. This collider was used to study stored beam parameters, injection, beam lifetime, and collisions. Injection of electrons and positrons was made by converting 1 GeV gamma rays in a small target inside the ring. The mean lifetime was consistent with gas bremsstrahlung. With the electrons stored, the whole magnet was rotated about a horizontal axis to inject positrons without losing electrons. A pulsed kicker was used to reduce the betatron oscillations of the injected particles. Luminosity was measured. The effects of Touschek scattering was discovered and studied.

10.3.3.2 VEP-1

VEP-1 at BINP was an early electron-electron collider used to study elastic scattering and double bremsstrahlung experiments [90]. This accelerator demonstrated the possibility to carry out colliding beam experiments for particle physics.

10.3.3.3 CBX

The Princeton-Stanford Colliding Beam Experiment CBX on the Stanford campus was an electron-electron collider used to study elastic scattering and double bremsstrahlung experiments [91]. G. O'Neill on this team proposed that radiation damping for a stored electron beam could help with injection and emittance reduction. The beam–beam tune shift limit was seen in this machine on the level of 0.02–0.05. CBX saw indications of the single beam resistive wall instability. Electron currents up to 1 A were stored. This accelerator also demonstrated the possibility to carry out colliding beam experiments for particle physics.

10.3.3.4 VEPP-2

VEPP-2 at BINP was one of the first electron-positron collider in the world [92, 93]. The rho and phi meson parameters were measured using annihilation into two pions followed later by omega and phi meson decay parameters in the vector meson region and the first observation of two-photon pair production. This collider was quickly rebuilt as VEPP-2M.

10.3.3.5 ACO

ACO at Orsay was an early electron–positron collider used to demonstrate collider principles [94]. This accelerator had six dipoles with quadrupoles in between. The ring was later used as a synchrotron light source demonstrating ring based FEL science. This accelerator gave way to the DCI collider a few years later.

10.3.3.6 ADONE

ADONE at Frascati was a one ring collider with a 50 MHz RF system [95]. ADONE had longitudinal phase feedback to damp beam instability modes, a 4 kG detector solenoidal field (MEA), magnet shunts to adjust the beta functions in the ring, and adjustable damping partition numbers generated by an RF frequency change. Injection was from a linac with an energy ramp after the fill. The measured luminosity increased a little over the fourth power of the energy. ADONE later confirmed the existence of the J/Psi.

10.3.3.7 CEA

The CEA Cambridge Electron Colliding Beam Facility used a linac for injection at 120 MeV (e^+) and 240 MeV (e^-) and then ramped both beams in the ring to full energy [96]. The accelerator technology advancements for the CEA are the first demonstration of a low beta insertion at the IP, single-turn pulsed orbit switching into an interaction point (IP) magnetic bypass, switching from energy cycling to dc operation at full energy and the addition of two damping magnets to redistribute the radial damping so that both betatron and synchrotron oscillations were damped. In particle physics the CEA measured that the R cross section ratio rose at higher center of mass energy, hinting at many discoveries to come later at other colliders.

10.3.3.8 SPEAR

SPEAR at SLAC was designed as a two ring collider but due to funding only one ring was built. Injection was at full energy from the SLAC 2-mile linac with a positron source in linac sector 11 [97]. The ring lattice was extremely flexible in the choice of operating tunes, dispersion, and beta values the IPs. Transverse horizontal and vertical instabilities (head-tail) were observed at about 0.5 mA per bunch which were cured by a positive chromaticity. The luminosity varied as the fourth power of the beam energy. SPEAR was very productive in particle physics with the discovery of the J/Psi and tau. SPEAR has been upgraded to a medium emittance light source.

10.3.3.9 VEPP-2M

VEPP-2M at BINP reached a luminosity 100 times VEPP-2. VEPP-2 served for a while as the injector [93, 98]. The ring operated with a superconducting wiggler with an 8 T field that was used to increase the radial emittance, decrease the damping time, increase the beam-beam tune shift for a higher luminosity, and also for suppression of intra-beam scattering. The collider could generate and use polarized beams. The particle physics results are many. Round beams have been studied at VEPP-2M to try to reduce beam-beam effects and, thus, increase the allowed beam-beam tune shift. Round beams required the use of solenoidal focusing at the IPs.

10.3.3.10 DORIS

DORIS at DESY started as a two ring collider with beams brought into collision with vertical dipoles in the IR and had a vertical crossing angle [99]. Injection was from a linac and a booster synchrotron. The single bunch currents in the collider were limited by higher parasitic modes in the RF cavities. The luminosity was limited by effects of the vertical crossing angle. DORIS was subsequently converted to a single ring collider with head-on collisions and went on to do many years of B meson physics.

10.3.3.11 DCI

DCI at Orsay was a two ring collider designed for high energy physics studies of charged and neutral particles and also of two photon physics [100]. The two rings were mounted one above the other and the beams were brought into collision with vertical bends near the IR. Both rings could store both e^- and e^+. The rings could operate with either two bunch collisions or four bunch collisions (opposite in both rings). The four bunch collisions aimed at charge cancelation of the beam-beam effects allowing higher beam-beam tune shifts and thus higher luminosity. The four beam scheme in reality partially worked but was strongly limited by incoherent and coherent beam-beam modes. This topic of four beam collisions has been theoretically studied for many years since. The DCI collider mostly operated in a one bunch mode in each ring (out of time) generating twice the luminosity. The peak luminosity scaled as the energy squared.

10.3.3.12 PETRA

PETRA at DESY had four interaction points and operated up to an energy of 24 GeV per beam with 2 × 2 bunches [101]. Additional seven cell RF cavities were installed over the years to achieve these energies. Second harmonic cavities (1 GHz) were installed to reduce the bunch length, cure a vertical single bunch instability and reduce several synchrotron-betatron resonances. The free space for the detector was reduced to ±4.45 m using mini-beta insertions allowing a vertical beta of 6 cm. The injected positrons were predamped in the accumulator storage ring PIA. PETRA is known for the discovery of the Gluon and for QCD studies. PETRA has been upgraded to a low emittance light source.

10.3.3.13 CESR

CESR at Cornell operated for B-meson studies for 20 years [102, 103]. The particle physics results included V_{ub} observations of "penguin" modes, $\mathrm{b} \rightarrow \mathrm{s}\gamma$ decays, CKM matrix constraining the unitarity triangle, and B mass and lifetime measurements. Injection was from a 200 MeV linac and a 12 GeV synchrotron. With electrostatic plates installed, CESR could collide up to 27 bunches separated in the accelerator arcs by what is now called a "Pretzel orbit" that was used to suppress parasitic beam-beam collisions and the related tune shifts. Several other colliders went on to use this technique to increase the number of bunches. It was discovered at CESR that a horizontal tune just above the half integer (<0.51) increased significantly the beam-beam limit allowing higher luminosity. Superconducting single-cell cavities with HOM damping were installed in CESR allowing up to 325 mA beams each of electrons and positron to the stored. The CESR interaction region used a combination of permanent-magnet and superconducting technologies for the vertically focusing quadrupoles. A ±2.5 mrad uncompensated IP crossing angle was ultimately used. Superconducting wigglers were later installed to allow operation at lower energy at the charm threshold. CESR is presently being used as a light source and as a test accelerator to study low emittance damping rings and electron cloud physics.

10.3.3.14 VEPP-4

VEPP-4 at BINP has been colliding beams for a long time including two modernizations [104]. Over the years, the vertical beta at the IP was reduced to 5 cm, superconducting wigglers were added to increase the luminosity, and a superconducting RF cavity was installed for bunch length reduction allowing the use of the lower IP beta. Bunch currents were limited by beam induced wakefields in the vacuum chambers at the level of 17 mA. Eight pairs of electrostatic separation plates allow two bunch operation in a Pretzel scheme. A transversely polarized beam could be injected into VEPP-4 from VEPP-3 for accurate measurements of the masses of the upsilon family particles. Other particle physics studied at VEPP-4M included precise tau and J/Psi mass measurements and two photon physics.

10.3.3.15 PEP

PEP at SLAC operated with three bunches per beam up to 14 GeV delivered to six interaction points [105]. PEP had aluminum vacuum chambers and aluminum RF cavities coated with TiN to suppress breakdowns. The head-tail microwave beam instability was studied extensively at PEP and cures investigated. Collisions with 1, 2, 4, and 6 IPs allowed studies of the beam-beam limits with different damping times per collision. The particle physics results at PEP included the first measurements of the tau lepton lifetime, the discovery that the B meson lifetime was unexpectedly long, analysis of jet structures, and the measurements of lifetimes and properties of charm and bottom hadrons.

10.3.3.16 Tristan

Tristan at KEK collided 2 $\times$ 2 bunches in four interaction points and was designed to search for high mass resonances [106]. The injector was a 2.5 GeV linac and a 377 m accumulator ring ramped to the 8 GeV injection energy. Tristan was the first large accelerator to use extensive superconducting RF technology.

10.3.3.17 SLC

The SLAC Linear Collider SLC was the first (and so far only) linear collider constructed. It was built to precisely measure the properties of the Z^0 meson. It was the first to measure the width of the Z^0 indicating only three families of light quarks and neutrinos. It also provided a precise indirect constraint on the Higgs mass [107]. The SLC collided single e^+ and e^- bunches at 120 Hz with 80% longitudinal e^- polarization at the IP coming from a polarized strained GaAs photo-gun. Other accelerator advances include BNS emittance damping in the linac, reduced emittances from e^- and e^+ damping rings, pulse-by-pulse IP position feedback, and a positron source and target with one-to-one e^- in to e^+ out conversion rate. About 10^{13} positrons were made and collided per second.

10.3.3.18 BEPC

BEPC at IHEP was built to produce tau and charm particle physics [108]. BEPC was a single ring collider with two collision points reaching 2.5 GeV per beam. The single bunch current was up to 22 mA and 140 mA in multi-bunch mode. Injection was from a full energy linac with two transport lines. A mini-beta optics at the IP using permanent magnet quadrupoles (0.5 m long) was installed reaching a vertical beta of 8.5 cm. Higher RF voltage was used to reduce the bunch length to match. BEPC measured precisely the tau lepton mass to 0.2 MeV out of a mass of 1777 MeV.

10.3.3.19 LEP

LEP at CERN has four collision points and is the largest (27 km) and highest energy (104.5 GeV per beam) collider built to date [109]. The particle physics completed at LEP included precise measurements of Z and W bosons, determination of the number of light neutrinos to be three, and exclusion of the Higgs mass below 114 GeV. To reduce power usage five cell RF cavities with attached spherical copper storage cavities were used to reach about 80 GeV. To reach 104.5 GeV additional superconducting RF cavities were added incorporating sputtered Nb on Cu surfaces. The bending dipoles in LEP needed only a low field so the steel laminations were spaced by concrete filler material. At the Z resonance the luminosity was limited by the beam-beam effect. At higher energies the luminosity limit was the available RF power. A Pretzel orbit scheme (8 and later 12 bunches) was used to increase the luminosity. At high energy, a low emittance optics was implemented. A beam-beam tune shift of about 0.083 as reached at 98 GeV.

10.3.3.20 DAFNE

DAFNE at Frascati was built to make precision measurements of K meson physics. Injection is made with a full energy linac and damping ring [110]. DAFNE operates with damping wigglers to decrease the emittances and shorten the damping time. A crab waist scheme was installed in the IP for increased luminosity and for tests of the SuperB IP concept [111]. Rolled permanent-magnet IP quadrupoles were used to compensate the detector solenoidal field.

10.3.3.21 PEP-II

PEP-II at SLAC was an asymmetric collider with two rings made to measure the properties of the b quark sector, the CP violation in the $B\overline{B}$ system, and to confirm the CKM matrix [112]. Injection was made at full energy at up to 30 Hz from the SLAC 3 km linac with the SLC e^- and e^+ damping rings and either e^- or e^+ accelerated on any given linac pulse. Accelerator advances at PEP-II include head-on asymmetric collisions at one IR, large bore permanent-magnet IP dipole and quadrupoles, local beta beats to correct chromaticity in the IP, fast IP position feedback, and bunch-by-bunch transverse and longitudinal feedbacks. The nearest final focus quadrupoles were inside the detector leaving a free space to the IP of about 0.5 m on each side. PEP-II holds the world's record of stored positrons at

3.2 A and for electrons at 2.1 A. PEP-II was the first collider to allow top-up injection (only a few Hz were needed) to keep the beam currents and luminosity constant, all with full continuous data collection by the particle physics detector.

10.3.3.22 KEKB

KEKB at KEK was an asymmetric two ring collider made to measure CP violation in the $B\overline{B}$ system and to confirm the CKM matrix [113]. Injection was made at full energy with the KEK J-shaped linac with either e^- or e^+ injected at 50 Hz with a several minute switch time between modes. KEKB had a lattice with a $5\pi/2$ phase advance to reduce the emittance well below that of a FODO cell. It had a crossing angle IP, used ARES RF copper cavities with an attached energy storage cell, and superconducting RF cavities to suppress longitudinal modes. Transverse bunch-by-bunch feedbacks were used to suppress instabilities. KEKB was the first collider to use superconducting crab cavities to reduce the effects of crossing angle collisions. KEKB also had top-up injection of both beams. KEKB holds the world's record for highest luminosity at 2.1×10^{34} cm^{-2} s^{-1}.

10.3.3.23 BEPC-II

BEPC-II at IHEP was built to provide tau and charm particle physics as a factory and also have synchrotron radiation production [114, 115]. BEPC-II is a double ring collider with one collision point reaching up to 2.1 GeV per beam with 93 bunches. The RF system has two superconducting single cavities at 500 MHz. The luminosity was recently limited by longitudinal instability from HOMs but was cured by a bunch-by-bunch longitudinal feedback system. BEPC-II has delivered several billion J/Psi events to the BES-III detector.

10.3.3.24 VEPP-2000

VEPP-2000 at BINP is a recent e+e− collider and has a single ring with two detectors and twofold symmetry [116, 117]. The particle physics being done at VEPP-2000 is tau and psi measurements and two gamma-physics. Injection comes from a 900 MeV booster synchrotron with one beam at a time. A round beam concept was applied in the ring design where a particle's angular momentum ($M = xz' - zx' =$ constant) is conserved yielding an enhancement of the beam's dynamic stability even with nonlinear effects of the beam-beam force included. This scheme requires equal emittances, equal small fractional tunes, equal betas at the IP, and no betatron coupling in the collider arcs. In practice with collisions for the detectors, only small adjustments in the tunes are needed to arrive at good luminosity conditions at the beginning of a fill. Observations show that a beam-beam parameter of 0.13 has been achieved and that round beams give a solid luminosity enhancement.

10.3.3.25 SUPERKEKB

SuperKEKB at KEK is an asymmetric two ring collider upgraded from KEKB and made to measure CP violation in the $B\overline{B}$ system to extend the understanding of the CKM matrix [118]. This collider has a 7.0 GeV High Energy Ring HER for e− and a 4.0 GeV Low Energy Ring LER for e+. The luminosity goal is 8×10^{35} cm^{-2} s^{-1}. The design beam currents will be doubled from KEK to 2.6 A on 3.6 A. The nano-beam scheme for increased luminosity decreases the overlapping length of colliding particles to about 0.25 mm with a 41.4 mrad half crossing angle [111]. The interaction region beta functions will be about 30/0.3 mm (h/v) with a bunch length of about 5.5 mm. The two rings without the interaction region were commissioning in 2016 where 1.01 A of e− were successfully stored in the HER and 0.87 A of e+ stored in the LER. The installation of the interaction region will be completed in early 2018 when luminosity commissioning will begin.

10.4 Asymmetric B-Factories

K. Oide

10.4.1 Physics Motivation

The idea of asymmetric B-factories was first introduced by P. Oddone in 1987 [119] to collide e^+e^- beams with different energies to measure the CP-asymmetry between the decay of B_0 and $\overline{B}_0$ mesons. The asymmetry of the energies of two beams boosts the generated particles longitudinally, then the difference of the decay time can be measured by the difference of the vertices, which was expected to be in about an order of 100 μm. The center-of-mass energy of the collision is set to the $\Upsilon(4S)$ resonance at 10.58 GeV. A very high luminosity around 10^{34} $cm^{-2}s^{-1}$ was required, which was more than 100 times higher than what had been achieved in colliders by that time.

10.4.2 Double Ring Collider

There may be several ways to realize the asymmetric collision. One way is to build a linear–linear or a ring–linear collider. Such a linear machine needs a very strong focusing $\beta_y^* \sim 100\ \mu$m to achieve the luminosity, then the bunch length must be as short as β_y^* to avoid the hour-glass effect. The bunch length itself can be obtained by bunch compressors, but the associated energy spread degrades the effective luminosity, since the width of the resonance $\Upsilon(4S)$ is only 20 MeV (2×10^{-4}). A huge damping ring would be necessary to realize such a short bunch length and a small energy spread simultaneously. Thus linear collision schemes seemed difficult.

As for the double-ring collision, a question is the sizes of the rings. If one can collide a large high energy ring (HER), for instance at 25 GeV, with a small low energy ring (LER) at 1.2 GeV, the total cost will be saved, assuming an existing tunnel for such a high energy ring. It was pointed out [120] that the collision of rings with different circumferences has somewhat fundamental difficulty: if two rings have the ratio of circumferences $m : n (m > n)$, the periodicity of the system becomes very long, *i.e.*, LCM$(m, n)/m$ times the revolution period of the larger ring. Then both rings will have dense resonance lines in the tune space which reduces the operable area, especially with a certain amount of the beam-beam tune shift. Thus collision of rings with different circumferences seemed difficult. Therefore only the double ring collider scheme with equal cicumferences remained.

Two projects of the asymmetric B-factories, PEP–II [121] at SLAC and KEKB [122] at KEK, were approved and started the construction by 1994. Both projects utilized the components and facilities of their previous generation colliders, PEP and TRISTAN, and built the BaBar and Belle detectors, respectively. Both machines started the collision experiments in 1999 and stopped the operation in April 2008 (PEP–II) and June 2010 (KEKB). Table 10.12 lists the main machine parameters corresponding to their best records [123, 124]. Both colliders achieved higher performance than their designs, and experimentally verified the Kobayashi–Maskawa model to bring the 2008 Nobel Prize in Physics.

10.4.3 Luminosity

The luminosity $\mathscr{L}$ of an asymmetric ring collider can be expressed by the following expression:

$$\mathcal{L} = \frac{\gamma_\pm}{2er_e}\left(1 + \frac{\sigma_y^*}{\sigma_x^*}\right)\left(\frac{I\xi_y}{\beta_y^*}\right)_\pm \left(\frac{R_\mathcal{L}}{R_y}\right) , \tag{10.1}$$

where γ, e, r_e, $\sigma_{x,y}^*$, I, $\beta_{x,y}^*$ are the Lorentz factor, electron charge, classical electron radius, beam sizes at the interaction point (IP), stored beam current in the ring, and the β-function at the IP, respectively. The suffix $\pm$ denotes each beam. The expression (10.1) is obtained from the beam-beam tune-shift parameter

$$\xi_{\pm x,y} = \frac{r_e}{2\pi\gamma_\pm}\frac{N_\mp\beta_{\pm x,y}^*}{\sigma_{x,y}^*\left(\sigma_x^* + \sigma_y^*\right)}R_{x,y} \tag{10.2}$$

Table 10.12 Progress of machine parameters of the PEP–II and KEKB B-factories

	PEP–II		KEKB (no crab)		KEKB (crab)	
	8/16/2006		11/15/2006		6/17/2009	
Date	LER	HER	LER	HER	LER	HER
Circumference [m]	2200		3016			
Beam energy [GeV]	3.1	9.0	3.5	8.0	3.5	8.0
Effective crossing angle [mrad]	0		22		0 (crab)	
Beam current [A]	2.90	1.88	1.65	1.33	1.64	1.19
Bunches	1722		1389		1584	
Bunch current [mA]	4.02	1.09	1.19	0.96	1.03	0.71
Bunch spacing [m]	1.2		1.8–2.4		1.8	
Horizontal emittance ε_x [nm]	30	50	18	24	18	24
RF frequency [MHz]	476		509			
Bunch length σ_z [mm]	10	10	8	6	8	6
β_x^* [cm]	30	30	59	56	120	120
β_y^* [cm]	0.9	1.1	0.65	0.59	0.59	0.59
Horizontal size @ IP [μm]	95	158	103	116	147	170
Vertical size @ IP [μm]	4.7	4.7	1.9	1.9	0.94	0.94
Beam-beam ξ_x	0.072	0.064	0.115	0.075	0.125	0.100
Beam-beam ξ_y	0.064	0.053	0.104	0.058	0.130	0.090
Luminosity [$\mathrm{nb}^{-1}\mathrm{s}^{-1}$]	12.1		17.6		21.1	
Integrated luminosity/day [pb^{-1}]	858		1260		1479	
Integrated luminosity/7 days [fb^{-1}]	5.41		7.82		8.43	
Integrated luminosity/30 days [fb^{-1}]	19.8		30.2		23.0	
Total integrated luminosity [fb^{-1}]	557		1040			

The left, center, right correspond to the highest performance of PEP–II, KEKB (no crab) and KEKB (crab), respectively. The integrated luminosities are the delivered numbers for PEP–II, and recorded for KEKB. 1 $\mathrm{nb}^{-1} = 10^{33}\ \mathrm{cm}^{-2}\mathrm{s}^{-1}$

and the definition of luminosity

$$\mathcal{L} = \frac{N_+ N_- f}{4\pi \sigma_x^* \sigma_y^*} R_{\mathcal{L}} \tag{10.3}$$

where N and f are the number of particles per bunch and the collision frequency ($I = Nef$), respectively, and we have assumed the beam sizes are common in two beams. The factors $R_{\mathcal{L},x,y}$ are the geometric reduction factors due to the hour-glass effect and the crossing angle.

While a round-beam scheme may have a merit of a factor of 2 on the luminosity according to Eq. (10.1), a flat beam scheme has been chosen in most e^+e^- colliders, as the round-beam focusing in both planes is more difficult for an extremely small β_*. For a flat beam, $\sigma_x^* \gg \sigma_y^*$, the luminosity is written as

$$\mathcal{L} \approx \frac{1}{2er_e} \left(\frac{\gamma I \xi_y}{\beta_y^*} \right)_{\pm} \left(\frac{R_{\mathcal{L}}}{R_y} \right) . \tag{10.4}$$

Then if there is no reason to differentiate ξ_y and β_y^* in the two rings,

$$\gamma_+ I_+ = \gamma_- I_- \tag{10.5}$$

is resulted. As the ratio of beam energies gets larger, the boost at the collision becomes larger, but the low energy ring must store higher beam current. Thus the energy ratio was a compromise between the physics merit and the accelerator difficulty. PEP–II chose 3.1 GeV and 9 GeV for positrons and electrons, while KEKB chose 3.5 GeV and 8 GeV. A larger ratio was more favored at PEP–II as it needs a magnetic separation of two beams at the IP as described later. The flavor of beams, the LER for positrons, was uniquely chosen at KEKB, where the positron acceleration for the HER was very difficult.

The actual operation of these machines, the condition (10.5) was not kept strictly, as shown in Table 10.12. One reason was that the natural size of each beam was not equal, for instance, the LER positron beam was relatively easy to be blown up due to the electron clouds at high current. Then there was a certain limit on the positron beam current and the HER current was increased beyond Eq. (10.5). This tendency was stronger in KEKB than PEP–II, as the former had stronger electron cloud effects than the latter as described later.

10.4.4 Crossing Angle

One of the design choices is the beam separation scheme near the IP. A crossing angle is a natural and easy solution of the separation, but the question was the experience at DORIS [125]. KEKB applied a horizontal crossing angle $2\theta_x = 22$ mrad, relying on simulation of the beam-beam effect. The corresponding Piwinski angle ($\equiv \theta_x \sigma_z / \sigma_x^*$) was 0.86. Their conclusion at the design stage was that the effect of the crossing angle on the beam-beam interaction would not be harmful up to their design beam-beam parameter 0.05, if the operating betatron tunes were carefully chosen. Their choice was right and verified the vertical beam-beam parameter of 0.06 in their luminosity marching. Crossing angles were also applied at CESR and DAΦNE colliders successfully in parallel with the KEKB operation. KEKB even prepared a crab-crossing scheme [126, 127] for the backup of the crossing angle. The ratio of the geometric reduction factors R_L/R_y in Eq. (10.1) does not drastically decrease for a large crossing angle as shown in [122].

PEP–II was much more nervous on the crossing angle, then they installed a magnetic separation scheme near the IP with permanent dipole magnets [128]. This scheme also worked, but their design around the IP had to be more complicated than with a crossing angle, and gave some limitations on the performance such as the detector background due to radiative Bhabha events [129], which was much less significant in Belle. As the space at the IP was limited, they could not install a compensation system for the detector solenoid field, which might have degraded the beam-optical performance. Another issue of the magnetic separation was the non-negligible effect due to the parasitic collision [130], which was never observed at KEKB.

10.4.5 Storing High Current

As described above, the luminosity is proportional to the stored current. To achieve the luminosity as high as $10^{34}\text{cm}^{-2}\text{s}^{-1}$, a stored current of near 3 A was required, which was one order higher than any high energy electron storage rings at that time. The first fundamental difficulty is to ensure the longitudinal stability of the beam.

The beam loading of the accelerating cavity is huge: a normal conducting cavity at the RF frequency $f_{\text{RF}} = 500$ MHz for the B-factories has a shunt impedance $R_s \approx 1.7\,\text{M}\Omega$. If the cavity is tuned at the harmonics, the 3 A beam generates 5.1 MV decelerating voltage at the cavity, which is even higher than the accelerating voltage V_c of the cavity, typically 0.5 MV. Thus the detuning of the cavity is necessary and the optimal amount of the detuning frequency is given by

$$\Delta f = -\frac{I \sin\phi_s}{2V_c}\frac{R_s}{Q} f_{\text{rf}} = -\frac{P_b \tan\phi_s}{4\pi U}\,, \tag{10.6}$$

where ϕ_s, Q, P_b, U are the synchronous phase, the Q-value, the beam power, and the stored energy of the cavity, respectively. If the magnitude of the detuning frequency becomes higher than or comparable to the revolution frequency, the cavity impedance hits the side bands of synchrotron motion to excite strong longitudinal coupled-bunch instabilities.

This issue of the beam-loading instability was solved in two B-factories in different ways. KEKB developed two types cavities with large stored energy, as Eq. (10.6) is inversely proportional to the stored energy. Both ARES [131] and superconducting [132] cavities could store electromagnetic energy 10 times larger than a conventional cavity. Then together with the HOM damping mechanism of them, the RF system of KEKB did not induce any beam instability up to the design current without a help of bunch-by-much feedback system. On the other hand, PEP–II took a different strategy to develop a sophisticated feedback system to reduce the effective impedance seen by the beam [133]. PEP–II applied a direct RF feedback system with newly developed sideband klystrons combing a longitudinal bunch-by-bunch feedback [134]. Both KEKB and PEP–II systems basically worked as expected nearly up to or even beyond their design currents.

Storing high currents caused a number of issues on the beam pipes, bellows, collimators, and even on the detectors. Direct hit of the beam of an ampere caused by beam instability or anything else easily melted down such components. The wakes at transitions resulted in discharge and heating to drive the catastrophe. A number of models have been developed and tried for the collimators, bellows, and HOM absorbers. Also machine protection system, loss monitors, and beam abort system had to evolve as the stored current increased.

10.4.6 Electron Cloud

Electron cloud was the one of the toughest issues for the asymmetric B-factories, specifically on the accumulation of the positron beam. The electron cloud had been known as a possible cause of beam instability in positive-charged beams since a long time ago such as the ISR era. Its observation [135] had been made at the Photon Factory (PF) of KEK and a theoretical explanation [136] had been done well before the start of the B-factories. What was new at the B-factories was the single-bunch instability induced by electron clouds [137]. The previous instability observed at the PF had been interpreted as a coupled-bunch instability, which was supposed to be cured by a bunch-by-bunch feedback. Thus at least KEKB was not well prepared for the single-bunch phenomena which have much higher frequency than the available feedback. Actually possibility of such a single-bunch effect had been suggested [138] before the construction of the B-factories, it had not been, however, well recognized. The single-bunch effect was experimentally confirmed at KEKB [139] as well as at CesrTA.

The electron cloud blowed up the vertical beam size drastically, and the threshold beam current was 0.4 A with four-bucket (2.4 m) bunch spacing at KEKB. The electron cloud appeared more severely in KEKB than in PEP–II, as the former had a round Cu beam pipe while the latter an Al antechamber with TiN coating. Thus the initial startup of the luminosity at KEKB was slower than PEP–II.

By applying weak magnetic field at the beam pipe, the electron cloud was removed at least in the free space. Either permanent magnets or solenoids were installed at KEKB and PEP–II to cover almost all straight sections and inside of some magnets such as quadrupoles and weak dipoles by 2004. The mitigation worked as expected and the blowup became unnoticeable at least for three-bucket spacing in the case of KEKB [140]. Beside the magnetic field, various mitigation techniques have been developed and tested at the B-factories, against the formation of the electron cloud, including antechambers [141], TiN or Diamond-like carbon coatings, grooved surface pipes [142], and clearing electrodes [143]. Those techniques will be effective for future super B-factories and damping rings of linear colliders. Also several measurements of the cloud density have been carried out.

Although the density of the electron cloud became below the instability threshold by magnetic field, the betatron tune shift due to the cloud still remained in the LER at KEKB to make the tune variation along the bunch train. A mitigation for the tune variation was making use of pulsed quadrupoles as done at KEKB [144].

10.4.7 Beam Optics

The luminosity of a ring collider is inversely proportional to the vertical β-function at the IP as shown in Eq. (10.4). The B-factories have used the smallest β_y^* as a ring collider so far. Generally speaking, a small β_y^* means higher chromaticity and higher nonlinearity arisen from sextupoles for the chromaticity correction. Thus the design of the ring lattice needs special care to ensure the dynamic aperture. One technique applied for KEKB was non-interleaved sextupole pairs separated by a $-I$ transformation that cancel the geometric nonlinearity of the sextupoles up to the second order [145]. Although the idea was very old but the reason why the application to a real ring had to wait until the B-factories was probably the necessary computer power to optimize the sextupole setting, as it requires a large number of sextupole families to extend the momentum acceptance. For instance, KEKB had 54 families of sextupole pairs. The relative betatron phase advance between the pairs became adequate by using the 2.5π cell structure in the case of KEKB arc section [146].

Another technique to enlarge the dynamic aperture was to place a special chromaticity correction section near the IP. The beam optics became somewhat similar to that for linear colliders in this case. KEKB designed such a section for the vertical correction, while PEP–II horizontal for their LERs.

These schemes worked as expected for the B-factories, and expected to work for future super B-factories and light sources. Once the chromaticity correction is solved, the next source of the nonlinearity is the fringe field of the final quadrupole and the geometric nonlinearity at the IP [147], which may be mitigated by additional octupoles placed at the final quadrupoles.

The x-y coupling and the residual vertical dispersion all over the ring were one of the keys to achieve a high luminosity by reducing the vertical emittance. Various techniques have been applied for such optics measurements and corrections [148–150]. A counter solenoid to the detector solenoid was also effective to reduce the coupling source in the case of KEKB. This was also important in the case of crab crossing where the luminosity performance was sensitive to the chromatic x-y coupling as described later.

10.4.8 Beam Diagnostics and Control

A number of beam diagnostic methods were developed and applied for the B-factories:

- Beam position monitors (BPMs) with a resolution better than 1 μm in the average mode. In some cases turn-by-turn or bunch-by-bunch electronics were equipped [151]. In the case of KEKB, the gain imbalance of the electrodes through the electronics was calibrated using a beam-mapping technique [152]. The design of the electrodes and the electronics were carefully done for the high-current operation.

- Beam-based alignment of BPMs was regularly carried out. Displacements of BPMs near sextupoles, caused by heating from the stored current, were monitored at KEKB [153].
- Bunch-by-bunch feedback systems were installed both in PEP–II and KEKB. Only PEP–II had the longitudinal system to suppress the beam-loading instability as described above. A collaboration including the DAΦNE team has been developed on the system for the present and future applications [154].
- Betatron tune monitor: controlling the betatron tune was extremely important to maximize the luminosity. The basic idea at KEKB was to monitor the tunes of pilot bunches in each ring that did not collide to the other beam. Tune feedback with these bunches was also applied to control them within an accuracy of $\Delta\nu \approx 10^{-4}$.
- Synchrotron radiation beam profile monitors. For the visible light, an interferometer was used especially for the vertical size measurement [155]. Special gated cameras were also used to observe the beam size of individual bunch, esp. to diagnose the electron-cloud effects [156].
- Beam loss monitors and beam abort system: both machines were very nervous to protect the machine against accidental beam losses caused by instabilities, RF trips, wrong injection, or whatever. The most sensitive and expensive loss monitor was the BaBar and Belle detectors, which generated beam abort signals if necessary. A number of beam loss monitors such as ionization chambers and PiN diodes were distributed around the ring, esp. near the collimators. The beam abort system consists of an abort kicker and a beam damp. The abort kicker had a rise time of 0.5 μs in the case of KEKB.
- The injectors had developed their own diagnostics including BPMs, wire scanners, streak cameras, etc.
- All accelerator components were controlled by computer control systems either by EPICS at KEKB [157] or a legacy system at PEP–II. An online modeling such as SAD for KEKB [149] was also important to achieve the luminosity.

10.4.9 Collision Tuning

Starting up the colliders after a period of long shut down, the following procedures were necessary to recover the luminosity:

- Global coupling/dispersion/β-function correction all over the ring. The global orbit was then locked to the "golden" orbit that was resulted by the optics correction.
- Locking the betatron tunes of the pilot bunches.
- The beam steering at the IP looking at the beam-beam deflection.

- In the case of the crossing angle at KEKB, the horizontal offset at the IP was controlled by looking at the vertical beam size measured by the interferometer [158].
- Tuning of the local coupling and dispersion at the IP by making offsets of orbits at sextupoles near the IP [149].
- Dithering technique was used at PEP–II to maximize the luminosity against the beam offsets [159].
- Skew sextupoles were introduced at KEKB to correct the local chromatic *x*-*y* coupling terms at the IP [124].

The horizontal tunes were chosen as close to a half integer as possible, to maximize the luminosity using the dynamic-β effect and expecting the reduction of the degree-of-freedom of the beam-beam interaction [160]. In the case of KEKB, the LER and the HER was operated at $\nu_x \approx 0.506$ and $\nu_x \approx 0.510$, respectively. Both the optics correction and the tune feedback were necessary to maintain a collision near the stop band.

10.4.10 Injector

The electron-positron injector must provide enough number of charges to the collider rings. PEP–II could fully utilized the injection system for SLC, which had more than enough performance for PEP–II, in the intensity, repetition, and the emittance, especially with the damping rings for both beams. On the other hand, the injector for KEKB was upgraded from that for TRISTAN, having only the minimum performance to satisfy the requirements of the injection to KEKB as shown in Table 10.13. In early days, it was thought that the performances of the two machines would be eventually limited by the performance of each injector. Actually such a situation did not happen. The key was the top-up operation applied for the both machines since 2004. Then the necessary strength of the injected beam became much smaller than the maximum performance even at KEKB [161]. Both machines were the first to have utilized the top-up for high-energy colliders, even earlier than the most of light sources. The 2-bunch per pulse acceleration and installing a C-band section in the linac [162] also contributed to make the gap between PEP–II

Table 10.13 Comparison of positron injection

	KEKB		PEP–II
	1999	2010	
Production energy [GeV]	4		30
Particles per pulse [10^{10}]	0.4	1.0	2
Repetition rate [Hz]	50		≤120
Invariant emittances H/V [μm]	~3000/3000		3/0.3
e^+/e^- switching time [s]	300	0.02	0 (simultaneous)

and KEKB smaller. KEKB has solved the conflict with the injection to the light sources by introducing a pulse-to-pulse switching of the linac.

10.4.11 Crab Crossing

KEKB operated with crab crossing from 2007 through 2010 using superconducting crab cavities [163] installed one cavity per ring [164]. KEKB had already achieved a luminosity 1.76×10^{34} cm^{-2}s^{-1} by then with the crossing angle, so the intension of the crab crossing was not simply the backup. Simulations of beam-beam effect indicated that a head-on or crab collision could increase the beam-beam parameter ξ_y even higher than 0.15 combining with a horizontal betatron tune close to a half integer [165]. Thus the hope was to experimentally verify the possibility of such a high beam-beam parameter, considering super B-factories.

The crab cavities were successfully installed and operated for more than 3 years. The luminosity was actually increased as shown in Table 10.12. The resulted beam-beam parameter was 0.09, which was indeed higher than the value with crossing angle, but much less than the simulation (0.15). It was not easy to single out the cause, but there were indications that remaining higher-order terms of the beam optics at the IP degraded the luminosity and ξ_y. One example was the chromatic x-y coupling terms at the IP [166]. By installing skew sextupoles in the arc as the tuning knobs, the luminosity was improved by up to 10% [124]. Then a speculation was that any higher order terms at the IP could degrade the performance. It was not easy to estimate how many terms were relevant and how to correct them, as there were almost no direct beam diagnostics at the IP except the luminosity. What was verified at KEKB was that the crab crossing itself should work for any colliders up to $\xi_{x,y} \lesssim 0.1$, including the Large Hadron Collider (LHC).

10.4.12 SuperKEKB

Although the crab crossing scheme tested at KEKB achieved then-highest luminosity, it also showed the difficulty to realize the very high beam-beam parameter $\gg$0.1. Thus the direction toward a higher luminosity asymmetric collider must change. An alternative idea, *Nano-beam scheme*, developed by P. Raimondi [167] saved the next generation of KEKB. The idea by Raimondi consists of:

- A large crossing angle, in terms of *Piwinski angle* $\theta_x\sigma_z/\sigma_x^* \gg 1$, where θ_x is the half horizontal crossing angle.
- A very short $\beta_y^* \sim \sigma_x^*/\theta_x \ll \sigma_z$.
- Small horizontal/vertical emittances.

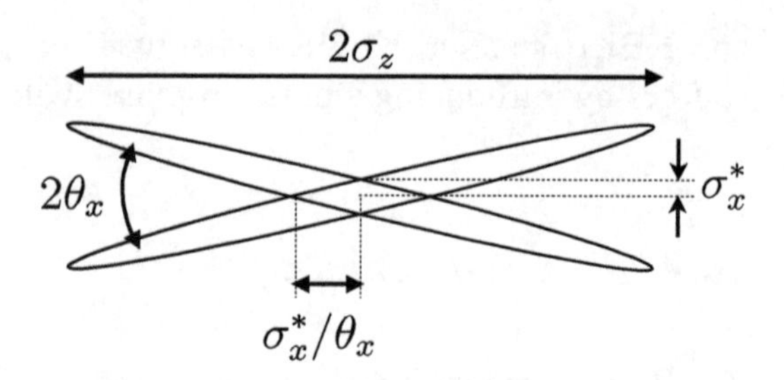

Fig. 10.13 Beam crossing in the nano-beam scheme

The formula for the luminosity, Eq. (10.1) is still valid in the nano-beam scheme. The major gain of the luminosity comes from a very short β_y^*. Unlike a head-on scheme, the bunches intersect one another only within σ_x^*/θ_x around the IP due to the large Piwinski angle as shown in Fig. 10.13.

As the length of the intersecting region is $\sigma_x^*/\theta_x \ll \sigma_z$, the condition to avoid an hour-glass effect becomes $\beta_y^* \lesssim \sigma_x^*/\theta_x \ll \sigma_z$. In the case of SuperKEKB, it is possible to choose $\beta_y^* \sim 0.3$ mm, which gives $\times 20$ gain compared to KEKB. Then the bunch length σ_z is not necessary to be very short, thus it is possible to avoid unfavorable effects such as the coherent synchrotron radiation. This scheme does not require a very high ξ_y, so $\xi_y \sim 0.09$, which has been achieved at KEKB, is assumed at SuperKEKB. The crossing angle itself can be larger than the previous machines, since it does not need crab crossing any more. As it does not require an operation very close to a half-integer tune, and also the horizontal beam-beam parameter is very small, the dynamic emittance effect is not an issue. The situation of the nano-beam crossing is similar to a collision with many micro-bunches which have a short bunch length $\sigma_x^*/\theta_x \ll \sigma_z$.

Actually Raimondi has proposed one more important idea, *crabbed waist scheme*, on top of the nano-beam scheme [167]. The crabbed waist aligns the vertical waist along the center line of the other beam, to reduce the dependence of the beam-beam effect on the horizontal displacement of a particle. This scheme should improve ξ_y by reducing the synchrotron-betatron coupling caused by the crossing angle. Although the crabbed waist scheme has merits on the collision itself, its realization needs further study. A simple way to introduce the crabbed waist is to install a pair of sextupoles in both sides of the IP. The nonlinearities of these sextupoles can be canceled by $-I$ or I transformation between the pair, but the unavoidable nonlinearities around the IP interfere the cancellation to reduce the dynamic aperture drastically in the case of SuperKEKB. Such nonlinearities include the fringe field of the final quadrupoles, geometric nonlinearities at the IP, and nonlinear fields in the quadrupoles and solenoids. As these terms increase for smaller β_y^*, the solution may be non-trivial and has not been found at least for SuperKEKB yet. Thus the crabbed waist scheme is not included in the base line design at SuperKEKB.

SuperKEKB started the beam operation in early 2016. The commissioning has been carried out in three phases: no collision (Phase 1, February–June 2016), collision without central vertex detector of Belle II (Phase 2, March–July 2018) [168], and collision with full Belle II (Phase 3, March 2019–). The design parameters [169]

Table 10.14 Parameters of SuperKEKB, design and typical values in phase 2, comparing to KEKB with crab crossing, where the effective crossing angle is zero

	SuperKEKB		SuperKEKB		KEKB (crab)		
	Design		June 2018		June 17, 2009		
Date	LER	HER	LER	HER	LER	HER	
Circumference	3016						m
Beam energy	4.0	7.0	4.0	7.0	3.5	8.0	GeV
Crossing angle	83				22/0 (crab)		mrad
Beam current	3.8	2.6	0.27	0.225	1.64	1.19	A
Bunches	2500		395		1584		
Bunch current	1.5	1.0	0.67	0.55	1.03	0.71	mA
Bunch spacing	1.2		~7.2		1.8		m
Hor. emittance	3.2	4.6	1.8	4.6	18	24	nm
RF frequency	509						MHz
Bunch length σ_z	6	5	4.6	5.3	8	6	mm
β_x^*	3.2	2.5	20	10	120	120	cm
β_y^*	0.27	0.3	3	3	5.9	5.9	mm
Hor. size @ IP	10	11	19	21	147	170	μm
Ver. size @ IP	0.048	0.048	0.56	0.56	0.94	0.94	μm
Piwinski Angle	24.9	18.9	10.0	10.5	(0.60)	(0.39)	
σ_x^*/θ_x	0.24	0.27	0.46	0.51	(13.3)	(15.4)	mm
Beam-beam ξ_x	0.0028	0.0012			0.125	0.100	
Beam-beam ξ_y	0.088	0.081	0.030	0.021	0.130	0.090	
Luminosity	800		2.3		21.1		/nb/s

1/nb = 10^{33} cm^{-2}s^{-1}

and a typical performance of Phase 2 operation [170] of SuperKEKB are listed in Table 10.14.

The commissioning in Phase 2 has achieved several milestones of the project:

- Verified the collision with nano beam scheme with Piwinski angle ~10. Although the luminosity was still much less than the design or achieved at KEKB, no fundamental limitations are found. The highest luminosity 56/nb/s was recorded during Phase 2 [170]. An experimental verification of coherent beam-beam instability [171] also assures the understanding of the beam-beam effect.
- The upgrade of the injector several new components: RF gun, positron target, and positron damping ring was basically successful [172].
- The mitigation of e-cloud in the LER has well suppressed the blowup of the vertical beam size up to 1 A of the stored beam [173].

The key toward the design luminosity is higher stored current with smaller β*s. Both the stored currents and β*s are still far from the design. Table 10.14 shows that the achieved β_y^* is much longer than the length of the interaction area, σ_x^*/θ_x. It means that the merit of nano-beam scheme has not been obtained yet. One possible obstacle for higher current and smaller β*s is the robustness of superconducting final focus quadrupoles [174] against quenches due to beam losses. Quenches of those magnets have been seen in the Phase 2 commissioning both for the stored and injected beams.

10.5 Tevatron—HERA—LHC

Karl-Hubert Mess · Peter Schmüser

10.5.1 Three Steps in the Evolution of Superconducting Accelerator Magnets

The first particle accelerator approaching the TeV energy range was the proton-antiproton collider Tevatron at the Fermi National Accelerator Laboratory. Much pioneering work was done at Fermilab, and many of the successful design and construction principles of the Tevatron dipole and quadrupole magnets have been adopted at the electron-proton collider HERA and the Large Hadron Collider LHC (CERN). The superconducting coils used in these accelerators show great similarities. They are wound with high precision from a multistrand cable containing many thousand fine niobium-titanium filaments in a copper matrix. The coils have a helium-transparent insulation and are confined by nonmagnetic clamps (often called *collars*). The clamps are assembled from precision-stamped stainless-steel or aluminum-alloy laminations and serve three purposes: they define the exact coil geometry, they exert a large prestress on the coils to prevent conductor motion during excitation of the magnet, and they take up the huge magnetic forces at large fields. Interesting differences, however, have evolved in the layout of the iron yoke. The Tevatron magnets are "warm-iron" magnets with a yoke at room temperature while the HERA and LHC magnets feature a "cold-iron" yoke inside the liquid helium cryostat.

In the beginning of the HERA project two different design lines were followed. Our group at DESY constructed and built 6 m long prototype dipoles of the "warm-iron" type which were basically copies of the Tevatron dipoles. These magnets were made with tooling suitable for series production and performed very well, the design field of 5.2 T was exceeded and excellent field quality was achieved. In parallel, an industrial company (Brown Bovery in Mannheim) designed and built two "cold-iron" prototype dipoles following a concept developed at Brookhaven National Laboratory. Here the coil was surrounded with epoxy-fibreglass form pieces and then directly mounted in an iron yoke which served as the clamping structure. Also the BBC dipoles showed remarkable performance. They exceeded the design field by an ample margin, however the field quality became poor for fields above 4 T owing to saturation effects in the nearby iron yoke.[2] With the aim in mind of combining the virtues of both design lines, while avoiding the relative drawbacks, the idea was conceived to take the well-proven aluminum-collared coil of the first design line and put an iron yoke immediately around the collars. Our main motivation was magnet safety in case on a quench (breakdown of superconductivity).

Quench protection is an important task in any large superconducting (sc) magnet system. Accelerator magnets have slim coils with a small copper-to-superconductor ratio. They are definitely not cryostable (cryostability means that the copper in the cable is able to carry the full current). If a HERA dipole quenches it is mandatory to ramp down the current of 5000–6000 A with a time constant of less than a second to prevent overheating and possible destruction of the coil. The protection of a single magnet is straightforward: if a quench is detected the coil is connected to a dump resistor via a thyristor switch and the current decays exponentially $I(t) = I_0 \exp(-t/\tau)$. The stored magnetic energy is dissipated in the dump resistor.

In an accelerator this simple solution is not possible since the magnets in a long string are connected in series. In HERA this string is a 45° arc comprising 52 dipoles and 26 quadrupoles. Current leads from the sc coils to the room-temperature environment are only installed at the ends of the string but not at individual magnets. Owing to the large inductivity of the 54-magnet string a long decay time constant $\tau \approx 20$ s must be chosen in order to limit the induced voltage against ground potential to less than 1000 V. The quenched coil would burn up during such a slow current decay, hence an electric bypass must be provided. Here the great advantage of a cold iron yoke comes in: it is easy to provide such a bypass by mounting a superconducting cable, reinforced with a copper bus bar, in a groove at the outer rim of the yoke. The bypass conductor is connected to the main current conductor via a "cold" silicon diode inside the liquid helium cryostat.[3] The diode has a threshold voltage of about 1 V at liquid helium temperature, hence the current flow through the bypass is zero as long as the magnet coil is in the superconducting state (during particle accceleration the ramp speed must be chosen so low that the inductive voltage stays well below the 1 V threshold). However, a quenched coil develops rapidly a resistive voltage exceeding 1 V, and then the main current switches automatically over to the bypass. The Tevatron possesses an active quench protection system which is more complicated and shall not be discussed here.

Potentially dangerous are quenches at localized spots in the coil. If the normal zone does not propagate fast enough along the coil it may happen that the stored magnetic energy is dissipated in the vicinity of the quench origin leading to local overheating. To avoid this, so-called "quench-heaters" are installed which are electrically heated when a quench has been detected. Their task is to spread the normal zone over the entire coil. Figure 10.14 shows that a warm-iron magnet may

---

[2]It has found out later in the LHC and RHIC projects that the detrimental effects of iron saturation can be alleviated by punching a suitable hole pattern into the iron yoke laminations.

[3]The cold diode concept was invented and thoroughly investigated at BNL.

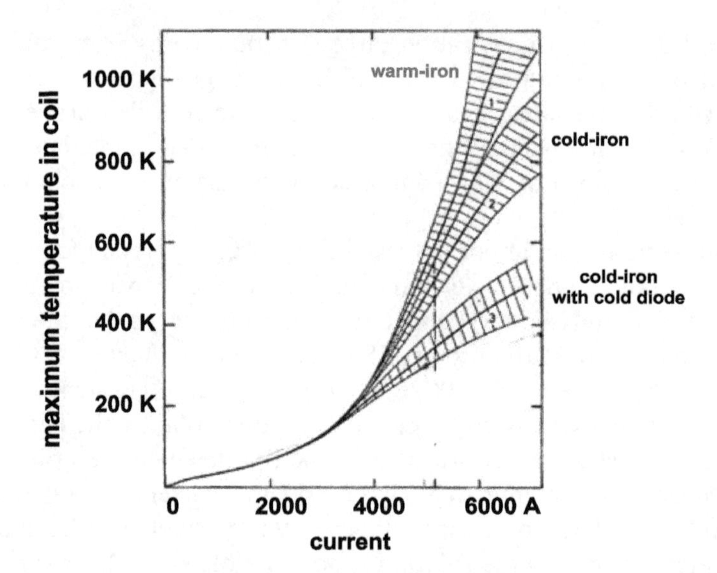

Fig. 10.14 Simulation of temperature rise in a dipole after a quench at a localized spot. Plotted is the maximum temperature in the coil as a function of coil current. It is assumed that the quench heaters fail and that the current is ramped down exponentially with a time constant of 17 s. Three magnets types are considered: the warm-iron HERA prototype dipole, the cold-iron BBC dipole without a cold diode, and the cold-iron BBC dipole with a cold diode. In the first two cases the bypass is provided by two curent leads from the superconducting cable in the liquid helium container to the room-temperature environment, a normal-conducting cable and a thyristor switch outside the cryostat. Adapted from a DESY Internal Note January 1984 by K.H. Mess and P. Schmüser

heat up to quite dangerous temperatures if the quench heaters fail, while the cold-iron magnet equipped with a cold diode and a superconducting bypass remains in a safe temperature regime.

The HERA-type magnet offers several more advantages. The cryostat can be a simple steel vessel which is easier to fabricate and less expensive than the very slim stainless-steel cryostat needed in the warm-iron magnet. The coil must be accurately centered with respect to the yoke to avoid field distortions and asymmetry forces. In the HERA design a precise centering is achieved by a tongue-groove combination in the collar and yoke laminations. In contrast to this, coil centering in a warm iron yoke requires many adjustable supports which constitute a considerable heat load on the cryogenic system.

The successful HERA magnet design has had a considerable impact on the layout of the LHC magnets.[4] The elegant twin-aperture magnet can be realized with this concept. The two counterrotating proton (or heavy ion) beams are guided and focused by two nearby sc coils of opposite polarity which are mounted in a common iron yoke. This yoke can be made rather slim since most the magnetic flux

[4]The evolution of the twin-aperture LHC magnets is described in an article by Lucio Rossi, CERN Courier October 2011.

of one coil returns through the aperture of the neigbouring coil and not through the iron. Another novel feature of the LHC magnets is the cooling by superfluid helium which extends the achievable field into the 9 T regime. A significant cost reduction is achieved by the fact that only one cryostat system is needed for both particle beams. The compactness of the magnets was very important for the installation in the tunnel, it would have been extremely demanding if not impossible to mount two separate proton storage rings in the narrow LEP tunnel.

The elegance and cost-effectiveness of LHC becomes obvious in comparison with the Superconducting Super Collider SSC project which was promoted in the USA in the 1980s. Here two counterrotating 20 TeV proton beams were planned to be accelerated and stored in two separate rings equipped with magnets of an improved Tevatron design. This "conservative" approach turned out far more costly than the innovative LHC solution. The SSC was cancelled in 1993 for budget reasons.

10.5.2 HERA Experience and the Design of Future Lepton-Hadron Colliders

The scientific results from HERA on quantum chromo-dynamics (QCD) are an essential pillar of our present understanding of the nature of strong interactions. These insights are of critical importance for the interpretation of the measurements of particle production processes at the Large Hadron Collider (LHC) at CERN. There are, however, very interesting phenomena and physics questions that were raised by the HERA experiments but remain unanswered as of today:

- How does QCD lead to the rich and complex structure we observe in nuclei?
- How are quarks and gluons and their spins distributed in nucleons and nuclei?
- Why do protons and neutrons have spin 1/2?
- How does the increase of gluon density with decreasing gluon fractional momentum x saturate?
- What is the distribution of partons that seed the new form of matter discovered at RHIC, the quark-gluon plasma?

The U.S. Nuclear physics community has formulated a physics program that addresses these questions in a White Paper [175]. This physics program is based on an Electron Ion Collider (EIC), which can provide collisions between polarized electrons and ions ranging from polarized proton to Uranium over a wide range of center of mass energies 29–140 GeV with a high luminosity in the order of 10^{34} cm^{-2} s^{-1}. There are two designs for an EIC under development in the U.S., one by Thomas Jefferson National Accelerator Facility, called JLEIC [176], and one by Brookhaven National Laboratory, called eRHIC [177]. Both designs are based on intersecting storage rings.

The experience gained at HERA with high luminosity lepton proton collisions, beam dynamics, various enabling technologies, the electron storage ring spin dynamics, and the control of collective effects have a tremendous impact on the design of these machines, in particular, the layout of the interaction regions.

HERA beam dynamics performance is used to benchmark computer codes needed to develop the EIC designs. HERA operational experience is extremely important for designing the complex interaction regions with unprecedented detector acceptance, assessing detector backgrounds induced by the beams and their mitigation. The design of the magnets which were developed for the HERA interaction regions have been carried over for the design of other colliders, such as BEPC-II, KEKB, ILC, LHeC, and eRHIC. Some beam dynamic phenomena observed in HERA directed the attention of the EIC designers to specific aspects of electron ion collisions.

The examples in the following sections illustrate the positive impact of the HERA design and operational experience on other colliders and in particular to the Electron Ion collider eRHIC.

10.5.2.1 Lepton-Hadron Beam-Beam Interactions

The concept of lepton proton collisions in HERA was to adjust the the strength of the beam-beam force for each beam such as this beam would collide with a beam of its own species. A typical beam-beam strength in terms of incoherent beam-beam tuneshift for $e^+ - e^-$ collisions was $x^e_{x,y} \simeq 0.03$. For Hadron, the strengths of beam beam interactions was typically below $x^p_{x,y} \simeq 0.01$. This succesful concept was used successfully in other colliders such as KEKB and is also adapted at the EIC design.

Occasionally, when the HERA beams were brought into collisions. an observation was made which is a concern for the electron ion collider. Under the influence of strong beam-beam interaction, the electron beam can develop coherent transverse oscillations. If this happens, the unstable motion of the electrons affects in turn the hadron beam via the beam-beam coupling. The hadrons will start to oscillate as well driven by the beam beam field of the unstable electron beam and the hadron beam emittance will then filament in phase space and become diluted that way. The effective beam size grows considerably and practically irreversely in the process which quickly renders the beam unusable.

This phenomenon was observed only occasionally at HERA and its rare occurrence made it impossible to perform systematic studies. Furthermore the impact on operations was very small and therefore, no resources have been invested to study the effect in detail.

The experience with lepton proton collisions in HERA, raised a concern that this effect would occur regularly in the EIC, because of significantly enhanced beam-beam parameters per collision in the EIC. This triggered realistic beam-beam interaction simulations for eRHIC. The beam-beam study was performed using the so-called strong, strong model, where each beam is described by a large number of super-particles and the thin lens beam-beam lenses are split into many beam-beam-lens slices. The individual particle's trajectories are impacted by the collective electro-magnetic forces induced by the particles of the opposite beam, respectively. This consitutes a fully dynamical model of the beam-beam interactions. The result of this study was that this effect indeed exists and would affect the performance of the EIC if not avoided by a careful choice of the machine parameters. The threshold in proton bunch intensity for this catastrophic instability is a factor of two above the chosen eRHIC design values of proton bunch intensity [177].

10.5.2.2 Beam-Gas Backgrounds of the Colliding Beam Detectors

The experience with beam induced backgrounds at HERA is very important for the design of the interaction regions and the collision detectors of the EIC or LHeC.

The combination of synchrotron radiation emitted by the electron beam in the separator- and the focusing fields when entering the interaction region and the corresponding strong sources of desorbed gases in the vicinity of the collision detectors caused initially strong hadron-beam-gas induced detector backgounds.

These backgrounds improved slowly by conditioning the surfaces of masks and absorbers by the presence of electron beam and scattered synchroton photons. Eventually a very high dynamical vacuum quality in the order of 1 nbar with full beam intensity was achieved. This required a installation of a large pumping speed and good vacuum conductance with integrated NEG pump in the IR quadrupole magnets and Titanium supplimation pumps. The cold beam pipe of the superconducting IR magnets (at a temperature of 80 K) acted as a cryopump and helped to improve the vacuum. But once the surface was saturated, the magnes needed to be warmed up and the accumulated gas gas molecules on the cold surfaces needed to be removed from the IR vacuum system using turbo pumps. Each warm-up and cooldown cycle marked a step in background improvement.

Another important lesson learned from HERA IR operations was that the vacuum system must be designed such that IR vacuum leaks are unlikely and that venting of the IR and detector vacuum for maintenance and repair purposes must be avoided by design.

The analysis of HERA backgrounds [178] shows the importance of extremely good vacuum quality in the beam vacuum chamber inside the central colliding beam detector. This makes high vacuum pumping speed and high vacuum conductance within the detector beam pipe mandatory. Techniques such as in situ bakeout and NEG coating of the detector beam chamber might be unavoidable to overcome this difficulty.

This experience is exploited in the design of the interaction regions of the EIC, in particular at eRHIC. First background simulations indicate that difficult initial background conditions as experienced in HERA can be avoided taking the lessons learned into account.

10.5.2.3 Hadron Beam Collimation

In lepton-hadron collisions with stong beam-beam interaction, the hadron beam will develop tails in the tranverse particle distribution. The controlled removal of these halo-particles is very important for low detector background and efficient data taking.

A sophisticated collimation system and collimator optimization scheme was implemented in HERA. It consisted of two stages of collimation, with one single primary collimator per oscillation plane and two secondary collimators per plane. A detailed and fully automated collimator adjustment procedure was developed which provided in general good background conditions routinely.

The HERA collimation system and its operational optimization techniques are being carried over to the eRHIC design version of the EIC.

Furthermore, high energy proton operation with the HERA-B [179] fixed target in HERA, revealed that the wire target could be integrated in the existing collimator system. It consitituted an excellent beam-edge spoiler target which significantly increased the efficiency of the two stage collimation system and led to a significant reduction of particle background in the detectors H1 and ZEUS.

Early simulations of halo particle backgrounds in HERA showed clearly, that the halo must be removed far away from the detectors and local shields or masks will only aggravate the background conditions for the colliding beam detectors [180].

Experience with HERA collimation system have been taken into account in the elaborated LHC collimation system and are an important input for the design of the collimation system of eRHIC.

10.5.2.4 Spin Polarization of the HERA Electron Beam

The good performance of the HERA electron spin in colliding beam operations at the North and the South interaction points with three pairs of spin rotators in the North, South, and East straight sections which provided a polarization of $\geq$50% [181] was vital for the HERMES experiment [182] and augmented the physics program of the H1 and ZEUS detectors [183]. This performance was achieved with uncompensated detector solenoids in the North (H1), South (ZEUS) interaction reagions. The spin tuning procedures developed for HERA which included the system of harmonic bumps, and spin-matching of the straigth sections between the rotator pairs, as well as the electron beam working point near the integer resonance were important factors in this success.

The HERA experience constitutes an important reference point for designing the EIC for high electron spin polarization. Spin physics is an important part of the EIC physics program. Collisions between highly polarized electron and Hadron beams with all combination of spin helicities enable an this physics program.

The spin matching techniques and the scheme of orbit optimization with harmonic bumps which contributed to the success of the HERA spin program, are being carried over to the eRHIC design to enable operation with highly polarized electron beams. As in HERA, an equilibrium polarization of at least 50% has been shown to be achievable in eRHIC at the highest electron beam operation energy of 18 GeV. This performance will enable to maintain a high level of average polarization of $\simeq$80% with the initial polization of the the electron beam of 85% if the stored electron bunches are replaced every 6 min on average [177].

10.5.2.5 Lessons Learned from HERA Dynamic Aperture

The dynamic aperture optimization in the HERA electron ring after the year 2000 was accomplished with only two sextupole families for the correction of chromaticity. Correction of higher order chromaticity which is due to the strong contribution of the interaction region quadrupole magnets to the chromaticity and the corresponding strong off-momentum distortion of the optical functions (beta-beat) is necessary to confine the beam tune foot print and to avoid destabilizing resonances.

The straight forward correction of off-momentum beta-beats with more families of sextupoles was avoided thereby avoiding large peak strengths of the sextupole magnets, which would have deteriorated the dynamic aperture. Instead, the horizontal and vertical betatron phase differences between the North and the South interaction points were chosen to be an odd integer of $\pi/2$. The off-momentum beta beat caused by the low beta quadrupoles of one IR was intrinsically canceled by the correponding beta beat of the other IR. This way, the nonlinear tuneshift with momentum was minimized.

The HERA chromatic correction scheme [184] also suppressed the generation of driving terms of the higher order nonlinear synchro-betatron resonances $Q_x + 3 \cot Q_s$=integer which otherwise would have affected the beam stability at the low tune values needed for high polarization operation.

Last but not least, the HERA chromatic correction scheme reduced the magnet currents of the sextupoles circuits which would have been required using the multi-family scheme. The HERA scheme is being carried over to eRHIC which is planned to be operated with two collidng beam detectors as well. The eRHIC chromatic correction scheme is an important constituent of the design which enables highest luminosity values.

10.5.2.6 HERA IR Magnet Design

In order to achieve hightest luminosity, the HERA interaction regions have been re-designed to allow for very small beta functions at the IP. The change was implemented in the year 2000 and HERA operated with about 3.5 times the design luminosity in the years 2004–2007.

This upgrade included accelerator magnets that were placed inside the colliding beam detectors H1 and ZEUS, and in particular inside the solenoidal detector fields of up to 5 T. Obviously, no magnetic steel was allowable inside the solenoid fields of the detectors. Furthermore, the accelerator magnets should have minimum mass to minimize the distorions of the trajectories of scattered particles, maximizong the detector acceptance this way and avoiding introduction of strong scattering targets insode the detector. Thus superconducting technology without steel collaring needed to be used. The available space for these magnets was very limited, as the detectors already existed and could not be modified. The magnets needed to be combinded function to separate the lepton and the hadron beam and to start focusing the electron beam as early as possible. The aperture of the magnet needed to be very large to allow the synchrotron beam to pass through the entire detector and low-beta quadrupole section. Furthermore, the aperture needed to provide space for the envelopes of the separated beams.

To meet these requirements, a new type of superconducting magnet was developed by a collaboration of scientists from Brookhaven National Laboratory and DESY [185]. The supercnducting magnet coil was produced by developing the 2-D direct wind method used for the RHIC corrector magnets to 3-D geometry such that the superconducting wire could be attached directly to the surface of a cylinder. The magnet coil was wound directly on the beam pipe using a seven-strand NbTi superconducting cable and the coil was stabilized with glass-fiber tape, a technique carried over from the HERA multipole corrector coils.

High field quality of superconducting magnets requires high precision in the manufacturing of the coil. For the HERA interaction region magnets, the precision was achieved by combining the RHIC and HERA corrector coil technologies.

The coil was wound on a precision-machined cylinder which was coated with epoxy. A computer controlled stylus placed superconducting wire with high precision on the cylinder. It also transmitted ultra-sound waves to the wire, which by means of friction melted the epoxy surface of the cylinder thereby fixing the superconducting wire in its position. After winding the first layer of the coil, it was complemented with GF spacers and fixed with glass-fiber tape under high tension before the coil was vacuum impregnated with epoxy. The field of the first layer was measured and compared with the calculated field for perfect wire position. Discrepancies were corrected by modifying the definition of the second layer wire positions correspondingly. After high precision machining the new surface, the magnet was ready for the second layer of coil to be wound. This way, the magnet achieved very high field quality of one unit of 10^{-4} at a radius of 25 mm with the center set off by up to 20 mm.

The HERA IR magnets were operated with 4.5 K supercritical He. The performance in routine operation was excellent.

The technique of the very successful HERA low beta quadrupoles was further developed by BNL scientists and the technique has been successfully applied to the low beta quadrupoles of the Beijing Electron-Positron Collider BEPC-II [186], the KEK b-factory SuperKEKB [187] and designs for the interaction regions of the International Linear Collider [188] have bee produced and prototyped. The Fig. 10.15 shows a photograph taken during the winding of the low-beta quadruole for superKEKB in September 2013. This technology is now quite mature and is used for many of the advanced eRHIC interaction region magnets [177].

Fig. 10.15 Winding of the superKEKB coil in September 2013 (Courtesy Brett Parker, BNL)

10.5.2.7 Conclusion

The accelerator and technical solutions as well as the operational proceedures which enabled the successful operation of the HERA Lepton-Hadron collider are an important resource for the design of future colliders, in particular for the design of ring-based electron-ion colliders.

10.6 LHC Layout and Performance to Date

R. Bailey · J. Wenninger

10.6.1 Introduction

The Large Hadron Collider (LHC) [189] is a two-ring superconducting accelerator and collider installed in a 27 km underground tunnel at CERN, on the Swiss-French border near to Geneva. The tunnel was originally constructed for the Large Electron Positron collider (LEP), which operated from 1989 to 2000.

The LHC has two high luminosity experiments, ATLAS and CMS, designed for a peak luminosity of 10^{34} cm^{-2} s^{-1}, one dedicated ion experiment, ALICE, and one

Table 10.15 LHC basic beam parameters

Parameter	Unit	Injection	Collision
Beam data			
Proton energy	GeV	450	7000
Relativistic gamma γ		479.6	7461
Number of particles per bunch		1.15×10^{11}	
Number of bunches		2808	
Longitudinal emittance (4σ)	eVs	1.0	2.5[a]
Transverse normalized emittance	μm rad	3.5[b]	3.75
Circulating beam current	A	0.582	
Stored energy per beam	MJ	23.3	362
Peak luminosity related data			
RMS bunch length[c]	cm	11.24	7.55
RMS beam size at the IP1 and IP5[d]	μm	375.2	16.7
RMS beam size at the IP2 and IP8[e]	μm	279.6	70.9
Geometric luminosity reduction factor F[f]		–	0.836
Peak luminosity in IP1 and IP5	$cm^{-2}\ s^{-1}$	–	1.0×10^{34}
Peak luminosity per bunch in IP1 and IP5	$cm^{-2}\ s^{-1}$	–	3.5×10^{30}

[a]The base line machine operation assumes that the longitudinal emittance is deliberately blown up in the middle of the ramp in order to reduce the intra beam scattering growth rates
[b]The emittance at injection energy refers to the emittance delivered to the LHC by the SPS without any increase due to injection errors and optics mis-match. The RMS beam sizes at injection assume the nominal emittance value quoted for top energy (including emittance blowup due to injection oscillations and mismatch)
[c]Dimensions are given for Gaussian distributions. The real beam will not follow a Gaussian distribution but more realistic distributions do not allow analytic estimates for the IBS growth rates
[d]The RMS beam sizes in IP1 and IP5 assume a β-function of 0.55 m
[e]The RMS beam sizes in IP2 and IP8 assume a β-function of 10 m
[f]The geometric luminosity reduction factor depends on the total crossing angle at the IP. The quoted number assumes a total crossing angle of 285 μrad in IR1 and IR5

experiment for B-physics, LHCb. The main parameters required to reach the high luminosity for ATLAS and CMS are given in Table 10.15.

The high beam intensities required for a luminosity of $L = 10^{34}\ cm^{-2}\ s^{-1}$ exclude the use of anti-proton beams and a single common vacuum and magnet system for both circulating beams (as was done in the SPPbarS and the TEVATRON) and implies the use of two proton beams. To collide two beams of equally charged particles requires opposite magnet dipole fields in both beams. The LHC is therefore designed as a proton-proton collider with separate magnet fields and vacuum chambers in the main arcs and with common sections only at the insertion regions where the experimental detectors are located.

There is not enough room for two separate rings of magnets in the LEP tunnel. Therefore the LHC uses twin bore magnets which consist of two sets of coils and beam channels within the same mechanical structure and cryostat.

The peak beam energy in a storage ring depends on the integrated dipole field along the storage ring circumference. Aiming at peak beam energies of up to 7 TeV inside the existing LEP tunnel implies a peak dipole field of 8.33 T and the use of superconducting magnet technology.

This presented quite a challenge; to make the most profitable use of the existing tunnel and to obtain the highest possible bending strength by exploiting the well-proven technology based on Nb-Ti Rutherford cables. To meet this challenge and to find the cheapest solution compatible with the required performance needed a substantial R&D program on magnets and associated technology. This was carried out between 1988 and 2001 by CERN in close collaboration with other European laboratories and European industry.

10.6.2 Layout

The basic layout of the LHC follows the LEP tunnel geometry and is depicted in Fig. 10.16. The LHC has eight arcs and straight sections, with dispersion suppressors in between.

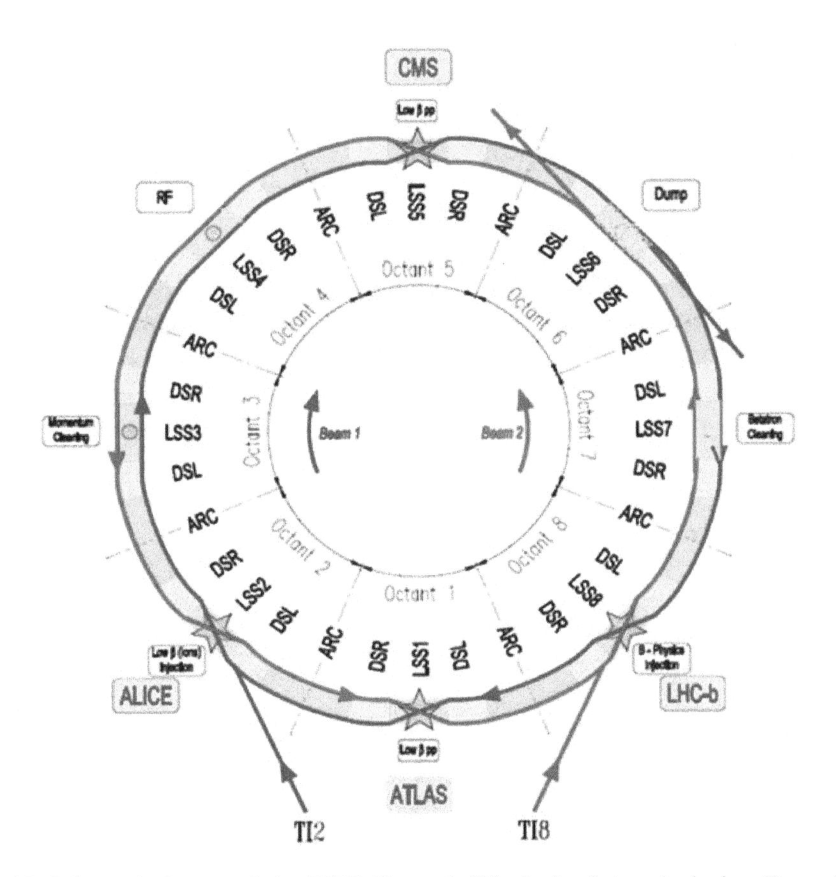

Fig. 10.16 Schematic layout of the LHC. Beam 1 (Blue) circulates clockwise. Beam 2 (Red) circulated counter-clockwise

10.6.2.1 The Straight Sections

Each straight section is approximately 528 m long and can serve as an experimental or utility insertion. The two high luminosity experimental insertions are located at diametrically opposite straight sections: the ATLAS experiment is located at point 1 and the CMS experiment at point 5. Two more experimental insertions, for ALICE and LHCb are located at point 2 and point 8 respectively. These insertion regions also contain the injection systems, for Beam 1 arriving from the SPS through the transfer line TI2, and for Beam 2 arriving from the SPS through the transfer line TI8. The injection kick occurs in the vertical plane with the two beams arriving at the LHC from below the LHC reference plane. The beams only cross from one magnet bore to the other at these four locations. The remaining four straight sections do not have beam crossings. Insertion 3 and 7 each contain two collimation systems, for momentum cleaning and betatron cleaning respectively. Insertion 4 contains two RF systems: one independent system for each LHC beam. The straight section at point 6 contains the beam dump insertion where the two beams are vertically extracted from the machine using a combination of horizontally deflecting fast-pulsed ('kicker') magnets and vertically-deflecting double steel septum magnets. Each beam features an independent abort system.

The two beams share an approximately 130 m long common beam pipe along the interaction regions (IR). The exact length is 126 m in IR2 and IR8 which feature superconducting separation dipole magnets next to the triplet assemblies and 140 m in IR1 and IR5 which feature normal conducting and therefore longer separation dipole magnets next to the triplet assemblies. Together with the large number of bunches (2808 for each proton beam), and a nominal bunch spacing of 25 ns, the long common beam pipe implies 34 parasitic collision points for each experimental insertion region (for four experimental IR's this implies a total of 136 unwanted collision points). Dedicated crossing angle orbit bumps separate the two LHC beams left and right from the central interaction point (IP) in order to avoid collisions at these parasitic collision points.

10.6.2.2 The Arcs

The arcs of the LHC are each made of 23 regular arc cells. The arc cells are 106.9 m long and are made out of two 53.45 m long half cells each of which contains one 5.355 m long cold mass (6.63 m long cryostat) short straight section (SSS) assembly and three 14.3 m long dipole magnets. The two apertures for Ring 1 and Ring 2 are separated by 194 mm. The two coils in the dipole magnets are powered in series and all dipole magnets of one arc form one electrical circuit. The quadrupoles of each arc form two electrical circuits: all focusing quadrupole magnets in Ring 1 and Ring 2 are powered in series and all defocusing quadrupole magnets of Beam 1 and Beam 2 are powered in series. The optics of Beam 1 and Beam 2 in the arc cells is therefore strictly coupled via the powering of the main magnetic elements.

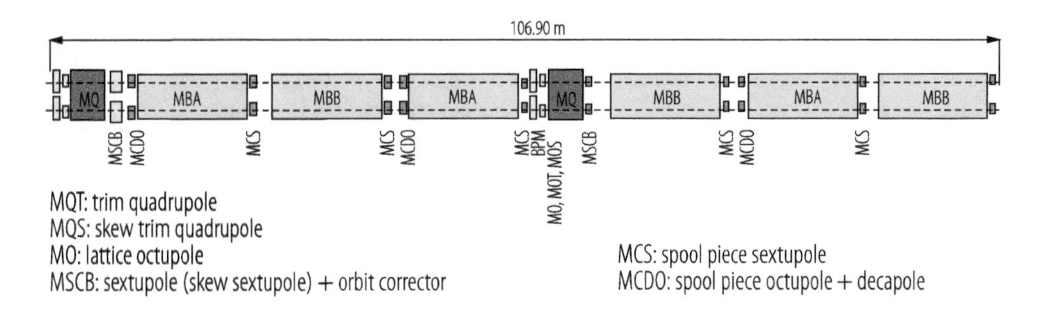

Fig. 10.17 The LHC arc cell layout

An LHC arc cell is also equipped with corrector magnets, which can be split into two distinct categories (see Fig. 10.17).

- The lattice corrector magnets attached on both sides of the main quadrupole magnets are installed in the Short Straight Section (SSS) cryostats.
- The spool-piece corrector magnets which are thin non-linear windings attached directly on the extremities of the main dipoles.

Contrary to the main dipole circuits and the two families (QF and QD) of lattice quadrupoles for which, in each sector, Ring 1 and Ring 2 are powered in series, the arc corrector magnets can be adjusted independently for the two beams.

10.6.2.3 The Dispersion Suppressors

A dispersion suppressor is located at the transition between an LHC arc and a straight section yielding a total of 16 dispersion suppressor sections. The aim of the dispersion suppressors is threefold:

- adapt the LHC reference orbit to the geometry of the LEP tunnel;
- cancel the horizontal dispersion arising in the arc and generated by the separation/recombination dipole magnets and the crossing angle bumps;
- help in matching the insertion optics to the periodic solution of the arc.

A generic design of a dispersion suppressor uses standard arc cells with missing dipole magnets. The LEP dispersion suppressor, which defines the geometry of the tunnel, was made of 3.5 cells with a 90° phase advance, optimized to suppress the dispersion. With the 2.5 times longer LHC dipole and quadrupole magnets, only two LHC cells can be fitted in the dispersion suppressor tunnel.

10.6.2.4 LSS1 and LSS5

IR1 and IR5 house the high luminosity experiments of the LHC and are identical in terms of hardware and optics (except for the crossing-angle scheme: the crossing

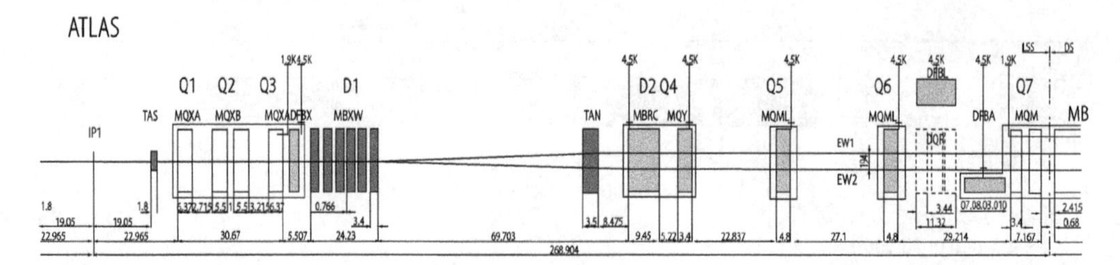

Fig. 10.18 Schematic layout of the right side of IR1

angle in IR1 is in the vertical plane and in IR5 in the horizontal plane). The small β-function values at the IP are generated with the help of a triplet quadrupole assembly. At the IP, the two rings share the same vacuum chamber, the same low-beta triplet magnets and the D1 separation dipole magnets. The remaining matching section (MS) and the dispersion suppressor (DS) consist of double-bore magnets with separate beam pipes for each ring.

Figure 10.18 shows the schematic layout of IR1.

Apart from the DS the insertions are comprised of the following sections, given in order from the interaction point:

- A 31 m long superconducting low-β triplet assembly operated at a temperature of 1.9 K and providing a nominal gradient of 205 T/m.
- A pair of separation/recombination dipoles separated by approximately 88 m. The D1 dipole located next to the triplet magnets has a single bore and consists of six 3.4 m long conventional warm magnet modules yielding a nominal field of 1.38 T. The following D2 dipole is a 9.45 m long, double bore, superconducting dipole magnet operating at a cryogenic temperature of 4.5 K with a nominal field of 3.8 T. The bore separation in the D2 magnet is 188 mm and is thus slightly smaller than the arc bore separation.
- Four matching quadrupole magnets. The first quadrupole following the separation dipole magnets, Q4, is a wide-aperture magnet operating at a cryogenic temperature of 4.5 K and yielding a nominal gradient of 160 T/m. The remaining three quadrupole magnets are normal-aperture quadrupole magnets operating at a cryogenic temperature of 1.9 K with a nominal gradient of 200 T/m.

The triplet assembly features two different quadrupole designs: the outer two quadrupole magnets are made by KEK and require a peak current of 6450 A to reach the nominal gradient of 205 T/m, whereas the inner quadrupole block consists of two quadrupole magnets made by FNAL and requires a peak current of 10,630 A.

The triplet quadrupoles are powered by two nested power converters: one 8 kA power converter powering all triplet quadrupole magnets in series and one 6 kA power converter supplying additional current only to the central two FNAL magnets. The Q1 quadrupole next to the IP features an additional 600 A trim power converter. The triplet quadrupoles are followed by the separation/recombination dipoles, D1 and D2, which guide the beams from the IP into two separated vacuum chambers.

Q4, Q5, Q6, Q7, Q8, Q9 and Q10 are individually powered magnets. The aperture of Q4 is larger to provide sufficient aperture for the crossing angle separation orbit. Two absorbers protect the cold magnets from particles leaving the IP. The TAS absorber protects the triplet quadrupole magnets and the TAN absorber, located in front of the D1 dipole magnet, protects the machine elements from neutral particles leaving the IP.

The matching section extends from Q4 to Q7 and the DS extends from Q8 to Q11. In addition to the DS, the first two trim quadrupoles of the first arc cell (QT12 and QT13) are also used for the matching procedure. All insertion and DS magnets are equipped with a beam screen [2]. The magnets left and right from the IP up to Q7 inclusive are placed symmetrically with respect to the IP. The positions of Q8, Q9 and Q10 left and right from the IP differ by approximately 0.5 m with respect to the IP due to the limited space in the DS.

10.6.2.5 LSS2

The straight section at IR2 houses the injection elements for Ring-1 as well as the ion beam experiment ALICE. During injection the optics must obey the special constraints imposed by the beam injection for Ring-1 and the geometrical acceptance in the interaction region (IR) must be large enough to accommodate both beams in the common part of the ring with a beam separation of at least 10 σ.

Figures 10.19 and 10.20 show the schematic layout of IR2.

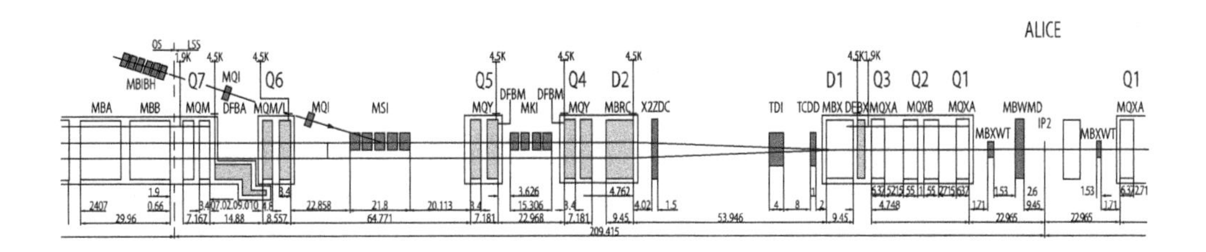

Fig. 10.19 Schematic layout of the left side of IR2

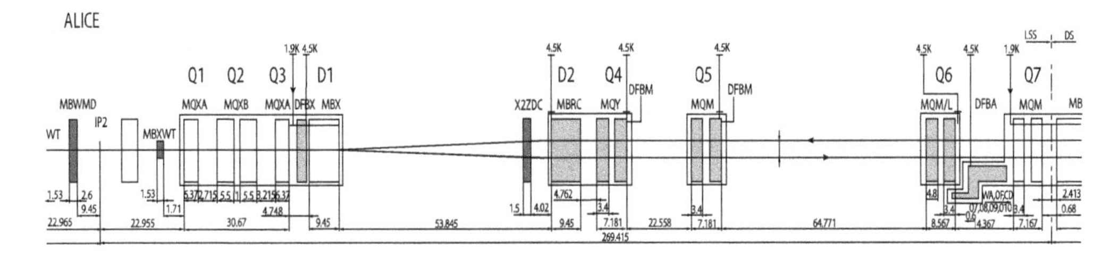

Fig. 10.20 Schematic layout of the right side of IR2

Apart from the DS the insertions comprise the following sections, given in order from the interaction point:

- A 31 m long superconducting low-β triplet assembly operated at 1.9 K and providing a nominal gradient of 215 T/m.
- A pair of 9.45 m long superconducting separation/recombination dipole magnets separated by approximately 66 m.
- Four matching quadrupole magnets. The first two quadrupole magnets following the separation dipole magnets, Q4 and Q5, are wide aperture magnets operating at 4.5 K and yielding a nominal gradient of 160 T/m. The remaining two quadrupole magnets are normal aperture quadrupole magnets operating at 1.9 K with a nominal gradient of 200 T/m.

The triplet quadrupoles are powered in series and are followed by the separation/recombination dipoles D1 and D2, which guide the beams from the IP into two separated vacuum chambers. Q4, Q5, Q6, Q7, Q8, Q9 and Q10 are individually powered magnets. The aperture of Q4 is increased to provide sufficient aperture for the crossing-angle separation orbit. The aperture of Q5 left of the IP is increased to provide sufficient aperture for the injected beam. The injection septum MSI is located between Q6 and Q5 on the left-side of the IP and kicks the injected beam in the horizontal plane towards the closed orbit of the circulating beam (positive deflection angle). The injection kicker MKI is located between Q5 and Q4 on the left-hand side of the IP and kicks the injected beam in the vertical plane towards the closed orbit of the circulating beam (negative deflection angle). In order to protect the cold elements in case of an injection failure a large absorber (TDI) is placed 15 m upstream from the D1 separation/recombination dipole left from the IP. The TDI absorber is complemented by an additional shielding element 3 m upstream of the D1 magnet and two additional collimators installed next to the Q6 quadrupole magnet. In order to obtain an optimum protection level in case of injection errors the vertical phase advance between MKI and TDI must be 90° and the vertical phase advance between the TDI and the two auxiliary collimators must be an integer multiple of $180° \pm 20°$.

The matching section extends from Q4 to Q7 and the DS extends from Q8 to Q11. In addition to the DS, the first two trim quadrupoles of the first arc cell (QT12 and QT13) are also used for the matching procedure. All magnets of the DS are equipped with a beam screen. The magnets left and right from the IP up to Q7 inclusive are placed symmetrically with respect to the IP. The positions of Q8, Q9 and Q10 left and right from the IP differ by approximately 0.5 m with respect to the IP due to the limited space in the DS.

10.6.2.6 LSS8

IR8 houses the LHCb experiment and the injection elements for Beam 2. The small β-function values at the IP are generated with the help of a triplet quadrupole assembly. At the IP, the two rings share the same vacuum chamber, the same low-

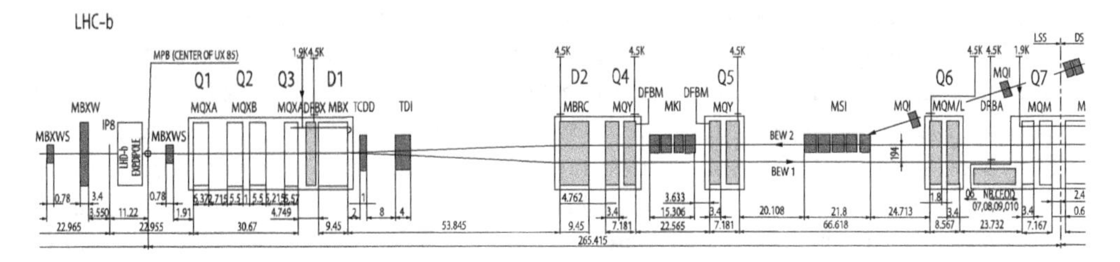

Fig. 10.21 Schematic layout of the right side of IR8

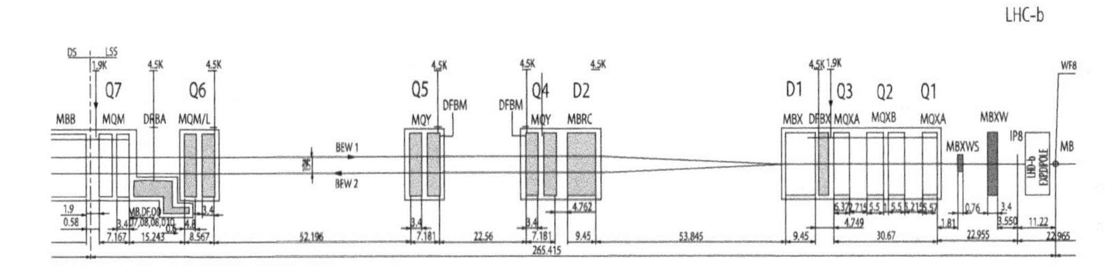

Fig. 10.22 Schematic layout of the left side of IR8

beta triplet magnets and the D1 separation dipole magnet. The remaining matching section (MS) and the DS consist of double-bore magnets with separate beam pipes for each ring.

Figures 10.21 and 10.22 show the schematic layout of IR8.

Apart from the DS the insertions contain the following sections, given in order from the interaction point:

Three warm dipole magnets compensate the deflection generated by the LHCb spectrometer magnet.

- A 31 m long superconducting low-β triplet assembly operated at 1.9 K and providing a nominal gradient of 205 T/m.
- A pair of separation/recombination dipole magnets separated by approximately 54 m. The D1 dipole located next to the triplet magnets is a 9.45 m long single-bore superconducting magnet. The following D2 dipole is a 9.45 m long, double bore, superconducting dipole magnet. Both magnets are operated at 4.5 K. The bore separation in the D2 magnet is 188 mm and is thus slightly smaller than the arc bore separation.
- Four matching quadrupole magnets. The first quadrupole following the separation dipole magnets, Q4, is a wide aperture magnet operating at 4.5 K and yielding a nominal gradient of 160 T/m. The remaining three matching section quadrupole magnets are normal aperture quadrupole magnets operating at 1.9 K with a nominal gradient of 200 T/m.
- The injection elements for Beam 2 on the right hand side of IP8. The 21.8 m long injection septum consists of five modules and is located between the Q6 and Q5 quadrupole magnets on the right-hand side of the IP. The 15 m long

injection kicker consists of four modules and is located between the Q5 and Q4 quadrupole magnets on the right-hand side of the IP. In order to protect the cold elements in case of injection failure a large absorber (TDI) is placed 15 m in front of the D1 separation/recombination dipole magnet right from the IP. The TDI is complemented by an additional shielding element between the TDI and D1 magnet (placed 3 m in front of D1) (TCDD) and by two additional collimators placed on the transition of the matching section left from the IP to the next DS section.

In order to provide sufficient space for the spectrometer magnet of the LHCb experiment, IP8 is shifted by 15 half RF wavelengths (3.5 times the nominal bunch spacing ~11.25 m) towards IR7. This shift of the IP has to be recuperated before the beam returns to the dispersion suppressor sections and implies a non-symmetric magnet layout in the matching section.

10.6.2.7 LSS3 and LSS7

The insertion IR3 houses the momentum cleaning systems of both beams, while IR7 houses the betatron cleaning systems of both beams. Particles with a large momentum offset are scattered by the primary jaw of IR3. Particles with a large H, V or combined H-V betatron amplitudes are scattered by the primary collimator jaws in IR7. In both cases the scattered particles are absorbed by secondary collimators.

Figure 10.23 shows the schematic layout of IR7.

The dispersion suppressor extends from Q8 to Q11. In addition to the DS, the first two trim quadrupoles of the first arc cell (QT12 and QT13) are also used for the matching procedure. All cryo-magnets are equipped with a beam screen. In IR3 and IR7, the underground galleries are not wide enough to house many high current power supplies. Therefore, contrary to the layout of the other IR's, the DS quadrupoles (Q7, Q8, Q9 and Q10) are made of a MQ+MQTL assembly (MQ + 2 MQTL at Q9) where the MQ's magnets are powered in series with the main arc quadrupoles. To avoid producing two kinds of MQ+MQTL assemblies, the dispersion suppressors left and right from the IP are not mirror symmetric with respect to each other. Instead, the DS quadrupole assemblies have the same orientation in the dispersion suppressors left and right from the IP and the MQ positions differ by approximately 0.5 m with respect to the IP in the two DS.

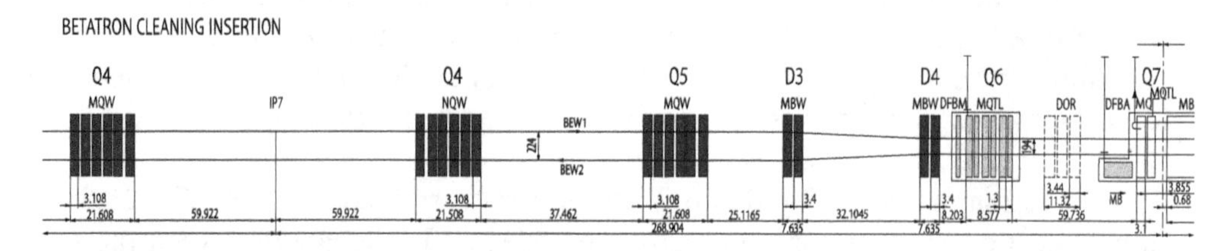

Fig. 10.23 Schematic layout of the right side of IR7

The layout of the Long Straight Section between Q7L and Q7R is mirror symmetric with respect to the IP. The right side of IR7 is shown in Fig. 10.23. This allows the symmetrical installation for the collimators of the two beams and minimizes the space conflicts in the insertion. Starting from Q7 left, the quadrupole Q6 (made of six superconducting MQTL modules) is followed by a dog-leg structure made of two sets of MBW warm single bore wide aperture dipole magnets (two warm modules each). The dogleg dipole magnets are labeled D3 and D4 in the LHC sequence with D3 being the dipole closer to the IP. The Primary Collimators are located between the D4 and D3 magnets, allowing neutral particles produced in the jaws to point out of the beam line, and most charged particles to be swept away. The inter-beam distance between the dogleg assemblies left and right from the IP is 224 mm, i.e. 30 mm larger than in the arc. This increased beam separation allows a substantially higher gradient in the Q4 and Q5 quadrupoles which are made out of six warm MQW modules. The space between Q5 left and right from the IP is used to house the secondary collimators at adequate phase advances with respect to the primary collimators.

The Q4 and Q5 quadrupoles left and right from the IP are powered in series. The warm dual-bore MQW quadrupole cannot be powered with different currents for each magnet aperture because the field quality is degraded to an unacceptable level even for a small imbalance in the field of the two apertures. The current must be equal or of opposite value in the bores to provide a good field quality. In order to obtain the required flexibility for the optics, two different kinds of powering schemes are used for the Q4 and Q5 quadrupole units. The magnets are identical, but in the MQWA type magnet the field is identical in both apertures, while in the MQWB type magnet, the field is opposite for both apertures. Each Q4 and Q5 assembly is made of five MQWA and one MQWB module. The nominal gradient of the MQWB unit is limited to 29.6 T/m while it can reach 35 T/m in the MQWA unit. This powering scheme breaks the exact antisymmetry by 29% providing enough flexibility to satisfy all the optics constraints. Again, Q5AL+Q5AR and Q5BL+Q5BR respectively are powered in series. As a by-product, this freedom in the straight section allows the trim strength needed in the DS to be limited so that regular MQTL's can be used.

In IR3, the most difficult constraint was to generate a large dispersion function in the straight section. Since the layout of the DS cannot be changed in IR3 this constraint means that the natural dispersion suppression generated in the DS is over compensated. To this end Q6 and Q5 were moved towards each other by a substantial amount, thus shrinking the space granted to the dog-leg structure D4-D3. It was therefore necessary to add a third MBW element to D3 and D4 in IR3. Apart from this IR3 and IR7 are identical.

10.6.2.8 LSS4

IR4 houses the RF and feed-back systems as well as some of the LHC beam instrumentation.

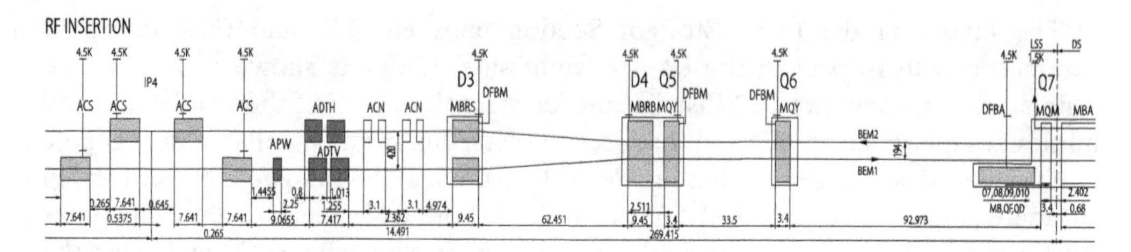

Fig. 10.24 Schematic layout of the right side of IR4

Figure 10.24 shows the schematic layout of IR4.

There are six superconducting quadrupole magnets in the straight section: Q5 Q6 Q7 on the left- and Q5 Q6 Q7 on the right-hand side of the IP. The outer dogleg dipoles, D4, sit next to the Q5 quadrupole magnets on each side of the IP. The RF cavities sit between the inner dogleg dipoles, D3. The layout of the DS, between Q7 and Q11, is identical to that in IR1 and IR5.

10.6.2.9 LSS6

IR6 houses the beam abort systems for Beam 1 and Beam 2. The beam extraction from the LHC is done by kicking the circulating beam horizontally into an iron septum magnet which deflects the beam in the vertical direction away from the machine components to absorbers in a separate tunnel. Each ring has its own system and both are installed in IR6.

Figure 10.25 shows the schematic layout of IR6.

In each of the dispersion suppressors up to six quadrupoles can be used for matching. The total of 16 quadrupoles is more than necessary to match both β-functions, the dispersion (both at the crossing point and in the arc) and adjust the phases. Although this number of parameters seems considerable, their variation is strongly limited by the aperture constraints which set limits on the β-functions and the dispersion inside the insertion. Special detection devices protect the extraction septum and the LHC machine against losses during the extraction process. The TCDS absorber is located in front of the extraction septum and the TCDQ in front of the Q4 quadrupole magnet downstream of the septum magnet.

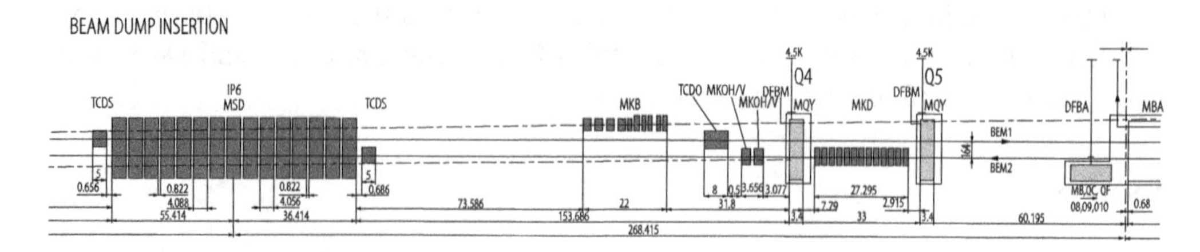

Fig. 10.25 Schematic layout of the right side of IR6

10.6.3 Performance

The LHC was first operated with beam for short periods in 2008 and 2009. In 2010 a first experience with the machine was gained at a beam energy of 3.5 TeV, and moderate beam intensity of up to around 200 bunches of 1.1×10^{11} protons per bunch (ppb). The reduced energy was chosen such as to minimize risks associated to the quality issue of the soldering of the busbar cables in the magnet interconnections. In 2011 the beam intensity was pushed to around 1400 bunches of 1.4×10^{11} ppb while 2012 was dedicated to luminosity production with higher bunch intensities 1.6×10^{11} ppb and a beam energy of 4 TeV. A bunch spacing of 50 ns was used in 2011 and 2012 to minimize the complication of electron clouds. The first operation period is commonly referred to as run 1. In early 2013 beam operation was stopped for a 2-year long shutdown (LS1) to consolidate the magnet interconnections in view of reaching the design beam energy. Many details on LHC operation during Run 1 may be found in [190].

Beam operation resumed in 2015 at 6.5 TeV following a dipole training campaign of 169 quenches at the end of Long Shutdown 1 (LS1). The LHC experiments expressed a strong preference for beams with 25 ns bunch spacing, as opposed to the 50 ns spacing used in 2011–2012, as this would result in a too high number of inelastic collisions per crossing (pile-up). On the machine side 25 ns beams pose additional challenges, e.g. the formation of electron clouds (e-clouds) in the vacuum chamber and a higher number of fast loss events, named Unidentified Falling Objects (UFOs). Given the number of new territories had to be explored, 2015 was considered a re-commissioning and a learning year, dedicated to preparing the machine for full luminosity production in 2016–2018. In 2016 the machine performance was pushed for the first time above the design luminosity of 10^{34} cm^{-2} s^{-1}, and by the end of that year the design had been exceeded by 40%. This excellent performance was possible thanks to beams of much lower emittance produced by the LHC injector chain, coupled to a reduced β^* of 40 cm as compared to the design value of 55 cm. In 2017 the performance could be pushed further with more than twice the design luminosity with a further reduction of the beam emittances and of β^* to 30 cm. The performance in 2017 was so exceptional that the luminosities of ATLAS and CMS had to levelled down to limit the event pile-up to 60. Figures 10.26 and 10.27 present the evolution of the peak and of the integrated luminosity between 2011 and 2017. A summary of the main parameters is presented in Table 10.16. Details on operation in LHC Run 2 up to and including 2016 can be found in [191].

The remarkable performance of LHC during run 2 between 2015 and 2017 was achieved despite unexpected limitations. In 2016 the SPS beam dump was damaged when high intensity LHC beams were dumped onto the block at 450 GeV, developing a vacuum leak. Because it was not practical to exchange the dump during

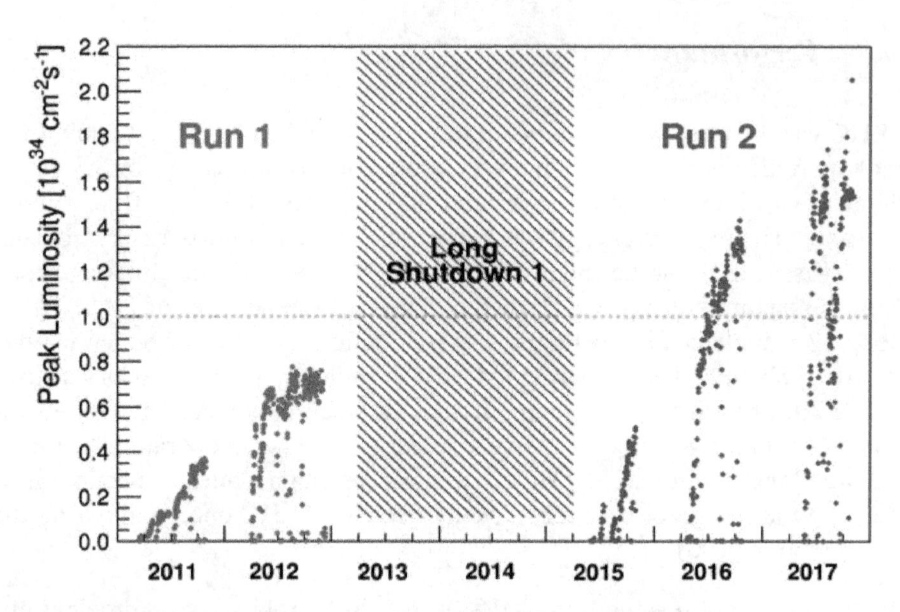

Fig. 10.26 Peak luminosity performance of the LHC in ATLAS and CMS between 2011 and 2017. In July 2016 the LHC reached the design luminosity of 1×10^{34} cm^{-2} s^{-1}

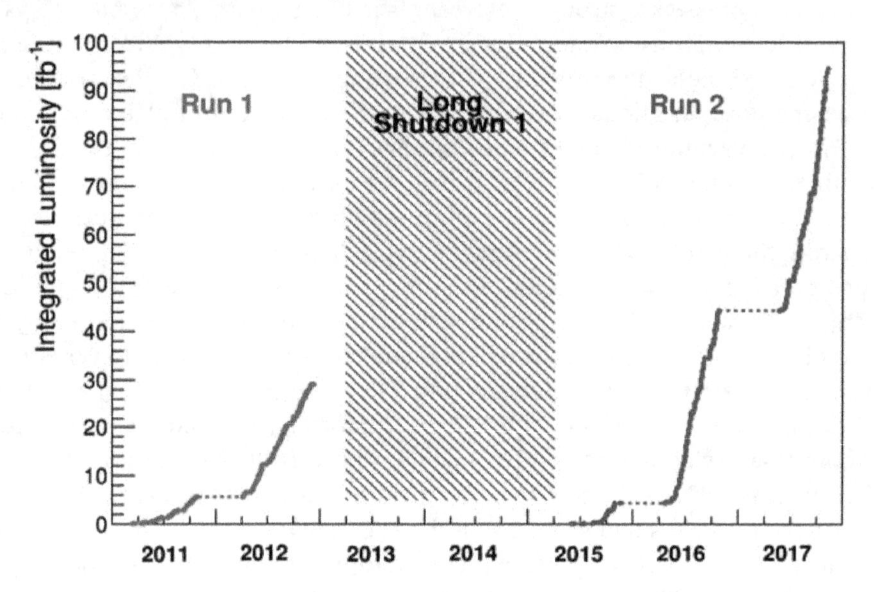

Fig. 10.27 Integrated luminosity delivered to the ATLAS and CMS experiments between 2011 and 2017

the run, the LHC beams were limited to 144 bunches during that year. This limitation was lifted in 2017 when a new dump was installed in the SPS. Unfortunately, an undetected vacuum pumping problem during the cool down of one LHC sector brought another limitation for the 2017 run. A few liters of Air introduced by a

Table 10.16 Performance reach in 2010 and 2011

Parameter	Unit	2012	2016	2017	Design
Energy	TeV	4	6.5	6.5	7
Bunch intensity	10^{10}	16	11	12.5	11.5
Bunches per beam		1380	2556	2556	2802
Emittance	μm	2.40	2.3	2.0	3.75
β_*	m	0.6	0.4	0.3	0.55
Luminosity 1 and 5	$cm^{-2}\ s^{-1}$	7×10^{33}	1.5×10^{34}	2×10^{34}	1×10^{34}

vacuum pump issue condensed as ice on the vacuum beam chamber. In the presence of very high intensity beams local losses would develop leading to beam dumps. Fortunately, it was possible to restore operation of the LHC with a low electron cloud variant of the 25 ns spacing LHC beam [192].

References

1. E. Regenstreif: Report CERN 62-03 (1962).
2. A. Hermann: History of CERN, Vol. I, North-Holland (1987).
3. U. Mersits, in: A. Hermann, et al.: History of CERN, Vol. II, North-Holland (1990)
4. S. Gilardone, D. Manglunki (eds.): Fifty Years of the CERN Proton Synchrotron, CERN-2011-004.
5. D. Simon, J.P. Riunaud: Proc. 17th Intern. Conf. High-Energy Accelerators, Accelerator Catalogue, I. Meshkov (ed.), Dubna (1998), p. 72
6. J. Coupard et al. (eds): LIU TDR Vol 1, CERN–ACC–2014-0337 15 December 2014
7. F. Gerigk, M. Vretenar (eds): LINAC4 TDR, CERN–AB–2006–084 ABP/RF
8. O. Brüning et al. (eds): LHC Design Report Vol. 1, CERN-2004-003
9. M. Giovanni (ed.): Multi Turn Eextraction design report, CERN-2006-011
10. J. Bernhard et al.: Proc. IPAC'18, IPAC2018, Vancouver, Canada doi:10.18429/JACoW-IPAC2018-TUPAF023
11. D. Simon: Proc. EPAC, Sitges (1996), p. 295.
12. N.C. Christofilos: unpublished manuscript (1950).
13. E.D. Courant, M.S. Livingston, H.S. Snyder: Phys. Rev. 88 (1952) 1190.
14. G.K. Green: Proc. Intern. Conf. High Energy Accelerators, M.H. Blewett (ed.), Brookhaven (1961), p. 39.
15. M. Plotkin: Brookhaven Nat. Lab. (1991) report-45058.
16. D.S. Barton: IEEE Trans. Nucl. Sci NS-30(4) (1983)2787.
17. L. Ahrens: HEACC'92, Intern. J. Mod. Phys. A (Proc. Suppl.) (1993) 109.
18. J.M. Brennan: IEEE Proc. Part. Accel. Conf., New York (1999), p. 614.
19. D.I. Lowenstein: Proc. 17th Intern. Conf. High-Energy Accelerators, Accelerator Catalogue, I. Meshkov (ed.), Dubna (1998), p. 72.
20. J.M. Brennan: IEEE Part. Accel. Conf., Dallas (1995), p. 1489.
21. K.A. Brown: Proc. IEEE Part. Accel. Conf., Portland (2003), p. 1545.
22. A. Zelenski: Proc. EPAC08, Genoa (2008), p. 1010.
23. Ya.S. Derbenev, A.M. Kondratenko: Sov. Phys. Dokl. 20 (1976) 562. Ya.S. Derbenev, et al.: Part. Accel. 8 (1978) 115.
24. T. Roser: 8th Intern. Symp. High-Energy Spin Physics, Particle and Fields Series 37, Minneapolis, MN (1988), p. 1442.

25. L.A. Huang: Proc. IEEE Part. Accel. Conf., Vancouver (2009), p. 4251.
26. V. Schoefer: Proc. IEEE Part. Accel. Conf., New York (2011), to be published.
27. C.J. Gardner: Proc. IEEE Part. Accel. Conf., Albuquerque (2007), p. 1862.
28. A. Pikin, et al.: Proc. Intern. Symp. Electron Beam Ion Sources and Traps, IOP Publishing, Stockholm (2010). http://iopscience.iop.org/1748-0221/5/09/C09003/pdf/jinst10_09_c09003.pdf
29. J. Alessi: Proc. IEEE Part. Accel. Conf., New York (2011), to be published.
30. V.V. Vladimirski, et al.: Proc. 4th Intern. Conf. High-Energy Accelerators, Dubna (1963), p. 233.
31. Institute of High Energy Physics, et al.: Proc. 6th Intern. Conf. High-Energy Accelerators, Cambridge (1967), p. 248.
32. E.F. Troyanov: Proc. 17th Intern. Conf. High-Energy Accelerators, Accelerator Catalogue, I. Meshkov (ed.), Dubna (1998), p. 29.
33. S. Ivanov: Proc. Russian Particle Accelerator Conf., Zvenigorod (2008), p. 130.
34. V.A. Teplyakov: Proc. 17th Intern. Conf. High-Energy Accelerators, Accelerator Catalogue, I. Meshkov (ed.), Dubna (1998), p. 31.
35. A.S. Gurevich: Proc. 17th Intern. Conf. High-Energy Accelerators, Accelerator Catalogue, I. Meshkov (ed.), Dubna (1998), p.30.
36. S. Ivanov. Proc. Russian Particle Accelerator Conf., Protvino (2010), p. 27.
37. A.G. Afonin: Proc. Russian Particle Accelerator Conf., Protvino (2010), session WECHX01.
38. S. Ivanov et al. Advances of Light-Ion Acceleration Program in the U70. Proc. 23rd Russian Particle Accelerators Conference RUPAC-2012, St.-Petersburg, 2012, pp. 100–102.
39. S. Ivanov, O. Lebedev. Transverse Noise Blow-up of the Beam in the U-70 Synchrotron. Instruments and Experimental Technique, Vol. 56, No. 3, 2013, pp. 249–255.
40. S. Ivanov, O. Lebedev. Attaining Square-Wave Stochastic Slow Extraction Spills from the U-70 Synchrotron. Instruments and Experimental Techniques, 2015, Vol. 58, No. 4, pp. 456–464.
41. D.W. Kerst: Symp. High-Energy Accel. and Pion Physics, CERN (1956), CERN Report 56-26, p. 36.
42. G.K. O'Neill: Symp. High-Energy Accel. and Pion Physics, CERN (1956), CERN Report 56-26, p. 64.
43. K.R. Symon, A.M. Sessler: Symp. High-Energy Accel. and Pion Physics, CERN (1956) CERN Report 56-26, p. 44.
44. CERN Study Group New Accelerators: Report CERN/542 (1964).
45. A. Hofmann: Proc. 11th Intern. Conf. High-Energy Accelerators, Accelerator Catalogue, J.H.B. Madsen, P.H. Standley (eds.), CERN (1980), p. 44.
46. K. Johnsen: CERN Report 84-13 (1984).
47. K. Johnsen: Proc. 5th Intern. Conf. High-Energy Accelerators, Frascati (1965), p. 3.
48. G. Plass: CERN Internal Report NPA/Int.61-8 (1961).
49. J.B. Adams: Proc. 8th Intern. Conf. High-Energy Accelerators, CERN (1971), p. 25.
50. J.B. Adams: Proc. 10th Intern. Conf. High-Energy Accelerators, Protvino (1977), p. 17.
51. K.-H. Kissler: Proc. 17th Intern. Conf. High-Energy Accelerators, Accelerator Catalogue, I. Meshkov (ed.), Dubna (1998), p. 8.
52. H. Haseroth: Phys. Reports 403-404 (2004) 27.
53. LEP Injector Study Group: Report CERN-LEP/TH/83-29 (1983).
54. M. Benedikt, et al. (eds.): LHC Design Report, Vol. III, CERN-2004-003.
55. C. Rubbia, et al.: Proc. Intern. Neutrino Conf., Aachen (1976), p. 683.
56. S. van der Meer: CERN Internal Report ISR-PO/72-31 (1972).
57. P. Bramham, et al.: Nucl. Instrum. Meth. 125 (1975) 201.
58. G. Carron, et al.: Proc. Part. Accel. Conf., San Francisco (1979), p. 3456.
59. H. Koziol, D. Möhl: Phys. Reports 403-404 (2004) 91.
60. G. Brianti: Proc. 14th Conf. High-Energy Accelerators, Accelerator Catalogue, S. Kurokawa (ed.), Tsukuba (1989), p. 56.
61. J. Gareyte: Proc. 11th Intern. Conf. High-Energy Accel., CERN (1980), p. 79.

62. H. Edwards: Annu. Rev. Nucl. Part. Sci. 35 (1985) 605.
63. M. Church: Proc. 17th Intern. Conf. High-Energy Accelerators, Accelerator Catalogue, I. Meshkov (ed.), Dubna (1998), p. 62.
64. F.R. Huson: Proc. 10th Intern. Conf. High-Energy Accelerators, Protvino (1977), p. 30.
65. S. Holmes (ed.): Report FNAL-TM-2484 (1998).
66. M. Church: Proc. Eur. Accelerator Conf., Paris (2002), p. 11.
67. A. Valishev, et al.: Proc. 23rd Particle Accelerator Conf., Vancouver (2009), p. 4230.
68. S.D. Holmes and V.D. Shiltsev, Annual Review of. Nuclear. and Particle Science, Vol.63, p.435.
69. M. Harrison, S. Peggs, T. Roser: Annu. Rev. Nucl. Part. Sci. 52 (2002) 425.
70. M. Harrison, et al.: Nucl. Instrum. Meth. A 499 (2003) 235.
71. M. Blaskiewicz, J.M. Brennan, K. Mernick: Phys. Rev. Lett. 105 (2010) 094801.
72. W. Fischer, et al.: Phys. Rev. ST Accel. Beams 11 (2008) 041002.
73. C. Montag, et al.: Phys. Rev. ST Accel. Beams 5 (2002) 084401.
74. W. Fischer, et al.: Proc. EPAC08, Genoa, Italy (2008), p. 1616.
75. A. Zelenski, J. Alessi, A. Kponou, D. Raparia: Proc. EPAC08, Genoa, Italy (2008) p. 1010.
76. H. Huang, et al.: Phys. Rev. ST Accel. Beams 7 (2004) 071001.
77. F. Lin, et al.: Phys. Rev. ST Accel. Beams 10 (2007) 044001.
78. Ya.S. Derbenev, A.M. Kondratenko: Part. Accel. 8 (1978) 115.
79. S.Y. Lee: Spin dynamcs and Snakes in Synchrotrons, World Scientific, Singapore (1997).
80. A. Bazilevsky, et al.: Proc. EPAC08, Genoa, Italy (2008), p. 1140.
81. M. Bai, et al.: Phys. Rev. Lett. 96 (2006) 174801.
82. W. Fischer, et al.: Phys. Rev. Lett. 115 (2015) 264801.
83. A. Fedotov et al.: Proc. NAPAC2016, Chicago, IL, USA (2016), p. 867
84. H. Blosser: Handbook of Accelerator Physics and Engineering, World Scientific (2006) 13.
85. E. Wilson: Handbook of Accelerator Physics and Engineering, World Scientific (2006) 57.
86. M. Tigner: Nuovo Cimento 37 (1965) 1228.
87. J. Seeman: *Nonlinear Dynamics Aspects of Particle Accelerators*, Springer-Verlag, Proc. 247 (1985) 121.
88. J. Rees: Handbook of Accelerator Physics and Engineering, World Scientific (2006) 11.
89. C. Bernardini, et al: High Energy Accelerator Conf. (1961) 256.
90. G. Budker, et al: J. Nucl. Energy C 8 (1966) 676.
91. G. O'Neill: High Energy Accelerator Conf. (1961) 247.
92. V. Auslender, et al: High Energy Accelerator Conf. (1965).
93. A. Skrinsky: Particle Accelerator Conf. (1995) 14.
94. G. Arzelier, et al: VIII High Energy Accelerator Conf. (1971) 127.
95. ADONE Group: Particle Accelerator Conf. (1971) 217.
96. J. Paterson, et al: Particle Accelerator Conf. (1971) 196.
97. J. Paterson, et al: Particle Accelerator Conf. (1975) 1366.
98. G. Tumaikin, et al: X High Energy Accelerator Conf. (1977) 443.
99. H. Nesemann, et al: Particle Accelerator Conf. (1983) 1998.
100. J. LeDuff, et al: XI High Energy Accelerator Conf. (1980) 566.
101. G. Voss, et al: XI High Energy Accelerator Conf. (1980) 748.
102. B. McDaniel, et al: Particle Accelerator Conf. (1981) 1984.
103. D. Rubin, et al: Particle Accelerator Conf. (1995) 481.
104. A. Blinoz, et al: XII High Energy Accelerator Conf. (1983) 183.
105. R. Helm, et al: Particle Accelerator Conf. (1983) 2001.
106. T. Nishikawa: XII High Energy Accelerator Conf. (1983) 143.
107. N. Phinney, et al: Particle Accelerator Conf. (1999) 3384.
108. J. Xu, et al: XII High Energy Accelerator Conf. (1983) 157.
109. S. Myers: Intern. Particle Accelerator Conf. (2010) 3663.
110. C. Milardi, et al: Particle Accelerator Conf. (2009) 80.
111. P. Raimondi: 2nd SuperB Meeting, Frascati (2006).
112. J. Seeman, et al: Eur. Particle Accelerator Conf. (2008) 946.

113. M. Tanaka, et al: Intern. Particle Accelerator Conf. (2011) 3735.
114. Q. Qin, et al: Intern. Particle Accelerator Conf. (2011) 3708.
115. Chinese Academy of Sciences: https://phys.org/news/2016-04-bepcii-luminosity-world-11033cm2s.html.
116. D. Shwartz, et al: Russ. Particle Accelerator Conf. (2010) 1.
117. Y. Rogovsky, et al: Physcis and Technique of Accelerators, Springer, 2014, Vol. 11, No. 5, pp. 651-655.
118. H. Koiso: Intern. Particle Accelerator Conf. (2017) 1275.
119. P. Oddone: Proc. UCLA Workshop Linear Collider BB Factory Conceptual Design, D. Stork (ed.), (1987), p. 243.
120. K. Hirata, E. Keil: Nucl. Instrum. Meth. A 292 (1990) 156.
121. *PEP-II: An Asymmetric B Factory. Conceptual Design Report*, SLAC–418, QCD183:S56:1993 (1993).
122. *KEKB B–Factory Design Report*, KEK Report 95–7 (1995).
123. J. Seeman: Conf. Proc. C 0806233 (2008) TUXG01.
124. Y. Funakoshi, T. Abe, K. Akai, Y. Cai, K. Ebihara, K. Egawa, A. Enomoto, J. Flanagan, et al.: Conf. Proc. C 100523 (2010) WEOAMH02.
125. A. Piwinski: IEEE Trans. Nucl. Sci. 24 (1977) 1408.
126. R.B. Palmer: Proc. 1988 DPF Summer Study on High-energy Physics in the 1990s (Snowmass 88), Snowmass, Colorado, 27 Jun – 15 Jul 1988, (1988), p. 613.
127. K. Oide, K. Yokoya: Phys. Rev. A 40 (1989) 315.
128. M. Sullivan, G. Bowden, H. DeStaebler, S. Ecklund, J. Hodgson, T. Mattison, M.E. Nordby, A. Ringwall, et al.: Conf. Proc. C 960610 (1996) 460.
129. B. Aubert, et al. (BABAR Collaboration): Nucl. Instrum. Meth. A 479 (2002) 1.
130. J. Seeman, M. Sullivan, M. Biagini, Y. Cai, F.J. Decker, M. Donald, S. Ecklund, A. Fisher, et al.: Proc. EPAC 2002, 3-7 Jun 2002, Paris, France, (2002), p. 434-436.
131. Y. Yamazaki, T. Kageyama: Part. Accel. 44 (1994) 107.
132. T. Furuya, et al.: Gif-sur-Yvette 1995, RF superconductivity, Vol. 2 (1995), p. 729.
133. H. Schwarz, R. Rimmer: Conf. Proc. C 940627 (1994) 1882.
134. J. Fox, T. Mastorides, C. Rivetta, D. Van Winkle, D. Teytelman: Phys. Rev. ST Accel. Beams 13 (2010) 052802.
135. M. Izawa, Y. Sato, T. Toyomasu: Phys. Rev. Lett. 74 (1995) 5044.
136. K. Ohmi: Phys. Rev. Lett. 75 (1995) 1526.
137. K. Ohmi, F. Zimmermann: Phys. Rev. Lett. 85 (2000) 3821.
138. T.O. Raubenheimer, F. Zimmermann (SLAC): Phys. Rev. E 52 (1995) 5487.
139. J.W. Flanagan, K. Ohmi, H. Fukuma, S. Hiramatsu, M. Tobiyama, E. Perevedentsev: Phys. Rev. Lett. 94 (2005) 054801.
140. H. Fukuma, J. Flanagan, K. Hosoyama, T. Ieiri, T. Kawamoto, T. Kubo, M. Suetake, S. Uno, et al.: AIP Conf. Proc. 642 (2003) 357.
141. Y. Suetsugu, K. Shibata, H. Hisamatsu, M. Shirai, K. Kanazawa: Vacuum 84 (2009) 694.
142. M.T.F. Pivi, F. King, R.E. Kirby, T. Markiewicz, T.O. Raubenheimer, J. Seeman, L. Wang: Conf. Proc. C 0806233 (2008) MOPP064.
143. Y. Suetsugu, H. Fukuma, L. Wang, M. Pivi, A. Morishige, Y. Suzuki, M. Tsukamoto, M. Tsuchiya: Nucl. Instrum. Meth. A 598 (2009) 372.
144. T. Mimashi, T. Ieiri, M. Kikuchi, A. Tokuchi, K. Tsuchida: Conf. Proc. C 0806233 (2008) TUPD011.
145. At least an application for a storage ring is seen in: R. Servranckx, K.L. Brown: IEEE Trans. Nucl. Sci. 26 (1979) 3598.
146. K. Oide, H. Koiso, K. Ohmi: AIP Conf. Proc. 391 (1997) 215.
147. K. Oide, H. Koiso: Phys. Rev. E 47 (1993) 2010.
148. J. Irwin, C.X. Wang, Y.T. Yan, K.L.F. Bane, Y. Cai, F.J. Decker, M.G. Minty, G.V. Stupakov, et al.: Phys. Rev. Lett. 82 (1999) 1684.
149. K. Akai, N. Akasaka, A. Enomoto, J. Flanagan, H. Fukuma, Y. Funakoshi, K. Furukawa, T. Furuya, et al.: Nucl. Instrum. Meth. A 499 (2003) 191.

150. Y.T. Yan, Y. Cai: Nucl. Instrum. Meth. A 558 (2006) 336.
151. T. Ieiri, K. Akai, H. Fukuma, M. Tobiyama: Nucl. Instrum. Meth. A 606 (2009) 248.
152. K. Satoh, M. Tejima: Conf. Proc. C 950501 (1995) 2482.
153. M. Tejima, M. Arinaga, T. Ieiri, H. Ishii, H. Fukuma, M. Tobiyama, S. Hiramatsu: Conf. Proc. C 0505161 (2005) 3253.
154. A. Drago, J.D. Fox, D. Teytelman, M. Tobiyama: Conf. Proc. C 0806233 (2008) THPC116.
155. T. Mitsuhashi, J.W. Flanagan, S. Hiramatsu: Proc. Seventh EPAC2000, 26-30 Jun 2000, Vienna, Austria, (2000), p. 1783-1785.
156. J.W. Flanagan, N. Akasaka, H. Fukuma, S. Hiramatsu, T. Mitsuhashi, T. Naito, K. Ohmi, K. Oide, et al.: Proc. Seventh EPAC2000, 26-30 Jun 2000, Vienna, Austria, (2000), p. 1119-1121.
157. N. Akasaka, A. Akiyama, S. Araki, K. Furukawa, T. Katoh, T. Kawamoto, I. Komada, K. Kudo, et al.: Nucl. Instrum. Meth. A 499 (2003) 138.
158. Y. Funakoshi, M. Masuzawa, K. Oide, J. Flanagan, M. Tawada, T. Ieiri, M. Tejima, M. Tobiyama, et al.: Phys. Rev. ST Accel. Beams 10 (2007) 101001.
159. L. Hendrickson, T. Gromme, P. Grossberg, T. Himel, D. Macnair, R. Sass, H. Smith, N. Spencer, et al.: Proc. Seventh EPAC2000, 26-30 Jun 2000, Vienna, Austria, (2000), p. 1897-1899.
160. K. Ohmi, K. Oide, E. Perevedentsev: Conf. Proc. C 060626 (2006) 616.
161. Y. Ogawa, A. Enomoto, K. Furukawa, T. Kamitani, M. Satoh, T. Sugimura, T. Suwada, Y. Yano, et al.: Conf. Proc. C 060626 (2006) 2700.
162. T. Kamitani, N. Delerue, M. Ikeda, K. Kakihara, S. Ohsawa, T. Oogoe, T. Sugimura, T. Takatomi, et al.: Conf. Proc. C 0505161 (2005) 1233.
163. K. Akai, J. Kirchgessner, D. Moffat, H. Padamsee, J. Sears, T. Stowe, M. Tigner: Proc. B factory workshop, 6-10 Apr 1992, Stanford, California, (1992).
164. K. Hosoyama, K. Akai, K. Ebihara, T. Furuya, K. Hara, T. Honma, A. Kabe, Y. Kojima, et al.: Conf. Proc. C 0806233 (2008) THXM02.
165. K. Ohmi, M. Tawada, Y. Cai, S. Kamada, K. Oide, J. Qiang: Phys. Rev. ST Accel. Beams 7 (2004) 104401.
166. Y. Ohnishi, K. Ohmi, H. Koiso, M. Masuzawa, A. Morita, K. Mori, K. Oide, Y. Seimiya, et al.: Phys. Rev. ST Accel. Beams 12 (2009) 091002.
167. P. Raimondi, D. Shatilov, M. Zobov, "Beam-Beam Issues for Colliding Schemes with Large Piwinski Angle and Crabbed Waist", arXiv:physics/0702033 [physics.acc-ph] (2007).
168. Y. Ohnishi, Proc.15th Annual Meeting of Particle Accelerator Society of Japan August 7-10, 2018, Nagaoka, Niigata, Japan (2018) WEOLP01.
169. *SuperKEKB Design Report*, https://kds.kek.jp/indico/event/15914/ (2014).
170. Y. Ohnishi, presentation at the 62nd ICFA Advanced Beam Dynamics Workshop on High Luminosity Circular e+e- Colliders (eeFACT2018), September 24-26 (2018), IAS, Hong Kong MOXAA02.
171. K. Ohmi, et al., "Coherent Beam-Beam Instability in Collisions with a Large Crossing Angle", Phys.Rev.Lett. 119 (2017) no.13, 134801.
172. K. Furukawa, et al., Proc. IPAC2018, Vancouver, BC, Canada, doi:10.18429/JACoW-IPAC2018-MOPMF073 (2018).
173. Y. Suetsugu, et al., "Mitigating the electron cloud effect in the SuperKEKB positron ring", Phys.Rev.Accel.Beams 22 (2019) no.2, 023201.
174. N. Ohuchi, et al., Proc. IPAC2018, Vancouver, BC, Canada, doi:10.18429/JACoW-IPAC2018-TUZGBE2 (2018).
175. A. Accardi et al, Electron Ion Collider: The Next QCD Frontier, arXiv:1212.1701 (2012)
176. https://www.jlab.org/jleic/index.html
177. J Beebe-Wang (Editor), eRHIC, Preconceptual Design Report, arXiv:xxxxxxxx
178. R. Yoshida, EIC Collaboration meeting Oct 2017 (BNL) unpublished, https://indico.bnl.gov/event/3492/contributions/10260/attachments/9195/11238/yoshida.pdf
179. Yu. Vassiliev et al, Multi-target operation at the HERA-B experiment, AIP Conference Proceedings 512, 359 (2000); https://doi.org/10.1063/1.1291460

180. R. Brinkmann. Simulation Of Background From Proton Losses In The Hera Straight Sections, Jul 1987. 22 pp. DESY-HERA-87-19 (1987) unpublished
181. http://www.desy.de/mpybar/psdump/eliana-cracow.paper.pdf
182. http://www-hermes.desy.de
183. Ziqing Zhang, Physics from Polarized ep Collisions at HERA, Proceeedings of the 25th Conference of Physics in Collisions, Prague (2005)
184. F. Willeke, Proceedings of the European Particle Accelerator Conference, Lausanne (2004).
185. B. Parker et al, TUOA02A, Proceedings of the European particle Conference 1998, Stockholm(1998); http://accelconf.web.cern.ch/AccelConf/e98/PAPERS/TUOA02A.PDF
186. B. Parker, Serpentine Coil Topology for BNL Direct Wind Superconducting Magnets, Particle Accelerator Conference Knoxville (2005) https://doi.org/10.1109/PAC.2005.1590546
187. N. Ouchi et al, DESIGN OF THE SUPERCONDUCTING MAGNET SYSTEM FOR THE SUPERKEKB INTERACTION REGION, Proceedings of PAC2013, Pasadena, CA USA WEODA1 (2013)
188. B. Parker, M. Anerella, J. Escallier, A. Ghosh, A. Jain, A. Marone, J. Muratore, P. Wanderer, BNL Direct Wind Superconducting Magnets, BNL-96547-2011-CP, Presented at the 22nd International Conference on Magnet Technology (MT-22) Marseille, France September 9-16, (2011) November 2011
189. The LHC Design Report, CERN-2004-003.
190. R. Alemany-Fernandez et al, Operation and Configuration of the LHC in Run 1, CERN-ACC-NOTE-2013-0041.
191. J. Wenninger, LHC towards nominal performance, Proceedings of IPAC17, Copenhagen, Dk (2017).
192. Proceedings of the 8th LHC operation workshop, Evian, France (2017), https://indico.cern.ch/event/663598.

11

Prominent Uses of Accelerators and Storage Rings

M. Dohlus, J. Rossbach, K. H. W. Bethge, J. Meijer, U. Amaldi, G. Magrin, M. Lindroos, S. Molloy, G. Rees, M. Seidel, N. Angert, and O. Boine-Frankenheim

11.1 Synchrotron Radiation and Free-Electron Lasers

M. Dohlus · J. Rossbach

11.1.1 Synchrotron Radiation

11.1.1.1 Basic Properties of Synchrotron Radiation

It is well known from Maxwell theory that electromagnetic radiation is emitted whenever electric charges are accelerated in free space. This radiation assumes quite extraordinary properties whenever the charged particles move at ultrarelativistic

M. Dohlus (✉) · J. Rossbach (✉)
DESY, Hamburg, Germany
e-mail: martin.dohlus@desy.de; joerg.rossbach@desy.de

K. H. W. Bethge · J. Meijer
University Leipzig, Leipzig, Germany
e-mail: jan.meijer@uni-leipzig.de

U. Amaldi · M. Lindroos
CERN (European Organization for Nuclear Research)Meyrin, Genève, Switzerland
e-mail: Ugo.Amaldi@cern.ch; Mats.Lindroos@cern.ch

G. Magrin
EBG MedAustron, Wiener Neustadt, Austria
e-mail: giulio.magrin@medaustron.at

S. Molloy
Accelerator Operations, MAX IV Laboratory, Lund, Sweden
e-mail: stephen.molloy@esss.se

speed: The radiation becomes very powerful and tightly collimated in space, and it may easily cover a rather wide spectrum ranging from the THz into the hard X-ray regime. When generation of such radiation is intended rather than being a side effect, the charged particles are normally electrons, thus kinetic energies are then typically in the multi-MeV range.

The theoretical treatment of synchrotron radiation starts traditionally from retarded Lienard-Wiechert potentials [1], allowing quantitative determination of radiation properties in detail:

$$\phi(0,t) = \frac{q}{4\pi\varepsilon_0}\left[\frac{1}{R\left(1-\vec{n}\cdot\vec{\beta}\right)}\right]_{t'=t-R/c}, \quad \vec{A}(0,t) = \frac{q}{4\pi\varepsilon_0 c}\left[\frac{\vec{\beta}}{R\left(1-\vec{n}\cdot\vec{\beta}\right)}\right]_{t'=t-R/c}. \tag{11.1}$$

Here, q is the accelerated point charge, ε_0 is the dielectric constant, and $\vec{R}$ is the distance vector from the charge to the observer, with the unit vector $\vec{n} = \vec{R}/R$ and $\vec{\beta} = \vec{v}/c$ the particle's velocity $\vec{v}$ normalized to the vacuum speed of light c, see Fig. 11.1. For simplicity, and without loss of generality, we assume the observer is located at the origin at $R = 0$.

All quantities in Eq. (11.1) must be taken at the retarded time $t' = t - R/c$, i.e. not at the time t of observation but at the time when a signal moving at speed c must

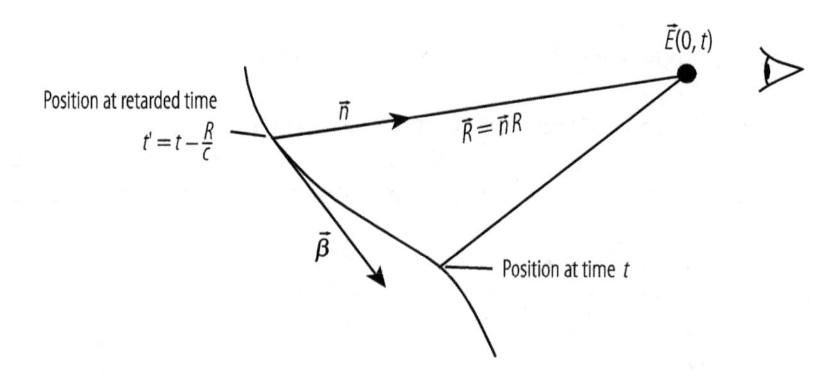

Fig. 11.1 Geometry of actual and retarded position of moving particle, and position of observer

G. Rees
Rutherford Appleton Laboratory, Didcot, UK

M. Seidel
Paul Scherrer Institut, Villigen, Switzerland
e-mail: mike.seidel@psi.ch

N. Angert · O. Boine-Frankenheim
GSI, Darmstadt, Germany
e-mail: N.Angert@gsi.de; O.Boine-Frankenheim@gsi.de

be emitted from the charge's position in order to reach the observer at time t. This latter fact is actually the origin for many peculiarities of synchrotron radiation.

The electric and magnetic fields are derived from Eq. (11.1) using $\gamma^2 = 1/(1-\beta^2)$, $\vec{E} = -\nabla\phi - \partial\vec{A}/\partial t$ and $\vec{B} = \nabla \times \vec{A}$:

$$\vec{E}(0,t) = \frac{q}{4\pi\varepsilon_0}\left[\frac{\vec{n}-\vec{\beta}}{\gamma^2\left(1-\vec{n}\cdot\vec{\beta}\right)^3 R^2}\right]_{t-R/c} + \frac{q}{4\pi\varepsilon_0 c}\left[\frac{\vec{n}\times\left\{\left(\vec{n}-\vec{\beta}\right)\times\dot{\vec{\beta}}\right\}}{\left(1-\vec{n}\cdot\vec{\beta}\right)^3 R}\right]_{t-R/c}. \tag{11.2}$$

and

$$\vec{B} = \frac{1}{c}\left[\vec{n}\times\vec{E}\right]_{t-R/c}. \tag{11.3}$$

While Eqs. (11.2) and (11.3) are well suited to calculate quantitatively the field of a charge moving on a well-known path, some general properties of the field are more conveniently determined from the equivalent expression Eq. (11.4) given in [2]:

$$\vec{E}(0,t) = \frac{q}{4\pi\varepsilon_0}\left[\frac{\vec{n}}{R^2} + \frac{R}{c}\frac{d}{dt}\left(\frac{\vec{n}}{R^2}\right) + \frac{1}{c^2}\frac{d^2}{dt^2}\vec{n}\right]. \tag{11.4}$$

The first term describes the static Coulomb field, scaling with R^{-2}. The second term modifies the field direction of the Coulomb field such that, in case of a charge moving at constant speed, the Coulomb field is NOT directed towards the retarded position of the particle (as it might be suggested by the first term) but rather to the instantaneous position of the charge just at the time t of observation. This will be shown below.

All radiation is described by the third term, and since it contains a contribution scaling with R^{-1}, it is the dominant field at large distances from the source. Thus, in order to consider properties of synchrotron radiation qualitatively, it is sufficient to understand the behaviour of

$$\vec{E}_{\text{rad}}(0,t) \propto \left[\frac{d^2}{dt^2}\vec{n}\right]. \tag{11.5}$$

Since the unit vector $\vec{n}$ cannot change its length, one can see [2] that $d^2\vec{n}/dt^2$ is always perpendicular to $\vec{n}$, i.e. to the retarded position of the particle if observed from a far distance. The fact that $\vec{E}_{\text{rad}}(0,t)$ is perpendicular to $\vec{n}$ is also seen from the second term of Eq. (11.2) (describing the radiation in far zone) which would yield zero if scalar multiplied by $\vec{n}$. Together with Eq. (11.3) it is seen that the Poynting vector $\vec{S} = \left(\vec{E} \times \vec{B}\right)/\mu_0$ is parallel to $\vec{n}$. It is thus immediately clear that the radiation detected by the observer always seems to be originating from the retarded position of the particle, i.e. the retarded position is the apparent origin of the radiation, as it is intuitively expected when the speed of light is taken into account.

Equation (11.5) suggests that one just has to inspect the acceleration of the charge transverse to its apparent line of sight in order to understand the behaviour of the electric field component. For example, it is thus easily seen that radiation from an electron moving on a circular orbit is linearly polarized in the plane of the circle if the observer is located in the same plane. If, however, the radiation is observed from a position elevated out of this (say, horizontal) plane, one expects also a vertical field component.

Since Eqs. (11.2) and (11.4) look so differently, it is useful to sketch how Eq. (11.2) can be derived from Eq. (11.4): a key point is that Eq. (11.2) requires all derivations to be taken with respect to the retarded time $t' = t - R/c$ while Eq. (11.5) contains derivatives with respect to the time of observation t. Explicitly, this reads

$$\vec{\beta} = -\frac{d\vec{R}}{cdt'} = -\frac{d\left(R\vec{n}\right)}{cdt'} = -\frac{dR}{cdt'}\vec{n} - R\frac{d\vec{n}}{cdt'}. \tag{11.6}$$

From $t' = t - R/c$ it follows: $dt'/dt = 1 - (dR(t)/cdt')(dt'/\mathrm{d}t)$. Solving for dt'/dt yields: $dt'/dt = 1/(1 + \mathrm{d}R/cdt')$.

Due to $\vec{n} \perp d\vec{n}/dt'$ we get from Eq. (11.6): $dR/cdt' = -\vec{n}\cdot\vec{\beta}$ and thus

$$\frac{dt'}{dt} = \frac{1}{1 - \vec{n}\cdot\vec{\beta}}. \tag{11.7}$$

For ultrarelativistic motion, and in particular in forward direction $\vec{n} \parallel \vec{\beta}$, dt' and dt differ by a really large factor

$$\frac{1}{1-\beta} = \frac{1+\beta}{1-\beta^2} \approx 2\gamma^2 \gg 1 \tag{11.8}$$

thus resulting in a huge compression of the time scale at which radiation properties are observed compared to the one at which electron motion takes place. It should be noted that this has nothing to do with any Lorentz transform but is rather a property of relativistic Doppler shift. The particle position at the retarded time is just a different point in space-time.

With Eq. (11.7) in mind, Eq. (11.4) can be rewritten (note we write $(n\beta)$ for $\left(\vec{n}\cdot\vec{\beta}\right)$ everywhere):

$$\vec{E}(0,t) = \frac{q}{4\pi\varepsilon_0}\frac{1}{R^2}\left\{\begin{array}{c} \vec{n}+\frac{1}{[1-(n\beta)]^3}\left[-\vec{\beta}+2\vec{\beta}(n\beta)-\vec{\beta}(n\beta)^2+3\vec{n}(n\beta)\right. \\ -6\vec{n}(n\beta)^2+3\vec{n}(n\beta)^3+3\vec{n}(n\beta)^2-\vec{n}\beta^2-2\vec{\beta}(n\beta) \\ \left.-2\vec{n}(n\beta)^3+\vec{\beta}\beta^2+\vec{\beta}(n\beta)^2\right] \end{array}\right\}$$
$$+\frac{q}{4\pi\varepsilon_0}\frac{1}{cR[1-(n\beta)]^3}\underbrace{\left[\vec{n}\left(n\dot{\beta}\right)-\dot{\vec{\beta}}+\dot{\vec{\beta}}(n\beta)-\vec{\beta}\left(n\dot{\beta}\right)\right]}_{\vec{n}\times\left[\left(\vec{n}-\vec{\beta}\right)\times\dot{\vec{\beta}}\right]}. \tag{11.9}$$

It is interesting noting that not only the last line but also all expressions after $3\vec{n}(n\beta)^3$ in the curly bracket stem from the "radiation term" $\frac{1}{c^2}\frac{d^2}{dt^2}\vec{n}$.

After expanding $\vec{n}$ by $\frac{[1-(n\beta)]^3}{[1-(n\beta)]^3}$, the curly bracket in Eq. (11.9) can be simplified into $\{\,\} = \frac{1}{[1-(n\beta)]^3}\frac{\vec{n}-\vec{\beta}}{\gamma^2}$, completing the proof that Eqs. (11.2) and (11.4) are equivalent.

In the following, we will restrict ourselves to properties of the radiation field $\vec{E}_{\text{rad}}$ described by the second term in Eq. (11.2) or by Eq. (11.5), which are equivalent if the observation is made at sufficiently large distance R from the charge ("far-zone approximation"), since in this case the contribution from the curly bracket in Eq. (11.9) to Eq. (11.5) can be neglected:

$$\vec{E}(0,t)_{\text{rad}} = \frac{q}{4\pi\varepsilon_0 c}\left[\frac{\vec{n}\times\left\{\left(\vec{n}-\vec{\beta}\right)\times\dot{\vec{\beta}}\right\}}{\left(1-\vec{n}\cdot\vec{\beta}\right)^3 R}\right]_{t-R/c}. \tag{11.10}$$

It should be emphasised that there are indeed practical cases where this approximation is not valid, e.g. if radiation from an undulator (see below) of length L_u is observed from a distance not much larger than L_u [3].

In most experimental cases the time evolution of the electric field vector is not observable but only the radiation power and its angular or spectral distribution.

In discussing properties of synchrotron radiation it is important to distinguish the instantaneously emitted power (and its angular distribution) from the time development of the power observed by an experimentalist fixed in the lab system. While the first is described in terms of the retarded time t', the latter is observed on the time scale t, which makes a big difference for ultrarelativistic motion, see Eq. (11.7). We point out again that this has nothing to do with a Lorentz transform.

Radiation Power and Its Angular Distribution

The radiation power density is described by the Poynting vector $\vec{S} = \left(\vec{E} \times \vec{B}\right)/\mu_0 = \varepsilon_0 c\left(\vec{E}_{\text{rad}}\right)^2 \vec{n}$, where $\vec{E}_{\text{rad}}$ is the far zone radiation field given in observer time t according to Eq. (11.10). In the present section, we want to refer the radiation power P in far zone to the motion of the electron (i.e. to an emission time interval dt'), so we need to consider

$$\frac{dP\left(t'\right)}{d\Omega} = R^2\left(\vec{S}\cdot\vec{n}\right)\frac{dt}{dt'} = R^2\left(\vec{S}\cdot\vec{n}\right)\left(1 - \vec{n}\cdot\vec{\beta}\right). \tag{11.11}$$

where $d\Omega$ is the solid angle element into which the power is emitted. Thus,

$$\frac{dP\left(t'\right)}{d\Omega} = \frac{q^2}{(4\pi)^2\varepsilon_0 c}\frac{\left[\vec{n} \times \left\{\left(\vec{n} - \vec{\beta}\right) \times \dot{\vec{\beta}}\right\}\right]^2}{\left[1 - \vec{n}\cdot\vec{\beta}\right]^5}. \tag{11.12}$$

This is the most general expression for the angular dependence of the energy loss into radiation of an accelerated point charge in far-field approximation.

There are essentially two different acceleration mechanisms which need to be distinguished: Acceleration $\dot{v}_\perp$ perpendicular to the direction of motion, usually provided by a magnetic field B_{ext}, and acceleration $\dot{v}_\parallel$ by an electric field E_{ext} parallel to the momentarily velocity.

The total radiation power due to $\dot{v}_\perp$ can be shown to be

$$P_{\text{rad},\perp} = \frac{q^2}{6\pi\varepsilon_0 c^3}\gamma^4\dot{v}_\perp^2 = \frac{q^2 c}{6\pi\varepsilon_0}\beta^4\frac{\gamma^4}{\rho^2}. \tag{11.13}$$

Here, $\varrho = \left|\vec{p}\right| / (q B_{\text{ext}})$ is the bending radius of the particle with momentum $\vec{p}$ and charge q in presence of a magnetic field B_{ext} directed perpendicular to $\vec{p}$.

The total radiation power due to $\dot{v}_\parallel$ is, on the other hand,

$$P_{\text{rad},\parallel} = \frac{q^2}{6\pi\varepsilon_0 c^3}\left(\gamma^3\dot{v}_\parallel\right)^2 = \frac{q^2 c}{6\pi\varepsilon_0}\left(\frac{d\gamma}{ds}\right)^2, \tag{11.14}$$

with $d\gamma/ds$ the acceleration due to the longitudinal electric field. In almost all practical cases this radiation power is much smaller than the one generated in a magnetic field of 1 T.

Thus, for the remainder of this paper, we restrict ourselves to acceleration due to magnetic fields.

For circular accelerators the total energy loss U_{rad} during one turn of the charged particle due to synchrotron radiation is an important quantity. From Eq. (11.13) one calculates

$$U_{\text{rad}} = \frac{1}{c}\int P_{\text{rad}} ds = \frac{2\pi\varrho}{c} P_{\text{rad}} = \frac{q^2}{3\varepsilon_0}\frac{\gamma^4}{\rho} \approx 88.5\,\text{keV}\,\frac{(E_0/\text{GeV})^4}{\rho/\text{m}}, \tag{11.15}$$

where the integral extends over all bending magnets. The last expression represents the radiation loss in practical units in the case of electrons or positrons, with E_0 the particle energy. For simplicity is has been assumed that ρ is constant in all bending magnets.

According to the γ^4-scaling of Eqs. (11.13) and (11.15), synchrotron radiation constitutes a massive challenge on the construction of electron/positron synchrotrons in the multi-GeV range. As an extreme example, in the electron-positron storage ring LEP at CERN each particle lost approximately $U_{\text{rad}} = 2850$ MeV per turn when running at its maximum particle energy of $E_0 = 100$ GeV, even though the bending radius was as large as $\rho = 3100$ m.

On the contrary, emission of synchrotron radiation is a negligible effect for hadrons in most practical cases.

In a magnetic field, the acceleration is always perpendicular to the velocity: $\dot{\vec{\beta}} \perp \vec{\beta}$. Then, Eq. (11.12) can be expressed in more practical units, reflecting the geometry illustrated in Fig. 11.2:

$$\frac{dP(\varphi,\theta,t')}{d\Omega} = \frac{q^2}{(4\pi)^2\varepsilon_0 c}\frac{\left|\dot{\vec{\beta}}\right|^2}{(1-\beta\cos\theta)^3}\left(1-\frac{\sin^2\theta\cos^2\varphi}{\gamma^2(1-\beta\cos\theta)^2}\right). \tag{11.16}$$

The direction of observation $\vec{n}$ is expressed here in terms of the angles φ,θ as illustrated in Fig. 11.2.

For ultrarelativistic particles, i.e. if $1-\beta \ll 1$, the denominators $1-\beta\cos\theta$ get very small around $\theta \approx 0$ such that the radiation is concentrated very much in the direction of $\vec{\beta}$.

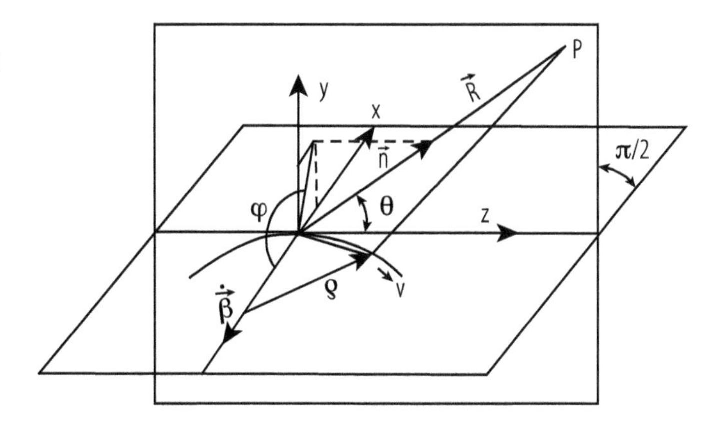

Fig. 11.2 Geometry and definition of parameters used in Eq. (11.16)

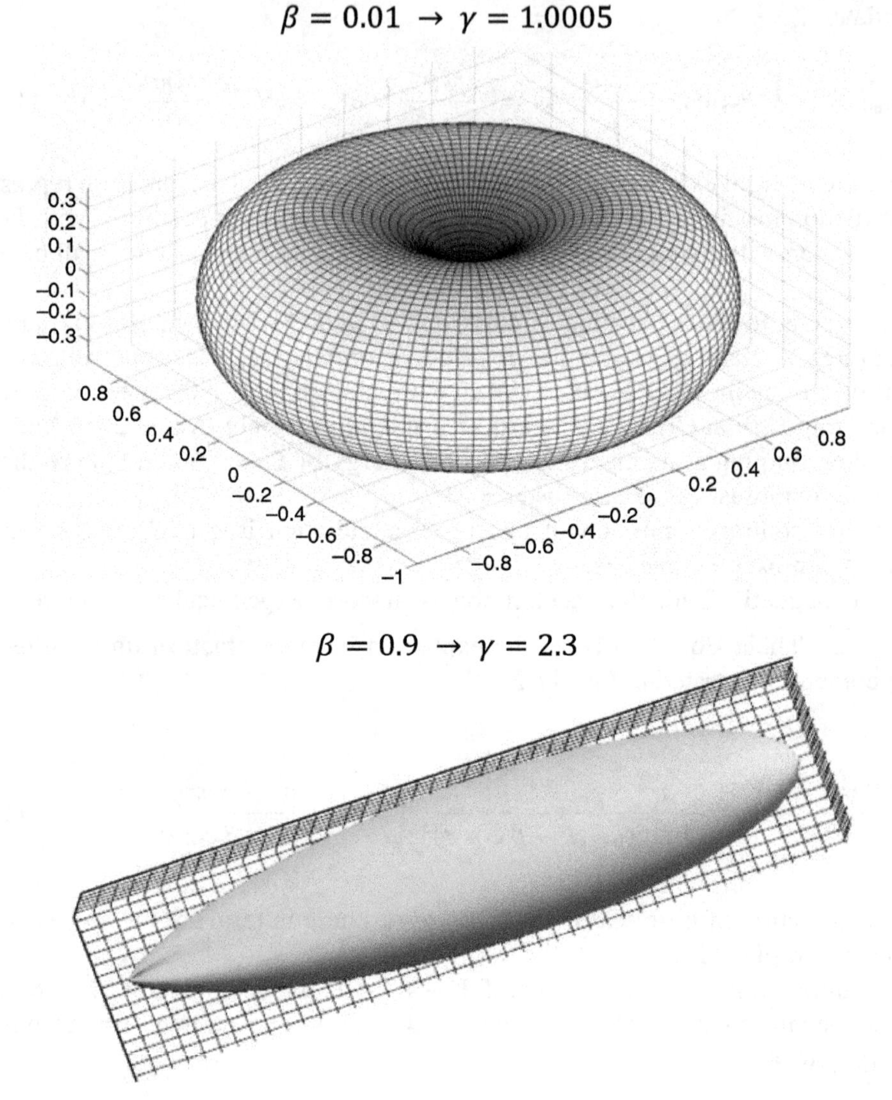

Fig. 11.3 Synchrotron radiation emitted by an ultrarelativistic charged particle is concentrated into a narrow cone with opening angle $1/\gamma$

Figure 11.3 illustrates the directivity of synchrotron radiation according to Eq. (11.16) for two different particle energies.

It should be pointed out that the directivity of synchrotron radiation described by Eq. (11.16) is an instantaneous property of emission, thus is depends only on the local magnetic field strength in the very moment of emission. An observer located in far distance may in fact observe field contributions stemming from several sections of the electron's trajectory. This will be discussed later.

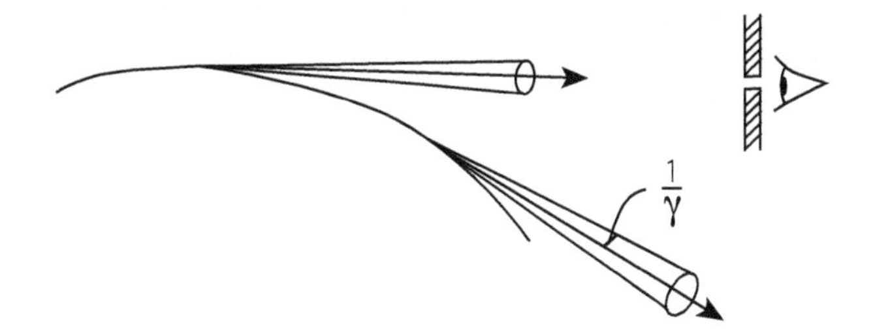

Fig. 11.4 Scenario of synchrotron radiation detection by a distant observer

For $\gamma \gg 1$ and $\theta \ll 1$ we get $1 - \beta\cos\theta \approx (1 + \gamma^2\theta^2)/(2\gamma^2)$ and thus:

$$\frac{dP(\varphi,\theta,t')}{d\Omega} = \frac{q^2}{2\pi^2\varepsilon_0 c}\frac{\left|\dot{\vec{\beta}}\right|^2\gamma^6}{\left(1+\gamma^2\theta^2\right)^3}\left(1 - \frac{4\gamma^2\theta^2\cos^2\varphi}{\left(1+\gamma^2\theta^2\right)^2}\right). \tag{11.17}$$

Equation (11.17) illustrates even more that the emission is concentrated into a cone of opening angle $\theta \approx 1/\gamma$ with respect to the forward direction. A rigorous calculation shows that the rms-opening angle is indeed exactly $1/\gamma$ [4].

For practical calculations it might be useful to replace $\left|\dot{\vec{\beta}}\right|$ by $\beta^2 c/\rho \approx c/\rho$ with $1/\rho = qB/p_0$ describing the bending radius ϱ.

An observer in far distance sees a radiation field only during the short time when the cone passes the observer's aperture, see Fig. 11.4.

For an observer in the plane of deflection, the radiation field has only a non-zero component $E_x(t)$ in this plane, as can be understood easily from Eq. (11.5). It is thus linearly polarized. Due to the strong retardation effect described by Eq. (11.7), this time duration is indeed much smaller than the time $\Delta t' \approx (2/\gamma)(\rho/c)$ the electron needs to cover an angle of $2/\gamma$ on its curved trajectory, namely shorter by a factor $2\gamma^2$. Thus one expects a radiation pulse duration $\Delta t \approx (2\rho)/(c\gamma^3)$. More precisely, the time profile (in the plane of deflection) is illustrated in Fig. 11.5 [4]. It consists essentially of a single spike with a characteristic duration $1/\omega_c$. The "critical frequency" ω_c is defined by

$$\omega_c = \frac{3}{2}\frac{\gamma^3 c}{\rho}. \tag{11.18}$$

11.1.1.2 Spectrum of Synchrotron Radiation from a Long Bending Magnet

In most practical cases, a user of synchrotron radiation is interested in the spectral properties rather than in time domain features. Since the user refers to radiation

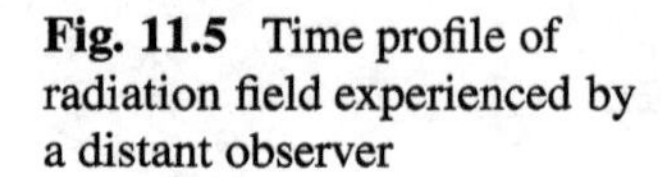
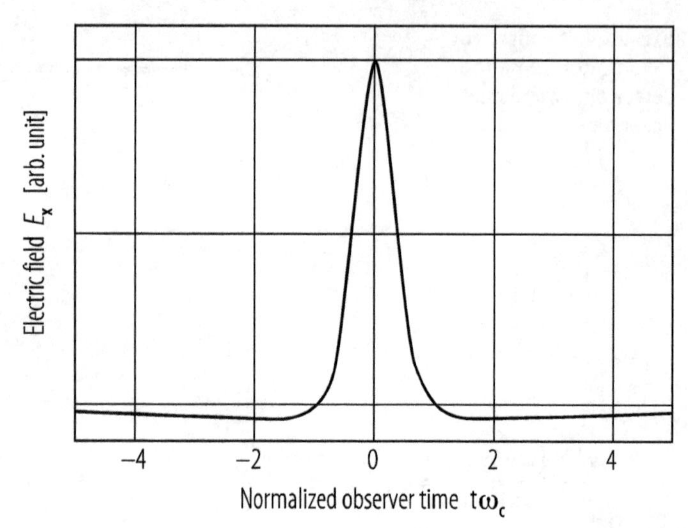

Fig. 11.5 Time profile of radiation field experienced by a distant observer

properties at the location of the observer, we have to investigate the expression

$$\frac{dP(\varphi,\theta,t)}{d\Omega} = R^2\left(\vec{S}\cdot\vec{n}\right) = R^2\varepsilon_0 c\left(\vec{E}(t)\right)^2 \tag{11.19}$$

instead of Eq. (11.11). $P(t) = \Delta W/\Delta t$ describes the amount of energy ΔW radiated within the observer's time interval Δt. The total energy radiated per passage into the solid angle $d\Omega$ is thus

$$\frac{dW}{d\Omega} = \varepsilon_0 c R^2 \int\limits_{-\infty}^{\infty} \left|\vec{E}(t)\right|^2 dt. \tag{11.20}$$

With the help of Parseval's theorem, the r.h.s. of Eq. (11.20) can be turned into frequency domain: $\int_{-\infty}^{\infty} \left|\vec{E}(t)\right|^2 dt = \int_{-\infty}^{\infty} \left|\vec{E}(\omega)\right|^2 d\omega$. Here, the Fourier transform of the electric field $\vec{E}(t)$ is used:

$$\vec{E}(\omega) = \frac{1}{\sqrt{2\pi}} \int\limits_{-\infty}^{\infty} \vec{E}(t)e^{i\omega t}dt. \tag{11.21}$$

The energy radiated into the solid angle $d\Omega$ and frequency interval $d\omega$ is thus given, in far field approximation, by

$$\frac{d^2W}{d\Omega d\omega} = 2\varepsilon_0 c R^2 \left|\vec{E}(\omega)\right|^2. \tag{11.22}$$

The factor of 2 appears since $\vec{E}(t)$ is a real quantity, such that negative frequencies are not considered. According to Eq. (11.10), this means that the expression

$$\vec{E}(\omega) = \frac{q}{4\pi\varepsilon_0 c\sqrt{2\pi}} \int_{-\infty}^{\infty} e^{i\omega(t-R/c)} \frac{\vec{n} \times \left\{\left(\vec{n} - \vec{\beta}\right) \times \dot{\vec{\beta}}\right\}}{\left(1 - \vec{n}\cdot\vec{\beta}\right)^3 R} dt \tag{11.23}$$

needs to be evaluated. The term $(t - R/c)$ appears in the exponent since all quantities $\vec{n}$, $\vec{\beta}$, $\dot{\vec{\beta}}$, R must be evaluated at the retarded time. It should be noted that, in order the solid angle $d\Omega$ to be well defined in Eq. (11.20), the relevant part of the trajectory should remain for all times within a volume of diameter much smaller than R, e.g. in a circular accelerator.

For motion on a circle in a constant magnetic field, the result reads [1]:

$$\frac{d^2W(\theta)}{d\Omega d\omega} = \frac{3q^2}{16\pi^3\varepsilon_0 c}\gamma^2\left(\frac{\omega}{\omega_c}\right)^2\left(1+\gamma^2\theta^2\right)^2\left[K_{2/3}^2(\xi) + \frac{\gamma^2\theta^2}{1+\gamma^2\theta^2}K_{1/3}^2(\xi)\right], \tag{11.24}$$

with the abbreviation

$$\xi = \frac{1}{2}\frac{\omega}{\omega_c}\left(1+\gamma^2\theta^2\right)^{3/2}. \tag{11.25}$$

$K_{2/3}^2$ and $K_{1/3}^2$ are Bessel functions of fractional order.

Equation (11.24) is of course not any more an instantaneous property but refers to the average over the entire passage of the electron. The angle θ describes the elevation of the observer with respect to that tangent to the circle of motion where $\vec{n}\cdot\vec{\beta}$ becomes maximum in Fig. 11.2 (corresponding to $\varphi = 90°$).

It should also be noted that, although the integral in Eq. (11.23) extends from minus to plus infinity, it does not consider the fact that the electron (normally) performs multiple revolutions in the synchrotron, such that the radiation cone passes the observer many times per second. This fact can be accounted for by multiplying Eq. (11.23) by the revolution frequency which would turn the radiated energy per passage into an average radiation power, and the spectrum would become a discrete one.

The term with $K_{2/3}$ describes the radiation polarized in the (horizontal) plane of deflection (σ-polarization), while the $K_{1/3}$ contribution is vertically polarized (π-polarization). As mentioned before and inferred from Eq. (11.5), the π-polarization has no intensity in the plane of deflection ($\theta = 0$), while the horizontal polarization has its maximum there. Both components have a rather broad spectral distribution with a maximum close to ω_c.

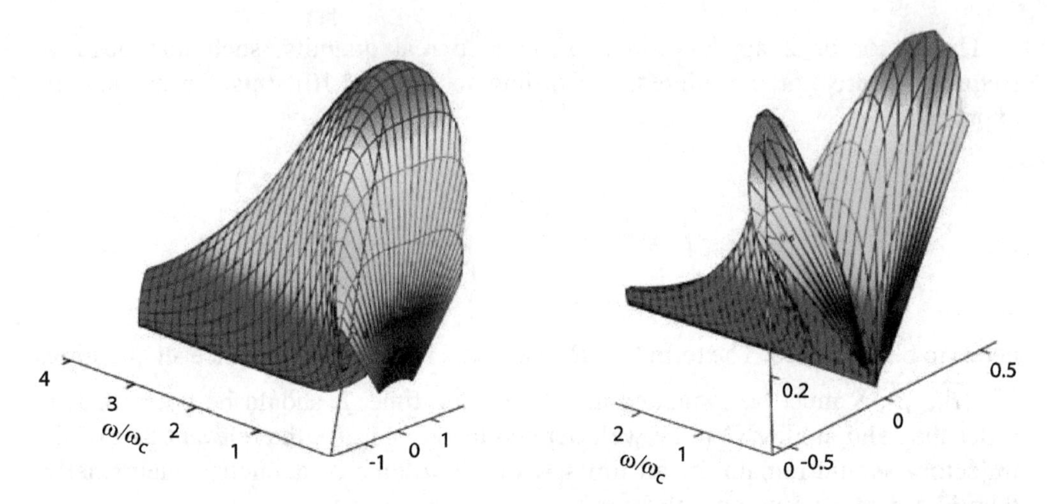

Fig. 11.6 σ-polarization (left) and π-polarization (right) component of synchrotron radiation from a bending magnet. The axis to the left describes the frequency normalized to the critical frequency ω_c, while in the plane perpendicular to ω/ω_c, the directivity of intensity is depicted, with the angle θ taken in normalized $\gamma\theta$ units. The intensity is in arbitrary units

According to Eq. (11.25), the functional dependence of $d^2W/(d\Omega\, d\omega)$ on ω and θ is a universal one if these quantities are normalized to ω_c and $1/\gamma$, respectively. Using such normalized variables, the frequency and angular dependence of the polarization components are illustrated separately in Fig. 11.6. The vertical opening angle decreases with rising frequency and is about $\pm 1/\gamma$ around $\omega = \omega_c$.

Since in most practical cases γ is very big and the opening angle is thus very small, the angular dependence is often not resolved by users. Integration of Eq. (11.24) over the vertical angle θ yields the spectral power density

$$\frac{dW}{d\omega} = \frac{\sqrt{3}q^2}{4\pi\varepsilon_0 c}\gamma\frac{\omega}{\omega_c}\int\limits_{\omega/\omega_c}^{\infty} K_{5/3}(x)dx = \frac{\sqrt{3}q^2}{4\pi\varepsilon_0 c}\gamma\left(\frac{dW}{d\omega}\right)_{\mathrm{norm}}. \tag{11.26}$$

Equation (11.26) summarizes both polarization components. Again, if the frequency is normalized to ω_c, the spectral power density can be expressed by the universal function

$$\left(\frac{dW}{d\omega}\right)_{\mathrm{norm}} = \frac{\omega}{\omega_c}\int\limits_{\omega/\omega_c}^{\infty} K_{5/3}(x)dx, \tag{11.27}$$

see Fig. 11.7 (note that, in contrast to Fig. 11.6, a double logarithmic scale is used). For a discussion of the individual σ- and π-components see, for instance, [4].

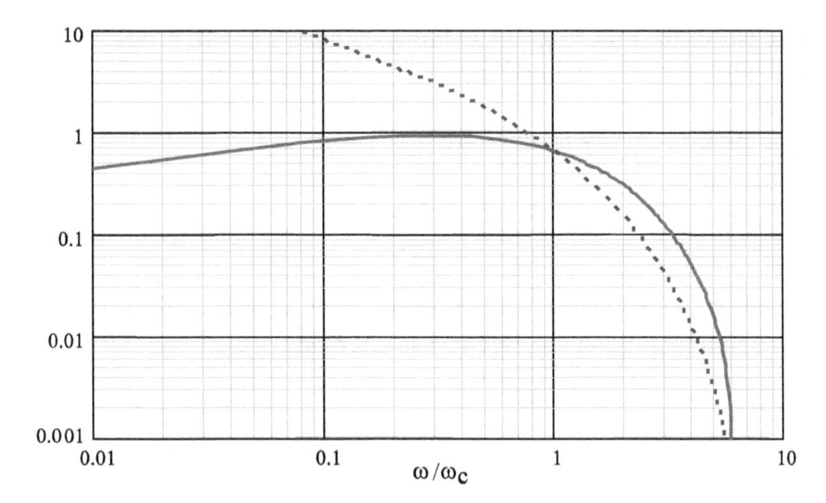

Fig. 11.7 Normalized power spectrum see Eq. (11.27) (solid line) and photon number spectrum (broken line) of synchrotron radiation

Photon Distribution

Sometimes it is necessary to pay attention to the fact that synchrotron radiation is emitted, as any electromagnetic radiation, in quanta (photons) of energy $\varepsilon_\gamma = \hbar\omega$, where $\hbar$ is Planck's constant. The spectral angular photon flux can be obtained from Eq. (11.24), dividing by $\hbar$:

$$\frac{d^2 N_\gamma}{d\Omega \, d\varepsilon_\gamma / \varepsilon_\gamma} = \frac{d^2 W}{d\Omega \, \hbar d\omega}. \tag{11.28}$$

Equation (11.28) calculates the number of photons N_γ emitted per unit solid angle into a relative photon energy interval $d\varepsilon_\gamma/\varepsilon_\gamma$. Again it is noted that this quantity refers to a single turn in the synchrotron.

The angular-integrated spectral photon spectrum corresponding to Eq. (11.26) can be expressed in the form

$$\frac{dN_\gamma}{d\omega} = \frac{\sqrt{3}\alpha\gamma}{\omega_c} \int\limits_{\omega/\omega_c}^{\infty} K_{5/3}(x)dx, \tag{11.29}$$

with the fine structure constant $\alpha = 1/137.036$. This spectrum is also depicted in Fig. 11.7.

Integrating Eq. (11.29) over all frequencies yields the total number N_γ of photons emitted per electron per turn in the synchrotron:

$$N_\gamma = \frac{5\pi}{\sqrt{3}}\alpha\gamma. \tag{11.30}$$

It is interesting to note that N_γ is typically about 100, i.e. it is a rather small number, although the photon number spectrum diverges for $\omega \to 0$, see Eq. (11.29) and Fig. 11.7.

The mean photon energy is given by

$$\langle \varepsilon_\gamma \rangle = \frac{8}{15\sqrt{3}} \hbar \omega_c. \tag{11.31}$$

The considerable granularity in the emission process has quite some impact on the electron beam parameters in electron storage rings [5].

11.1.1.3 Simple Means of Changing the Emission Spectrum

Users often don't appreciate the spectrum of synchrotron radiation, either because of its large frequency width or because it might not contain sufficiently high frequencies for the particular application. In the latter case, the most straight forward solution would be making use of the strong γ-dependence of Eqs. (11.24) and (11.26) and increasing the electron energy. However, there are often considerable technical limitations in this respect, in particular at electron storage rings operating in the GeV regime.

Wavelength Shifters

As the synchrotron radiation spectrum is normalized to the critical frequency ω_c, according to Eq. (11.18) the spectrum can also be hardened by increasing the magnetic field strength, thus reducing the bending radius ρ. To this end, often superconducting magnets are applied. In order to restrict the subsequent modification of the electron beam's design orbit to a small section of the storage ring, a sequence of dipole magnets is frequently used with zero net deflection angle. Such an arrangement is called *wavelength shifter*. The radiation properties are determined in the same way as for ordinary synchrotron radiation. Beyond hardening the spectrum and increasing the flux, there is also some advantage in terms of flexibility in the geometrical arrangement of radiation beam lines.

Short-Magnet Radiation, Edge Radiation

When discussing the characteristic time profile of synchrotron radiation it was assumed in the context of Figs. 11.4 and 11.5 that the electron would propagate in the magnetic field for sufficiently long time such that the radiation cone passes the observer's aperture in its entire angular extension of $\pm 1/\gamma$. To this end, the dipole magnet must have a length of at least

$$\Delta l_c \approx c\Delta t' \approx \frac{2\rho}{\gamma} \approx \frac{2m_0 c}{eB}. \tag{11.32}$$

For electrons and magnetic fields in the $B \approx 1$ T range, this results in a few millimeters, which would be of little relevance in most cases. However, if B is much weaker, or if protons are considered (e.g. at LHC/CERN), the resulting time profile of the radiation field is shorter than shown in Fig. 11.5 and assumed for the calculation of the synchrotron radiation spectrum. As the time profile gets shorter, its Fourier transform extends to higher frequencies, thus the spectrum gets "harder". In contrast to increasing B or γ, this hardening is, however, not accompanied by increased flux as the instantaneous properties of emission depend only on the local magnetic field strength which does not change by shortening the magnet.

A spectrum hardening effect similar to short-magnet radiation takes place if the magnetic field rises at the entry face of the magnet (or drops at the exit, respectively) over a distance comparable to or smaller than given by Eq. (11.32). In such case, the time profile of the radiation field exhibits a rising (or falling, respectively) edge steeper than that seen in Fig. 11.5. As a consequence, the Fourier transform extends to rather high frequencies. At high-energy proton synchrotrons this is being used to extend the spectrum towards wavelengths which are easy to observe.

11.1.1.4 Wigglers and Undulators

Definitions

Undulators and wigglers provide a periodic magnetic field over a part of the synchrotron's circumference. In most cases, the magnetic field perpendicular to the electron beam's design orbit can be described by a pure sinusoidal—at least within the small spatial area where the electron beam is present. If the field acting on the electron beam has only one non-zero Cartesian component, the device is called a planar undulator (or wiggler, respectively). The field close to the axis can then be described by

$$B_y(z) = B_0 \sin(k_u z)\,. \tag{11.33}$$

Here, B_0 is the field amplitude and $k_u = 2\pi/\lambda_u$ is the undulator wave number, with λ_u the undulator's period. The z-axis is along the electron beam's initial design momentum, and we have chosen arbitrarily the magnetic field to be in the vertical y-direction.

Typically, the length L_u of these devices is a few meters, and the field integral

$$I_1 = \int_0^{L_u} B_y(s)ds \tag{11.34}$$

is made zero such that there is no over-all deflection of the electron beam's trajectory.

A key parameter is the undulator parameter

$$K = \frac{eB_0\lambda_u}{2\pi m_0 c} \approx (0.934 B_0/\mathrm{T}) \cdot (\lambda_u/\mathrm{cm}) \,. \tag{11.35}$$

As this quantity refers to electrons (or positrons), we have assumed m_0 to be the electron rest mass m_e in the last part of the equation. A device with $K \leq 1$ is called undulator, while wigglers exhibit values $K > 1$. Sometimes devices with $K > 1$ are also called undulators if they are used in terms of radiation properties typical of undulators, e.g. when observing the line spectrum in forward direction (see below).

Undulators and wigglers are realized in basically three varieties of technology: iron-based electromagnets, permanent magnets and superconducting magnets. For details, see e.g. [6].

Particle Trajectory

The equation of motion of an electron (with elementary charge $-e$) moving in presence of the field of Eq. (11.33) reads in Cartesian components

$$\ddot{x} = \frac{e}{\gamma m_e} B_y(z) \cdot \dot{z}, \text{ and}$$

$$\ddot{z} = -\frac{e}{\gamma m_e} B_y(z) \cdot \dot{x}. \tag{11.36}$$

To first-order approximation, the periodic solution reads

$$x(t) \cong \frac{K}{\beta\gamma k_u} \sin(k_u \beta c t), \text{ or } x(z) \cong \frac{K}{\beta\gamma k_u} \sin(k_u z), \text{ and}$$

$$z(t) \cong \beta c t. \tag{11.37}$$

This motion is illustrated in Fig. 11.8.

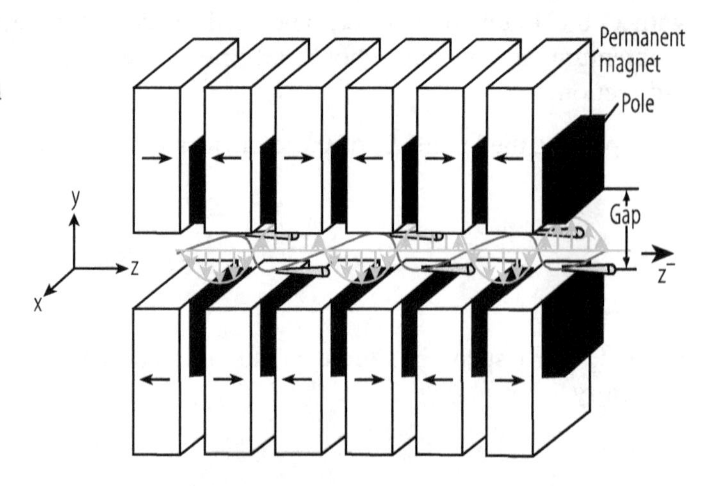

Fig. 11.8 Periodic electron motion (red) in a planar undulator fabricated in hybrid permanent magnet technology. The magnet field is indicated by green arrows

It should be noted that, in order this periodic trajectory to happen, the electron beam must enter the device on an orbit representing the appropriate initial conditions. In addition, the undulator field must begin and end with a quarter-period undulator section.

From Eq. (11.37) the maximum deflection angle can be calculated:

$$\vartheta_{\max} \approx \left(\frac{dx}{dz}\right)_{\max} = K \cdot \frac{1}{\gamma\beta} \approx \frac{K}{\gamma}. \tag{11.38}$$

The maximum orbit excursion of the electron is

$$x_{\max} \cong \frac{K}{\gamma k_u}. \tag{11.39}$$

Under many typical conditions, this results in only a few micrometers and is thus much smaller than the typical electron beam diameter.

In Eq. (11.37) the longitudinal motion $z(t)$was described only to first order, not taking into account its coupling to the transverse motion. A more precise, second order calculation results in

$$z(t) = \overline{v}_z t - \frac{K^2}{8\gamma^2 k_u} \sin(2\omega_u t), \tag{11.40}$$

where the abbreviation $\omega_u = \overline{\beta} c k_u$ and the average longitudinal velocity

$$\overline{v}_z \approx c\left[1 - \frac{1}{2\gamma^2}\left(1 + \frac{K^2}{2}\right)\right] \equiv \overline{\beta} c \tag{11.41}$$

have been introduced. According to Eq. (11.40), the oscillatory motion in the transverse plane translates into a small modulation of the longitudinal velocity around $\overline{v}_z$. This average longitudinal velocity differs, as described by Eq. (11.41), from c due to two effects: the factor $1 - 1/(2\gamma^2) \approx \beta$ describes by how much the electron's total speed differs from c. The factor $1 + K^2/2$ describes the *mean* additional longitudinal retardation due to the transverse velocity components.

Radiation Properties

Calculation of the radiation spectrum starts again from Eqs. (11.22) and (11.23), however, now the oscillatory motion in the undulator field must be considered when evaluating Eq. (11.23). Qualitatively, it should be clear from Figs. 11.4 and 11.9 that the time profile of the wiggler radiation field ($K > 1$) should resemble the standard dipole field case with the only difference that there will be a series of radiation spikes, see Fig. 11.10. The radiation spectrum differs thus not very much from the synchrotron radiation case but will be N times more intense with N being the number of wiggler periods (see Fig. 11.11).

On the contrary, in the undulator case $K \leq 1$, an observer will only see radiation field within a small angle of approximately $\pm 1/\gamma$ with respect to the forward

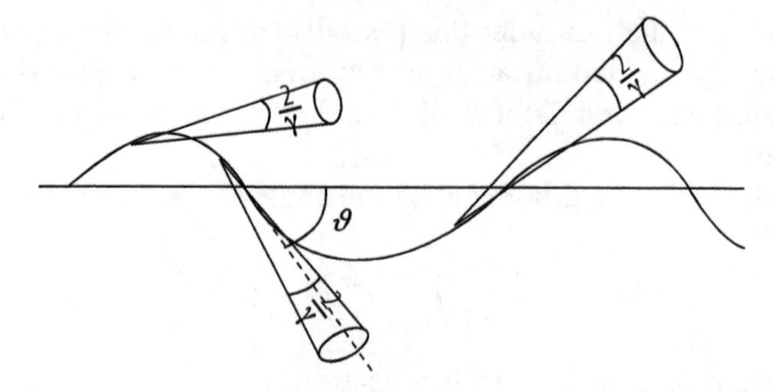

Fig. 11.9 Emission of radiation cones along the oscillatory trajectory in an undulator or wiggler. In this schematic, the wiggler case $K > 1$ is illustrated, where the deflection angles $\vartheta = K/\gamma$ are larger than $1/\gamma$

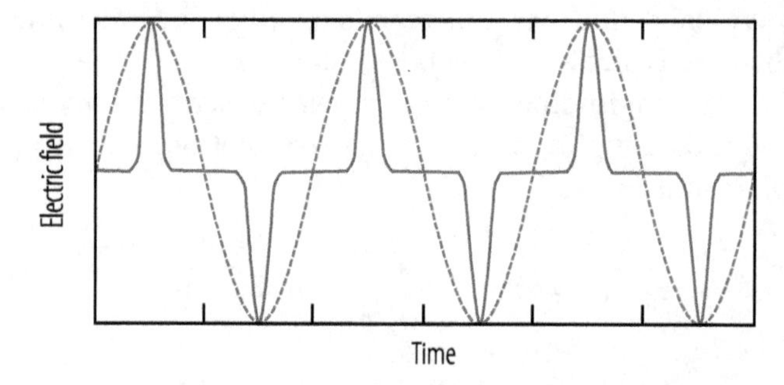

Fig. 11.10 Time profile of radiation field for wiggler (solid line) and undulator radiation (broken line). Note that these two time profiles do not belong to the same undulator period, although they have obviously the same fundamental period in time. The reason is that the slippage between the longitudinal electron velocity and c is larger in the wiggler than in the undulator case

direction, and within this angle he will observe a continuous oscillatory field with N periods. Fourier transform of this time profile results in a fundamental wavelength

$$\lambda_1 = \frac{\lambda_u}{2\gamma^2}\left(1 + \frac{K^2}{2} + \gamma^2\theta^2\right), \tag{11.42}$$

and higher harmonics $\lambda_k = \lambda_1/k$. From symmetry consideration one can immediately understand that in forward direction $\theta = 0$ there will be only odd harmonics.

The angular spectral energy distribution in forward direction (see Fig. 11.11) is given by [4]

$$\begin{aligned}\left(\frac{d^2W}{d\Omega d\omega}\right)_{\theta=0} \cong \frac{q^2N^2K^2\gamma^2}{4\pi\varepsilon_0 c\left(1+\frac{K^2}{2}\right)^2} \sum\nolimits_{k=2n+1} k^2\left(\frac{\sin\left(\pi N\frac{\omega-k\omega_1}{\omega_1}\right)}{\pi N\frac{\omega-k\omega_1}{\omega_1}}\right)^2 \\ \times\left| J_{\frac{k-1}{2}}\left(\frac{kK^2}{4\left(1+\frac{K^2}{2}\right)}\right) - J_{\frac{k+1}{2}}\left(\frac{kK^2}{4\left(1+\frac{K^2}{2}\right)}\right)\right|^2 .\end{aligned} \tag{11.43}$$

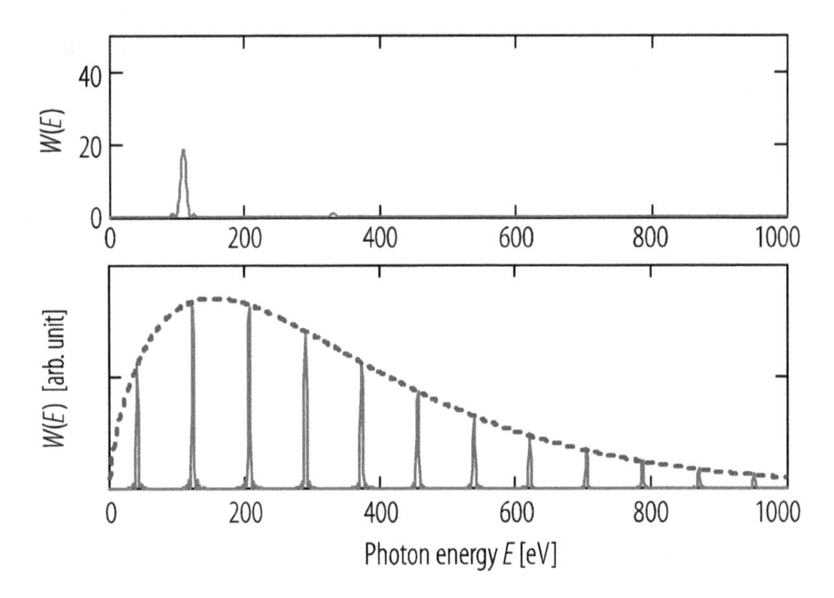

Fig. 11.11 Undulator spectrum in forward direction for $\gamma = 1000$ and $N = 10$ undulator periods. The undulator parameter is $K = 0.5$ (top) and $K = 2$ (bottom). The undulator period is $\lambda_u = 2$ cm. The vertical scales are in arbitrary units. The broken line in the lower diagram shows the spectrum of synchrotron radiation for a magnetic dipole field equal to the peak field of the undulator, with the vertical scale matched to the undulator spectrum arbitrarily

The sinc-function with argument $\pi N(\omega - k\omega_1)/\omega_1$ describes the spectral distribution of the individual undulator harmonics. The total line width of harmonics k is

$$\frac{\Delta\omega}{\omega_k} \approx \frac{1}{kN}. \tag{11.44}$$

The photon number spectrum radiated into the frequency interval $\Delta\omega/\omega_k$ in forward direction is given by

$$\frac{dN_\gamma}{d\Omega} = \alpha N^2 \gamma^2 \frac{\Delta\omega}{\omega_k} k^2 \frac{K^2}{\left(1+\frac{K^2}{2}\right)^2} \left| J_{\frac{k-1}{2}} \left(\frac{kK^2}{4\left(1+\frac{K^2}{2}\right)} \right) - J_{\frac{k+1}{2}} \left(\frac{kK^2}{4\left(1+\frac{K^2}{2}\right)} \right) \right|^2. \tag{11.45}$$

α is again the fine structure constant.

As compared to the synchrotron radiation spectrum Eq. (11.24), undulator radiation provides a spectral density in forward direction larger by some factor N^2. One factor N results from having a number of N undulator periods. The other factor N stems from the fact that, due to interference, the radiation spectrum is concentrated into narrow resonance lines. Thus, in forward direction, the field amplitudes add up coherently, not the intensities. The total radiation energy increases with N as expected.

Sometimes, e.g. for spectroscopy applications, only the spectral contribution of the undulator harmonics in forward direction is of interest. This radiation has a very narrow opening angle:

$$\sigma_\theta \approx \sqrt{\frac{1+K^2/2}{2\gamma^2 kN}} = \sqrt{\frac{\lambda_1}{\lambda_u kN}} = \sqrt{\frac{\lambda_k}{N\lambda_u}} = \sqrt{\frac{\lambda_k}{L_u}}. \tag{11.46}$$

It is some factor of $1/\sqrt{kN}$ narrower than with synchrotron radiation.

Helical Undulators

A helical undulator generates a magnetic field vector following a helical orientation like

$$\vec{B}_{\text{hel}} \cong B_0 \begin{pmatrix} -\sin k_u z \\ \cos k_u z \\ 0 \end{pmatrix}. \tag{11.47}$$

The resulting electron orbit is (appropriate initial conditions given) also a helix, with velocity components

$$\frac{v_\perp}{c} = \frac{1}{c}\sqrt{v_x^2 + v_y^2} = \frac{K}{\gamma}, \text{ and}$$

$$\beta_z = \frac{1}{c}\sqrt{v^2 - v_x^2 - v_y^2} = \sqrt{\beta^2 - \left(\frac{K}{\gamma}\right)^2} \approx 1 - \frac{1}{2\gamma^2}\left(1 + K^2\right), \tag{11.48}$$

which are both constant.

The major differences compared to planar undulators are:

(i) For calculation of the fundamental wavelength according to Eq. (11.42) one needs to replace $K^2/2$ by K^2.
(ii) According to Eq. (11.5), the electric field vector of undulator radiation in forward direction will also move on a perfect helix, thus this radiation will be circular polarized, and it will consist of only the fundamental harmonics, independent of the magnitude of K.

11.1.1.5 Radiation from Many Electrons

Brilliance

The emittance of a synchrotron radiation photon beam, i.e. the product of rms opening angle σ'_γ and rms source size σ_γ is limited due to diffraction. For a perfect

Gaussian optical mode, the minimum-emittance is achieved, given by [4, 7]

$$\varepsilon_\gamma = \sigma'_\gamma \sigma_\gamma = \frac{\lambda}{4\pi}. \tag{11.49}$$

with the wavelength of radiation λ.

As the emittance of the photon beam can never be smaller than indicated by Eq. (11.49), the very narrow opening angle of undulator radiation, Eq. (11.46), means that photons radiated from an undulator have a rather large apparent source size.

If a radiating electron beam has an emittance ε_e smaller than given by Eq. (11.49), the resulting radiation will be indistinguishable from radiation of a point source, if the opening angle σ'_e of the electron beam is matched to the opening angle of the radiation. A crude estimate for σ'_e is $\sigma'_e \approx \sqrt{\varepsilon_e/\beta}$, where β is the magnet optics beta function. Thus, a comparison with Eq. (11.46) would result in an estimated matched beta function $\approx L_u/4\pi$. However, such a small value cannot be kept constant within the undulator length if there are no further magnetic focusing elements. A more appropriate value is thus somewhat larger [4]:

$$\beta_{\text{match}} \approx \frac{L_u}{2}. \tag{11.50}$$

It should be noted that these considerations apply for both transverse directions x/y independently.

An electron storage ring providing an electron emittance

$$\varepsilon_e \leq \frac{\lambda}{4\pi} \tag{11.51}$$

in both x and y is said to work at the diffraction limit. Designing electron storage rings operating at (or close to) the diffraction limit is a difficult task, in particular if the wavelengths of interest are very short, e.g. in the hard X-ray regime. In tendency, this requires large ring diameters and sophisticated electron beam optics arrangements. Quite some efforts in this direction are under way at several big laboratories at the time of writing the present article.

In practice, Eq. (11.51) can often not be fulfilled. In this case, since both the photon and electron distributions are Gaussian in good approximation, the effective, combined distribution is given by a convolution and results in

$$\Sigma^2 \approx \sigma_\gamma^2 + \sigma_e^2. \tag{11.52}$$

This holds for both, transverse dimensions and opening angles, and applies again for both transverse directions x/y independently.

Many experiments using synchrotron (or undulator) radiation rely on the possibility to focus a small spectral fraction $\Delta\omega/\omega$ of the photon beam, selected e.g. by a monochromator, onto a small spot at the experiment. The figure of merit for such experiments is the brilliance B:

$$B = (dn/dt) / \left(4\pi^2 \Sigma_x \Sigma_y \Sigma_{x'} \Sigma_{y'} d\omega/\omega\right). \tag{11.53}$$

This quantity is also called brightness by some authors. dn/dt is the number of photons per unit time, and Σ_x, etc., are the rms photon beam extensions in x,y dimensions, convoluted in the spirit of Eq. (11.52).

In RF accelerators, electrons are arranged within bunches such that the peak beam current is much larger than the average current. Thus, the instantaneous value of brilliance during the very moment of bunch passage is much larger that its time average.

Undulator beam lines at modern storage rings reach peak brilliance values up to $B_{\text{peak}} \approx 10^{25}$ mm^{-2} mrad^{-2} s^{-1} (0.1%)$^{-1}$, see Fig. 11.12.

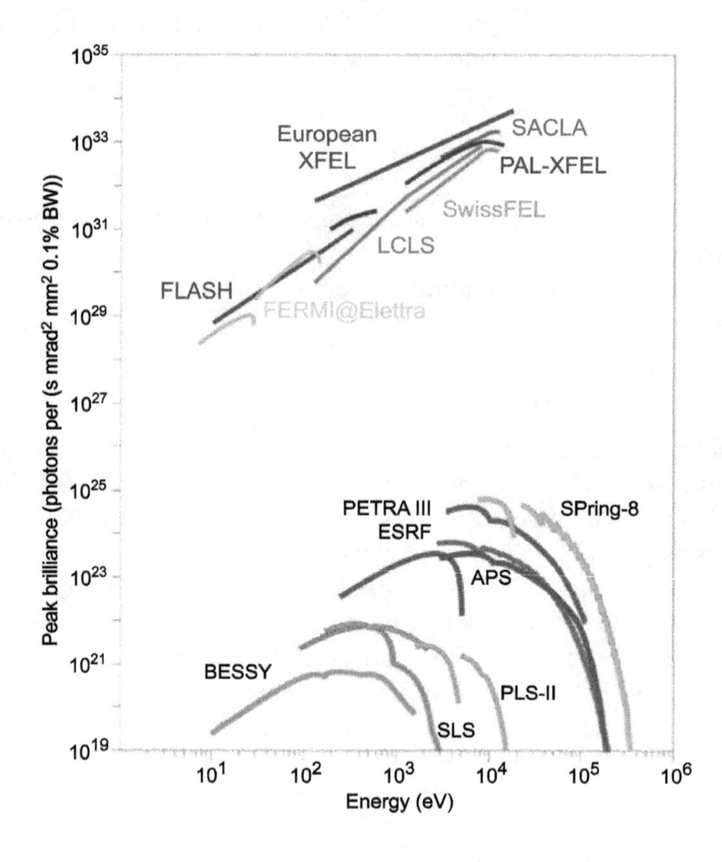

Fig. 11.12 Peak brilliance values delivered by typical undulator beamlines at some state-of-the-art synchrotron radiation storage rings. Values for FEL facilities are also shown. It should be noted that some of the FEL facilities (FLASH, European XFEL) are driven by a superconducting linear accelerator providing an electron bunch repetition rate larger by several orders of magnitude than normal conducting ones, resulting in a correspondingly larger rate of photon pulses. This difference is not visible in this figure of peak values but may be beneficial for several scientific applications

Average brilliance values can exceed values $B_{avg} \approx 10^{21}$ mm^{-2} mrad^{-2} s^{-1} $(0.1\%)^{-1}$.

In case of a perfectly diffraction limited beam in both *x/y* directions, it is concluded from Eqs. (11.49), (11.51), and (11.52) that Eq. (11.53) simplifies to:

$$B = 4\,(dn/dt)\,/\left(\lambda^2 d\omega/\omega\right). \tag{11.54}$$

Coherent Synchrotron Radiation

In sections "Basic Properties of Synchrotron Radiation" through "Wigglers and Undulators", emission by a point charge has been considered. In this case, the intensity scales with the radiating charge q squared. In other words, if this quasi-point charge consists of N_e electrons, the total power would be N_e^2 larger than the power P_0 radiated by a single electron.

In general, if radiation of many electrons is considered, we need to add up first the electric field vectors $\vec{E}_k$ of all electrons coherently before calculating the intensity. In the following we restrict ourselves to the one-dimensional case, where all electrons follow the same path, but just at different times ("pencil beam"). In most cases, the transverse extension of the bunch does not alter the results significantly.

In case of a pencil beam, the electric field contributed by any of the electrons differs from the field of the others only in terms of the phase factor $\exp\{i\omega(t - R_k/c)\}$ in Eq. (11.23), since, at time t, the electron with index k is located at a retarded distance R_k while others will be located at a different distance. Calling these individual phases differences φ_k, the power radiated by the ensemble scales like

$$\begin{aligned} P(\omega) \propto \left|E^2\right| &= \left(\vec{E}_1 + \vec{E}_2 + \ldots\right)\left(\vec{E}_1 + \vec{E}_2 + \ldots\right)^* \\ &= \sum_{k,j}^{N_e} \vec{E}_k \vec{E}_j^* \propto \sum_{k,j}^{N_e} e^{-i(\varphi_k - \varphi_j)} \\ &\propto N_e + \sum_{k \neq j}^{N_e} e^{i(\varphi_k - \varphi_j)}. \end{aligned} \tag{11.55}$$

How much of phase difference is introduced by a difference in longitudinal position z_k depends, of course, on the wavelength λ considered:

$$\varphi_k - \varphi_j = \frac{2\pi}{\lambda}\left(z_k - z_j\right) = k\left(z_k - z_j\right) = \frac{\omega}{c}\left(z_k - z_j\right). \tag{11.56}$$

If the longitudinal positions of electrons are random, the last term in Eq. (11.55) cancels and the resulting power is given by

$$P_{\text{inc}}(\omega) = N_e P_0(\omega), \tag{11.57}$$

with P_0 the power of a single electron. This contribution is called incoherent radiation.

If, however, the charge distribution is non-random, we describe the longitudinal charge distribution by a normalized density function $\rho(z)$. This function determines, when evaluating the double sum of phase factors in Eq. (11.55), how often each phase difference value appears, so the double sum can be translated into a Fourier transform of $\rho(z)$. The expectation value of the power radiated by the ensemble is then

$$P(\omega) = N_e P_0 + N_e (N_e - 1) \left| F_{\text{long}}(\omega) \right|^2 P_0(\omega) . \tag{11.58}$$

Here,

$$F_{\text{long}}(\omega) = \int_{-\infty}^{\infty} \rho(z) e^{i \frac{\omega}{c} z} dz \tag{11.59}$$

is the longitudinal form factor of the charge distribution inside the electron bunch. The contribution

$$P_{\text{coh}}(\omega) = N_e (N_e - 1) \left| F_{\text{long}}(\omega) \right|^2 P_0(\omega) . \tag{11.60}$$

to the total radiation power is called coherent synchrotron (or undulator, respectively) radiation. Due to the large number N_e of electrons in the bunch, P_{coh} can exceed P_{inc} by many orders of magnitude, even if the longitudinal charge profile has only a small Fourier content at the wavelength of interest.

At storage rings, where the charge distribution can be described very well by a Gaussian, with typical rms bunch length values of a few millimetres, P_{coh} becomes noticeable only at wavelengths in the far infrared. On the other hand, emission of radiation is suppressed very effectively by shielding due to the vaccum chamber. This happens, if in the vacuum chamber, which can be regarded as a curved waveguide, is no propagating mode whose phase velocity matches with the particle velocity. A typical parameter is the shielding wavelength $2\pi\sqrt{h^3/R}$ that can be calculated for a flat chamber of height h and curvature radius R [8]. Thus, coherent synchrotron radiation is not observed at most storage rings, but it has been provoked by arranging the storage ring parameters to achieve very short bunches [9].

The dramatic increase of radiation power can also be achieved if the longitudinal charge density is modulated at the wavelength of interest. This is the physical basics of the free-electron laser (FEL).

At linear accelerators much shorter bunches can be realized than at storage rings, and longitudinal bunch profiles may exhibit a very rich internal structure at scales down to the micrometer range, in particular at high-gain FEL facilities, where bunch lengths in the few micrometer range are needed, see Sect. 11.1.2. At such accelerators, infrared spectroscopy of coherent synchrotron radiation (or of optical transition radiation) represents a powerful tool for electron beam diagnostics [10].

11.1.2 Free-Electron Lasers

11.1.2.1 One Dimensional FEL Theory

The interaction of electrons with electromagnetic waves in an undulator of an FEL is sketched in Fig. 11.8. Important aspects of the FEL process can be described in a model, where electromagnetic fields do not depend on the transverse coordinates x,y and the trajectories of particles with different transverse initial conditions are just transversely shifted. This does not exclude transverse motion of particles.

For the description of the interaction of electrons with waves we need only three types of state quantities. Two of them are particle coordinates in longitudinal phase space: the ponderomotive phase ψ, see below and the relative energy offset η. A third quantity $\hat{E}_x$ stands for the complex amplitude of the electric field of the plane electromagnetic wave, and z, the length along the undulator axis, is the independent coordinate. To simplify the FEL equations, we are not interested in oscillations with the undulator period λ_u, but in the variation of our quantities from period to period or in average versus one period. Therefore the FEL equations for a bunch with N particles (of index ν) per period are in principle of the following type

$$\begin{aligned} \tfrac{d}{dz}\psi_\nu &= f_\psi\left(\psi_1\cdots\psi_N, \eta_1\cdots\eta_N, \hat{E}_x, z\right) \\ \tfrac{d}{dz}\eta_\nu &= f_\eta\left(\psi_1\cdots\psi_N, \eta_1\cdots\eta_N, \hat{E}_x, z\right) \\ \tfrac{d}{dz}\hat{E}_x &= f_E\left(\psi_1\cdots\psi_N, \eta_1\cdots\eta_N, \hat{E}_x, z\right). \end{aligned} \tag{11.61}$$

If the undulator parameters are z independent, and with some approximations they can be written as

$$\begin{aligned} \tfrac{d}{dz}\psi_\nu &\sim \eta_\nu \\ \tfrac{d}{dz}\eta_\nu &\sim \mathrm{Re}\left\{\hat{E}_x \exp\left(i\psi_\nu\right)\right\} \\ \tfrac{d}{dz}\hat{E}_x &\sim b = N^{-1}\sum\exp\left(-i\psi_\nu\right). \end{aligned} \tag{11.62}$$

The first equation relates the relative position in the bunch (ponderomotive phase) to the energy offset. This is just a linearized version of the equation of motion, that describes that a particle with more energy (higher η) is deflected less by the undulator, performs oscillations with smaller amplitude, has a shorter trajectory through one period and increases its phase (longitudinal position) relative to a particle without deviation from reference energy.

The second equation represents the change of particle energy caused by the electromagnetic wave. This depends on the phase of the particle ψ_ν relative to the phase of the wave $\arg\left(\hat{E}_x\right)$ and on the absolute amplitude $\left|\hat{E}_x\right|$. This clarifies the definition of the ponderomotive phase: ψ_ν is $\Delta z_\nu 2\pi/\lambda_l$ with Δz_ν the length shift relative to a reference particle of reference energy ($\eta_{ref} = 0$), and wavelength λ_l into forward direction according to Eq. (11.42).

The third equation assumes that only one spectral line (usually the fundamental wavelength) is excited and is characterized by the amplitude $\hat{E}_x$. This amplitude is driven by the microbunching of the particle distribution with the same wavelength. Therefore the Fourier coefficient of that wavelength, the bunching factor b, determines the increase of $\hat{E}_x$. This approach of a resonant and slowly varying amplitude (SVA) is used in most FEL programs (as in ALICE or GENESIS [11, 12]).

For a systematic presentation of FEL theory see [7, 13–16] and the literature quoted therein.

11.1.2.1.1 Low-Gain FEL Theory

The low-gain approximation assumes that the electromagnetic wave amplitude does not change during the passage of one electron bunch through the undulator. This assumption may sound like a contradiction to the purpose of the FEL of amplifying the intensity of a radiation field. The approximation makes it possible, however, to estimate in a simple way the amount of amplification as a cumulated effect during a single passage of the undulators. Therefore only the change of the ponderomotive phase and of the particle energy are considered. This causes microbunching and a net gain or loss of the particle energy. The microbunching can reach saturation if the electromagnetic wave is strong enough or the undulator sufficiently long. The net change of particle energy is an indirect method to calculate the change of field energy by utilizing energy conservation. The gain function $G = \Delta I/I_0$ is the ratio of the intensity change ΔI to I_0, the intensity of the electromagnetic wave assumed for low-gain theory. Strong bunching is not in contradiction to low-gain operation if G is small, but it requires presence of a very large initial intensity. This is normally accumulated over many round trips of the radiation within an optical cavity, with low-gain amplification at each round trip. The technical realization of such a FEL "oscillator" is thus based on the existence of an optical cavity consisting of low-loss mirrors.

The light wave co-propagating with the electron beam is taken as a plane wave $E_x(z,t) = E_0 \cos(k_l z - \omega_l t + \psi_0)$ with wavelength λ_l and wave number $k_l = \frac{2\pi}{\lambda_l}$. The motion of a particle in a planar undulator is described by Eqs. (11.37) and (11.40):

$$x(z(t)) \approx \frac{K}{\gamma k_u} \sin(k_u z(t))$$
$$z(t) \approx \overline{v}_z t - \frac{K^2}{8\gamma^2 k_u} \sin(2\omega_u t) .$$

The transverse velocity is

$$v_x(t) \approx \frac{K\overline{v}_z}{\gamma} \cos(k_u z(t)) , \tag{11.63}$$

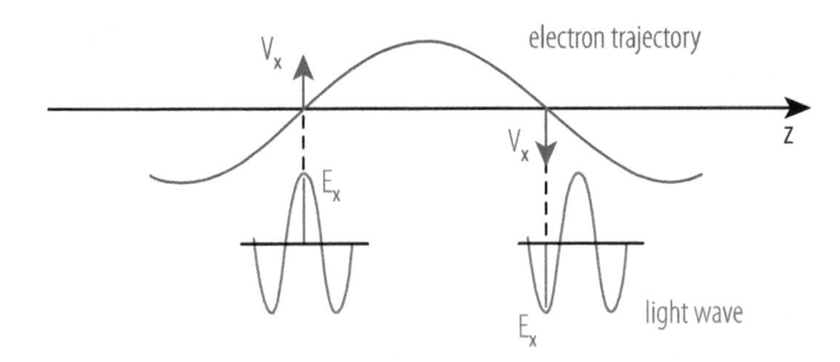

Fig. 11.13 Condition for sustained energy transfer from electron to light wave

and the time derivative of the electron energy is

$$m_e c^2 d\gamma/dt = -e\mathbf{v}\cdot\mathbf{E} = -eE_x\left(z(t),t\right)v_x(t). \tag{11.64}$$

Sustained energy transfer from electron to light wave requires that the light wave slips forward with respect to the electron by $\lambda_l/2$ per half period of the electron trajectory, see Fig. 11.13. This is fulfilled if $\overline{v}_z T_u + \lambda_l = cT_u$, with $T_u = 2\pi/\omega_u \approx \lambda_u/c$. This is a condition for the average longitudinal velocity Eq. (11.41) that relates the particle energy γ and undulator properties λ_u, K to the photon wavelength

$$\lambda_l = \frac{\lambda_u}{2\gamma^2}\left(1 + \frac{K^2}{2}\right). \tag{11.65}$$

It is the same wavelength as for the radiation of a single electron in forward direction. (Compare Eq. (11.42) with $\theta = 0$.) Slippages by $3\lambda_l/2$, $5\lambda_l/2$ ... are also permitted, leading to odd higher harmonics ($\lambda_l/3$,$\lambda_l/5$ ...) of the FEL radiation. However slippages of $2\lambda_l/2$, $4\lambda_l/2$ ... yield zero net energy transfer, hence even harmonics are absent in FEL radiation.

To calculate the average derivative of electron energy, as it is required for the FEL Eq. (11.62), we have to calculate the mean value of Eq. (11.64) in a time interval of the length T_u. Supposed the resonance condition Eq. (11.65) is fulfilled, the derivative of energy is a periodic function in time (with period T_u) and the mean value is

$$\langle d\gamma/dt\rangle_{T_u} = -e\frac{E_0\overline{v}_z}{2\gamma m_e c^2}\hat{K}\cos\psi_0, \tag{11.66}$$

with

$$\hat{K} = K\left[J_0\left(\frac{K^2}{4+2K^2}\right) - J_1\left(\frac{K^2}{4+2K^2}\right)\right]. \tag{11.67}$$

We rewrite Eq. (11.66) for $\eta(z) = (\gamma - \gamma_r)/\gamma_r$ with $z = \overline{v}t$ as independent coordinate to get an equation of the required type:

$$\frac{d\eta}{dz} = -e\frac{\hat{K}}{2\gamma^2 m_e c^2} \text{Re}\left\{E_0 \exp\left(i\psi_0\right)\exp\left(i\psi\right)\right\}. \tag{11.68}$$

Additional to the wave phase ψ_0 we consider the ponderomotive phase ψ which is individual per particle. This reflects that an arbitrary particle can be shifted in time relative to the "reference" particle with the trajectory $x(t)$, $z(t)$.

The slip $\Delta\psi = k_l\left(\overline{v}_z\left((1+\eta)\,\gamma_r\right) - \overline{v}_z\left(\gamma_r\right)\right)T_u$ of the ponderomotive phase in one undulator period due to η is caused by the energy dependency of the average longitudinal velocity Eq. (11.41). Therefore the longitudinal dispersion is in linear approximation

$$\frac{d\psi}{dz} = 2k_u\eta. \tag{11.69}$$

FEL Pendulum Equations and Gain Function

The particle dynamic in longitudinal phase space (ψ,η) is fully determined by Eqs. (11.68) and (11.69). They are formally equivalent to that of a mathematical pendulum. The phase space trajectories of few particles is illustrated in Fig. 11.14. The region of bounded motion is separated from the region of unbounded motion by a curve called the separatrix. Initially the particles are evenly distributed on the black line, but the endpoints are, in average, closer to the phase $\pi/2$. This illustrates microbunching. In the right diagram for the initial condition $\eta > 0$, we see a loss of the energy averaged over all particles. This net change of particle energy is an indirect method to calculate the change of field energy and light intensity by utilizing energy conservation.

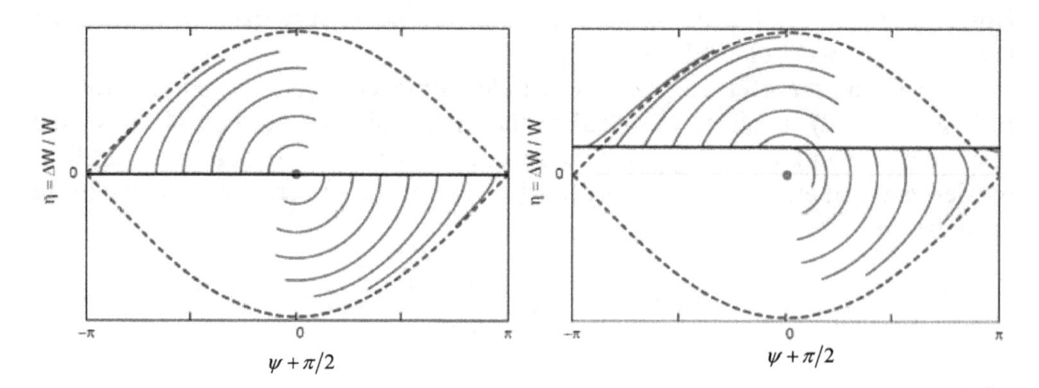

Fig. 11.14 Phase space (ψ,η) trajectories for 15 electrons of different initial phase (red) and separatrix (blue). Left picture: electrons are initially on resonance energy. Right picture: electron energy is initially above resonance energy

The FEL gain function is defined as the relative increase in light intensity during one passage of the undulator: $G = (I - I_0)/I_0 = \left|\tilde{E}_x/E_0\right|^2 - 1$. The gain is proportional to the negative derivative of the line-shape curve of undulator radiation (Madey theorem) [17]

$$G(\xi) = -\frac{\pi e^2 \hat{K}^2 N_u^2 \lambda_u^2 n_e}{4\varepsilon_0 m_e c^2 \gamma_r^3} \cdot \frac{d}{d\xi}\left(\frac{\sin^2 \xi}{\xi^2}\right), \tag{11.70}$$

with the detuning $\xi = \pi N_u \eta$, n_e the number of electrons per unit volume and N_u the number of undulator periods.

Low Gain FELs
Low gain FELs like the infrared FEL at JLAB consist of a short undulator in an optical cavity fed by a multi bunch source. Upon each bunch passage through the undulator the light intensity grows by only a few per cent, but after very many round trips a large average FEL beam power can be achieved, e.g. more than 10 kW in the infrared FEL at JLAB [16, 18].

11.1.2.1.2 High-Gain FEL Theory

For wavelengths far below the visible it is not possible to build optical cavities. An option is then making the undulator much longer, such that the gain during a single passage becomes attractive and an optical cavity arrangement becomes obsolete. In this case, the low-gain assumption $|G| < < 1$ cannot be made any more, meaning that the stimulation and propagation of electromagnetic waves must be taken into account. Results from LCLS (Linac Coherent Light Source, SLAC, Stanford, USA) are shown in Fig. 11.15 [19].

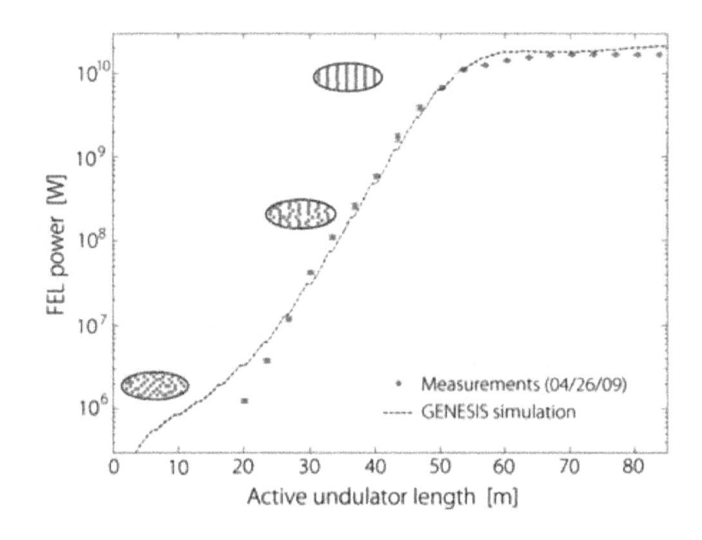

Fig. 11.15 Exponential growth and saturation of the FEL power in LCLS at $\lambda = 0.15$ nm as function of active undulator length [19]. The progressing microbunching is indicated schematically

Some important effects as microbunching, exponential growth and saturation of the FEL power and even the start-up-from-shot-noise can be studied with help of the one dimensional periodic model Eq. (11.62). The equations for $\frac{d}{dz}\psi_\nu$ and $\frac{d}{dz}\eta_\nu$ are the same as for the low gain case (Eqs. 11.68 and 11.69). The stimulation of the electric field amplitude is calculated with help of the wave equation.

Wave Equation

In one dimensional theory the electromagnetic field is described by a plane wave. According to this approach the finite beam cross-section is extended to infinity and each electron gets an infinite number of doubles in the expanded volume. The point particles are replaced by 1D charge sheets. All quantities as charge density, bunch current and electromagnetic fields are independent on the transverse coordinates x, y. The radiation field obeys the 1D inhomogeneous wave equation

$$\left[\frac{\partial^2}{\partial z^2} - \frac{1}{c^2}\frac{\partial^2}{\partial t^2}\right] E_x(z,t) = \mu_0 \frac{\partial j_x}{\partial t}, \tag{11.71}$$

where j_x is the transverse current density resulting from the sinusoidal motion. We make the ansatz

$$E_x(z,t) = \hat{E}_x(z,t) \exp\left[i\left(k_l z - \omega_l t\right)\right] \tag{11.72}$$

with a complex amplitude function $\hat{E}_x(z,t)$. The microbunching effect is anticipated by assuming a small periodic modulation $\hat{j}_1(z,t)$ of the longitudinal current density

$$j_z(z,t) = j_0(z - \overline{v}t) + \hat{j}_1(z,t) \cdot \exp\left(ik_l\left(\frac{z + \hat{z}\sin(2k_u z)}{\overline{\beta}} - ct\right)\right), \tag{11.73}$$

with $j_0(z - \overline{v}t)$ the current density without microbunching and $\hat{z} = K^2/\left(8\gamma^2 k_u\right)$. Note that the exponential term describes the fast oscillation in time and space, while $\hat{E}_x(z,t)$, $\hat{j}_1(z,t)$ and $j_0(\overline{v}t)$ are slowly varying amplitudes (SVA). The variations of SVA quantities in z and t are slowly compared to the λ_u respectively λ_l/c. Therefore the longitudinal oscillation of the bunch (compare Eq. (11.40)) can be neglected for SVA quantities but not for the exponential function. The transverse current density is

$$j_x(z,t) \approx \frac{v_x(z)}{\overline{v}_z} j_z(z,t), \tag{11.74}$$

with $v_x(z)$ from Eq. (11.63). Combining Eqs. (11.71), (11.72), and (11.74) and neglecting derivatives of SVA quantities compared to fast terms yields

$$\left(\tfrac{\partial}{\partial z}+\tfrac{1}{c}\tfrac{\partial}{\partial t}\right)\hat{E}_x\left(z,t\right)=-\tfrac{c\mu_0 K}{2\gamma_r}\hat{j}_1\left(z,t\right)\cos\left(k_u z\right)\exp\left(ik_u z\right)\exp\left(ik_l\hat{z}\sin\left(2k_u z\right)\right).$$

On the right hand side appears the product of three functions that are z periodic in λ_u respectively $\lambda_u/2$, but for the SVA approach we are only interested in variations large compared to the undulator period. Therefore we average this product along one undulator period and get

$$\left(\frac{\partial}{\partial z}+\frac{1}{c}\frac{\partial}{\partial t}\right)\hat{E}_x\left(z,t\right)=-\frac{c\mu_0\hat{K}}{4\gamma_r}\hat{j}_1\left(z,t\right). \tag{11.75}$$

Again the longitudinal oscillation is regarded by the modified undulator parameter $\hat{K}$, Eq. (11.67).

Periodic Approach

The periodic approach is applicable if the initial bunch and electromagnetic stimulation are sufficiently long in time, or more precisely if time variations are slowly compared to coherence time L_{coh}/c, with L_{coh} the coherence length defined below in Eq. (11.85). The time dependency in Eq. (11.75) is neglected $\hat{E}_x\left(z,t\right)\rightarrow\tilde{E}_x(z)$, $\hat{j}_1\left(z,t\right)\rightarrow\tilde{j}_1(z)$, $j_0\left(z-\overline{v}t\right)\rightarrow j_0$ and only particles in one micro-period are considered. We choose N electrons with start phases ψ_ν in the range $0\leq\psi<2\pi$. Then the modulation amplitude follows from Fourier series expansion

$$\tilde{j}_1=2j_0\sum_{\nu=1}^{N}\exp\left(-i\psi_\nu\right)/N. \tag{11.76}$$

The sum $\sum\limits_{\nu=1}^{N}\exp\left(-i\psi_\nu\right)/N$ is called bunching factor.

Coupled First-Order Equations

Combining Eqs. (11.68), (11.69) and (11.75) one obtains a set of coupled first-order equations

$$\begin{aligned}\tfrac{d}{dz}\psi_\nu&=2k_u\eta_\nu\\ \tfrac{d}{dz}\eta_\nu&=-\tfrac{e\hat{K}}{2m_e c^2\gamma^2}\,\mathrm{Re}\left\{\tilde{E}_x\exp\left(i\psi_\nu\right)\right\}\\ \tfrac{d}{dz}\tilde{E}_x&=-\tfrac{\mu_0 c\hat{K}}{4\gamma}\tilde{j}_1\end{aligned} \tag{11.77}$$

which, together with Eq. (11.76), describe the evolution of the phases ψ_n and energy deviations η_n of the N electrons, as well as the growth of $\tilde{E}_x(z)$ and $\tilde{j}_1(z)$. Longitudinal Coulomb forces ("space charge forces") are of minor importance in short-wavelength FELs [14] and are neglected here and in Eq. (11.78) below.

Third Order Equation
The main physics of the high-gain FEL is contained in the first-order Eq. (11.77) but these can only be solved numerically. If the modulation current $\tilde{j}_1$ remains small a linear third order differential equation for the electric field can be derived (see e.g. [7, 14, 20, 21]):

$$\tilde{E}_x''' + 4ik_u\eta\tilde{E}_x'' - 4k_u^2\eta^2\tilde{E}_x' - \frac{i}{\left(\sqrt{3}L_{g0}\right)^3}\tilde{E}_x = 0 \tag{11.78}$$

with the 1D power gain length

$$L_{g0} = \frac{1}{\sqrt{3}}\left[\frac{4\gamma_r^3 m_e}{\mu_0\hat{K}^2e^2k_u n_e}\right]^{1/3}. \tag{11.79}$$

Exponential Gain and Saturation
The solution of Eq. (11.78) is of the form $\tilde{E}_x(z) = \sum_{j=1}^{3} A_j \exp\left(\alpha_j z\right)$. For the special case $\eta = 0$ (electrons are on resonance energy) one finds $\alpha_{1,2} = \left(\pm 1 + i/\sqrt{3}\right)/\left(2L_{g0}\right)$, $\alpha_3 = -i/\left(\sqrt{3}L_{g0}\right)$. In case of laser seeding (see below) with an initial field E_0, all amplitudes are equal, $A_j = E_0/3$. The light power stays almost constant in the "lethargy regime", $0 \le z < \sim 2L_{g0}$, but then it grows exponentially (see Fig. 11.16)

$$P(z) \propto \exp\left(2\,\mathrm{Re}\left[\alpha_1\right]z\right) \equiv \exp\left(z/L_{g0}\right). \tag{11.80}$$

The Eq. (11.77) yield the same result as Eq. (11.78) in the lethargy and exponential gain regimes but describe FEL saturation in addition. The saturation

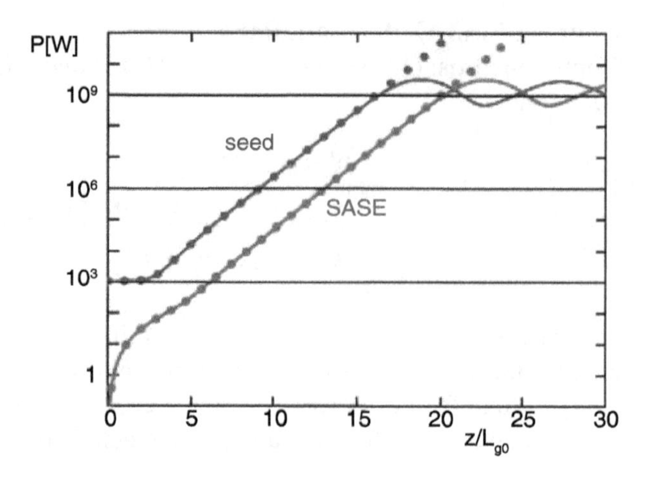

Fig. 11.16 FEL power as a function of z/L_{g0} in a seeded FEL (blue) and a SASE FEL (red). Solid curves: numerical integration of coupled first-order Eq. (11.77). Dots: analytic solution of third-order Eq. (11.78)

power is

$$P_{sat} \approx \rho P_b, \tag{11.81}$$

where P_b the electron beam power, and ρ is the dimensionless FEL (Pierce) parameter [21]

$$\rho = \frac{\lambda_u}{4\pi\sqrt{3}L_{g0}} = \left[\frac{\pi}{8}\frac{I_0}{I_A}\frac{\hat{K}^2}{\gamma_r^3 A_b k_u^2}\right]^{1/3}. \tag{11.82}$$

(I_0 peak current, $I_A \approx 17$ kA Alfven current, A_b beam cross section). For short-wavelength FELs ρ is typically of the order $10^{-4} \cdots 10^{-3}$.

FEL Gain-Function and Bandwidth

For a short undulator (length $\leq L_{g0}$), the high-gain FEL theory agrees with the low-gain theory, but in long undulators strong differences are seen: the gain is much larger and the gain-function approaches a Gaussian (Fig. 11.17). The high-gain FEL acts as a narrow-band amplifier with an rms bandwidth [7]

$$\sigma_\omega/\omega = \sqrt{\frac{\rho\lambda_u}{z}\frac{9}{2\pi\sqrt{3}}}. \tag{11.83}$$

Self-Amplified Spontaneous Emission (SASE)

SASE [20, 21] permits the startup of lasing without seed radiation. Intuitively speaking, spontaneous undulator radiation produced in the first section of a long undulator serves as seed radiation in the remaining section. More precisely speaking, because of the random electron distribution, the current contains a noise term which has a

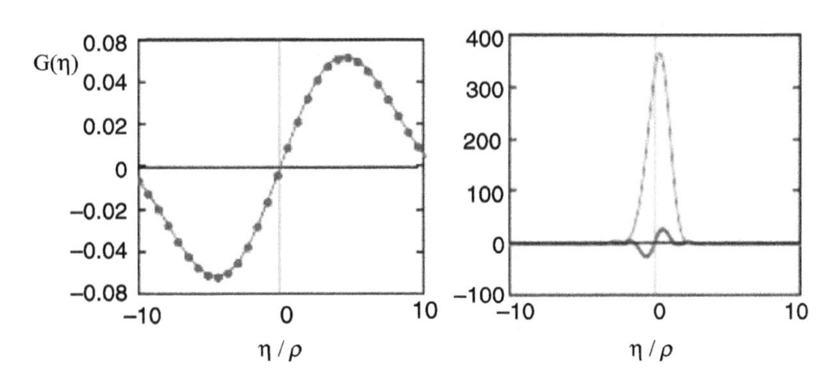

Fig. 11.17 FEL gain function $G(\eta)$ plotted vs. η/ρ at two positions in a long undulator: left $z = 1L_{g0}$, right $z = 8L_{g0}$. Red curves: high-gain theory. Blue dots/curve: low-gain theory (Madey theorem)

spectral component within the FEL bandwidth. The effective shot-noise power and modulated current density are [7, 22]

$$P_n = \rho\gamma m_e c^2 \sigma_\omega / (2\pi), \quad \tilde{j}_1(0) \approx \sqrt{e I_0 \sigma_\omega} / A_b. \tag{11.84}$$

The computed power rise for typical parameters of the soft x-ray FEL FLASH (see e.g. [7]) is shown in Fig. 11.16. Saturation is achieved at an undulator length $L_u \approx 20L_{g0}$. The SASE bandwidth at saturation is $\sigma_\omega^s/\omega \approx \rho$. SASE radiation exhibits shot-to-shot fluctuations in its output spectrum. The coherence length at saturation is

$$L_{coh} \approx \sqrt{\pi} c / \sigma_\omega^s = \lambda_l / \left(2\sqrt{\pi}\rho\right) \approx 11 \frac{\lambda_l}{\lambda_u} L_{g0}. \tag{11.85}$$

For a bunch length $L_b > L_{coh}$, the average number of spikes in the wavelength spectra is $M = L_b/L_{coh}$ (assuming full transverse coherence). M can be interpreted as the number of coherent modes of the FEL pulse. In the exponential gain regime the normalized radiation pulse energy $u = U_{rad}/\langle U_{rad}\rangle$ fluctuates according to the gamma distribution [14]

$$\rho_M(u) = \frac{M^M u^{M-1}}{\Gamma(M)} e^{-Mu}, \quad \sigma_u^2 = 1/M. \tag{11.86}$$

Phase Space and Simulation of Microbunching
The FEL dynamics resembles the synchrotron oscillations of a proton in a synchrotron or storage ring. In the (ψ, η) phase space the particles rotate clockwise, hence particles in the right half of an FEL bucket transfer energy to the light wave, while those in the left half withdraw energy, see Fig. 11.18 and compare Fig. 11.14. Equation (11.77) are well suited for modelling the microbunching. For $z \geq 12L_{g0}$ pronounced microbunches evolve in the right halves of the FEL buckets and increase the light intensity, while beyond $18L_{g0}$ they move into the left halves and reduce it. The FEL power oscillations in Fig. 11.16 are caused by this rotation in phase space.

Higher Harmonics
Close to saturation, the periodic sequence of narrow microbunches (see Fig. 11.18) corresponds to a modulation current with rich harmonics contents. In a planar undulator, odd higher harmonics will be amplified. The third (fifth) harmonic can reach 1% (0.1%) of the fundamental power.

11.1.2.2 Three Dimensional Effects

The realistic description of high-gain FELs has to be based on a three-dimensional (3D) theory, taking into account electron beam emittance and energy spread, and optical diffraction. In idealized cases, e.g. a round beam with uniform longitudinal charge density, an FEL eigenmode equation including all these effects can be

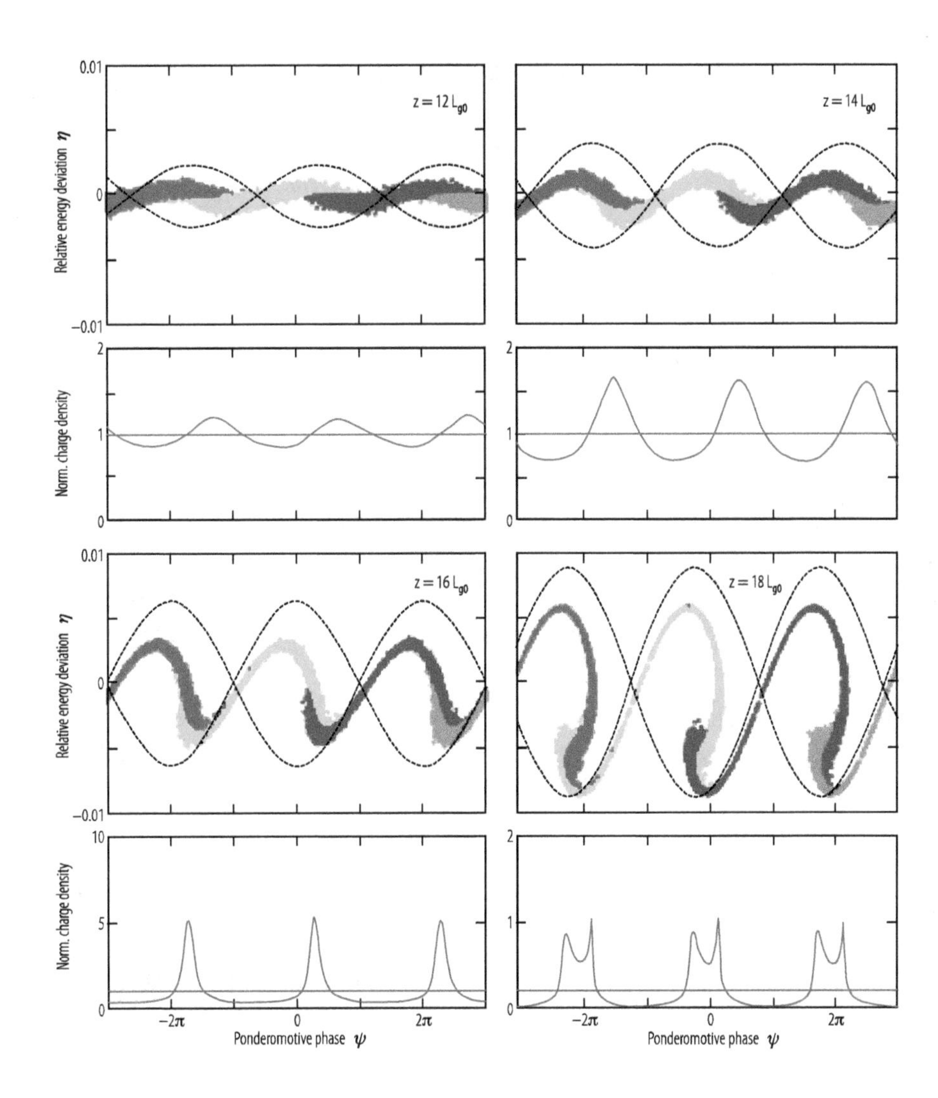

Fig. 11.18 Evolution of the microbunch structure at $z = 12L_{g0}$, $14L_{g0}$, $16L_{g0}$, $18L_{g0}$. Upper subplots: distribution of particles in (ψ, η) phase space. Three FEL buckets are indicated by dashed curves. Lower subplots: normalized charge density as function of ψ

developed [14, 23]. More realistic cases require sophisticated simulation codes such as FAST [24], GENESIS [11] or GINGER [25]. These are indispensable for the design of short-wavelength FELs.

3D Gain Length

The 3D gain length L_g is typically 30–50% longer than the 1D gain length L_{g0}. According to [23] L_g can be expressed in terms of three dimensionless parameters: $X_\gamma = L_{g0}4\pi\sigma_\eta/\lambda_u$ (energy spread parameter), $X_d = L_{g0}\lambda_l/\left(4\pi\sigma_r^2\right)$ (diffraction parameter, σ_r rms beam radius) and $X_\varepsilon = L_{g0}4\pi\varepsilon/(\beta_{av}\lambda_l)$ (angular spread parameter, ε emittance, β_{av} average beta function).

$$L_g = L_{g0}(1 + \Lambda) \tag{11.87}$$

$$\Lambda = a_1 X_d^{a_2} + a_3 X_\varepsilon^{a_4} + a_5 X_\gamma^{a_6} + a_7 X_\varepsilon^{a_8} X_\gamma^{a_9} + \\ + a_{10} X_d^{a_{11}} X_\gamma^{a_{12}} + a_{13} X_d^{a_{14}} X_\varepsilon^{a_{15}} + a_{16} X_d^{a_{17}} X_\varepsilon^{a_{18}} X_\gamma^{a_{19}}$$

$a_1 = 0.45, a_2 = 0.57, a_3 = 0.55, a_4 = 1, 6,$
$a_5 = 3.0, a_6 = 2.0, a_7 = 0.35, a_8 = 2.9,$
$a_9 = 2.4, a_{10} = 51, a_{11} = 0.95, a_{12} = 3.0,$
$a_{13} = 5.4, a_{14} = 0.7, a_{15} = 1.9, a_{16} = 1140,$
$a_{17} = 2.2, a_{18} = 2.9, a_{19} = 3.2.$

Gain Guiding

Gain guiding counteracts the diffractive widening of the FEL beam since most of the light is generated in the central core of the electron beam [26]. Gain guiding permits the FEL beam to follow slow, "adiabatic" motions of the electron beam and is thus crucial for the tolerable deviation of the electron beam orbit from a perfectly straight line in the long undulator of an x-ray FEL.

Transverse Coherence

The fundamental Gaussian mode TEM00 has its highest intensity on the beam axis while higher modes extend to larger radial distances. The TEM00 mode grows fastest along the undulator, owing to its superior overlap with the electron beam. Near saturation it dominates and the FEL radiation possesses a high degree of transverse coherence, as verified by double-slit diffraction experiments [27].

Seeding

FEL "Seeding" means to provide an initial electromagnetic wave at the entrance of the undulator with the help of an external laser pulse of adequate wavelength. Various seeding methods have been proposed to improve the longitudinal coherence properties of SASE radiation, and to reduce the relative timing jitter between pump and probe signals in time domain experiments aiming at femtosecond level resolution. Direct seeding requires a coherent signal well above the shot-noise level. In the VUV such signals may be obtained by high harmonic generation (HHG) in a gas [28, 29]. At shorter wavelengths, self-seeding [30] may be applied: a SASE signal, produced in a short undulator, is passed through a monochromator and serves as narrow-band seed radiation in the main undulators following further downstream. In a high-gain harmonic generation (HGHG) FEL [31], the electron beam is energy-modulated in an undulator by interaction with a powerful laser. A magnetic chicane converts the energy modulation to a density modulation. A second undulator causes the density-modulated beam to emit coherent radiation at a higher harmonic frequency. In an echo-enabled harmonic generation (EEHG) FEL [32], a second modulator followed by a second chicane are inserted before the radiator. The electron beam interacts twice with two laser pulses in the two modulators. The longitudinal phase space distribution becomes highly nonlinear, leading to density modulations at a very high harmonic number initiated by a modest energy modulation.

11.1.2.3 Technical Requirements

Very bright electron beams are required to drive ultraviolet and x-ray FELs. Higher peak current and smaller cross sections reduce the gain length (see Eq. (11.82)). High peak currents require longitudinal bunch compression, but the energy spread is increased by this process (which affects the gain length through X_γ in Eq. (11.87)). Very low-emittance beams can be generated by specially designed particle sources with photocathode or with thermionic emission. The beam cross section in the undulator can be reduced by stronger focusing (i.e., smaller β_{av}), but the increased angular spread will eventually degrade the FEL gain (through X_ε in Eq. (11.87)). The FEL design optimization is therefore multi-dimensional and beyond our scope here. Typical requirements on electron beams are

$$I_0 \geq 1\ \mathrm{kA}, \quad \sigma_\eta < \rho/2, \quad \varepsilon \sim \lambda_l/(4\pi)\,. \tag{11.88}$$

These requirements apply to the "slice" beam qualities defined on the scale of the coherence length (see Eq. (11.85)). For harmonic generation FELs, the slice energy spread should be much smaller than the ρ-parameter of the final amplifier because the additional energy modulation imposed on the beam becomes the effective energy spread there. Beam current, slice emittance and energy spread should be "flat" along the bunch in order not to increase the final radiation bandwidth. However, there are proposals for introducing, on purpose, some longitudinal variation of the electron energy within the bunch to generate a controlled frequency chirp in the FEL radiation pulse [33].

High-quality electron beams as described can be produced with linear accelerators but not with storage rings, mainly due to synchrotron radiation effects.

11.2 Accelerators in Industry

K. H. W. Bethge · J. Meijer

11.2.1 Introduction

Accelerators in their earliest stages of development served exclusively as tools of fundamental research; today they find application over a broad range of technical, industrial and medical areas.

One driving force of Ernest Lawrence for the development of the cyclotron was a cancer illness of his mother. Thus the possibility of treating tumors by radiation stands at the beginning of accelerator development.

Initially the application of accelerators in industry [34, 35] was, a by-product of fundamental research. This was because it was not needed continuously and because radiation safety requirements were in many cases too high to make industrial application economical.

On the other hand the installation of accelerators in special industrial laboratories or production lines does not find much publicity and therefore information about these accelerators is hard to obtain.

Two main lines of machine installations have to be considered (a) electron accelerators and (b) ion accelerators.

Additionally, the accelerators can be sub-divided into machines that analyze or modify materials. Electron microscopes are a simple example related to analysis of materials. In addition, there is Accelerator Mass Spectrometry (AMS) which is used to analyse samples of specific species for their applicability particularly in the pharmaceutical industry.

Some applications belong to different fields like the production of radioactive probes for medical application, e.g., PET (positron emission tomography) or SPECT (single photon emission computer tomography). Such probes are industrially produced in close consultation with medical consumers.

The industrial application of synchrotron radiation for photolithographic processes is in general part of the cooperation between large research institutes with suitable installations and relevant industrial partners: the beamtime is paid for by the relevant industries.

11.2.2 Electron Accelerators

Electron accelerators for industrial applications [36] are classified by their energy, focusing capability as well as by the obtainable dose. The large number of possible applications in different industries are listed in Table 11.1 [35]. The electron microscope is the classical low energy electron accelerator for analyzing samples. Unfortunately, the Scherzer theorem forbids a defocussing symmetric lenses with constant voltage. Electron microscopes with achromatic and aspherical lenses are very challenging. In the case of electron beam processing, the incident energy determines the maximum material thickness and the electron beam current and power determine the maximum processing rate.

The purpose of electron irradiation is the transfer of energy doses to produce neutral or positive charged radicals which react very rapidly with other chemical compounds. Spin correlation plays the important role particularly for neutral radicals. Thus the processes listed in Table 11.1 become production processes in the different branches of industry [36]. The energy transfer rate is nearly constant over a large energy range between 500 keV and several MeV, thus a homogeneous

Table 11.1 Possible applications of electron accelerators in industries [35]

Industries	Processes	Products
Chemical	Crosslinking	Polyethylene
Petrochemical	Depolymerisation	Polypropylene
	Grafting	Copolymers
	Polymerisation	Lubricants
		Alcohol
Electrical	Crosslinking	Building
	Heat-shrink	Instrument
	Memory	Telephone wires
	Semiconductor	Power cables
	Modification	Insulating tapes
		Shielded cable splices
		Zener diodes
		ICs, SCRs
Coatings	Curing	Adhesive tapes
Adhesives	Grafting	Coating paper products
	Polymerization	Wood/plastic composites
		Veneered panels
		Thermal barriers
Plastics	Crosslinking	Food shrink wrap
Polymers	Foaming	Plastic tubing and pipes
	Heat shrink memory	Molded packing forms
		Flexible packing laminates
Rubber	Vulcanization	Tire components
	Green strength	Battery separators
	Graded cure	Roofing membrane
Health	Sterilization	Medical disposals
Pharmaceutical	Polymer modification	Membranes
		Powders and ointments
		Ethic drugs
Pollution	Disinfection	Agricultural fertilizers
Control	Precipitation	Safe stack gas emission
	Manomer entrapment	Ocean-life nutrients form
Sludge		OSHA and EPA compliances
Pulp	Depolymerization	Rayon
Textiles	Grafting	Permanent-press textiles
	Curing	Soil-release textiles
		Flocked and printed fabrics
Aerospace	Curing, repair	Composite structures

IC integrated circuit, *SCR* silicon controlled rectifier (thyristor), *OSHA* Occupational Safety & Health Administration, *EPA* Environmental Protection Agency production of defects like point defects in semiconductors is possible. In comparison to ion beam related defects, radiation damages produced byelectrons are loosely connected, thus avoiding extended defects.

The electron irradiation of polymers produces a substantial fraction of radicals (charged or neutral) which react strongly with other polymers, monomers or additional substances. The processing of polymers by electron bombardment fulfills the demands of modern industrial production e.g. the compatibility with environmental requirements. This is particularly the case for all processes which include the curing of adhesive coatings. Furniture, cloth production and paper industries have adopted electron beam processes. Electron beams of lateral dimensions of 1 m or more are incorporated into production lines [36]. In many of these processes the need to remove dangerous fumes can be avoided.

The vulcanization process in the rubber industry proceeds under electron beam treatment without the addition of sulfur. The applied doses range from 30 to 50 kGy.

For curing of coatings, adhesives, ink on paper, plastic and metal substrates low energy accelerators (75–300 keV) are used. Such materials consist of oligomers (acrylated urethane polyesters, acrylated epoxies and polyethers) and monomers (trimethylolpropane triacrylate) to provide fluidity before curing. The radiation technique avoids the use of volatile solvents, thus helping to reduce pollution. Comparatively low doses of less than 50 kGy are used.

High energy accelerators (up to 10 MeV) are used to cure fiber-reinforced composite materials reducing processing time and the costs. Doses of 150–250 kGy are needed to obtain a combination of polymerization and crosslinking. These composite parts are now being used in automobiles and aircrafts.

The insulation of electrical wires as well as the production of jackets on multiconductor cables by radiation crosslinking was one of the first commercial applications. Materials used in this application are, e.g., polyethylene, polyvenylchloride, ethylene-propylene rubber, polyvenylidene fluoride and ethylene tetrafluoroethylene copolymer. Besides the creation of radicals, the electron beam can be used to create defects in crystals. One of the applications is the change or creation of the color of gemstones e.g. to create fancy diamonds. The electron beam in combination with a heat treatment changes the diamond color from yellow over green to red. These modified diamonds must be declared as "treated" to distinguish them from very rare colored natural diamonds with extreme high worth [37]. In the semiconductor industry, electron radiation is used to produce defects. These defects induce a life time reduction of charge carriers and allow a fast switching of electron high power devices [38]. This procedure is known as life-time killing and became very important for the production of high power devices e.g. for automotive or energy industrial products.

Radiation crosslinking stabilizes the initial dimensions of products and imparts the so-called "memory" effect. If the material is heated the original resistivity is maintained. Examples of commercial products using this effect are encapsulations for electronic components or exterior telephone cable connectors.

Low doses (30–50 kGy) of electron beam radiation are applied to automobile tires before assembling the different components.

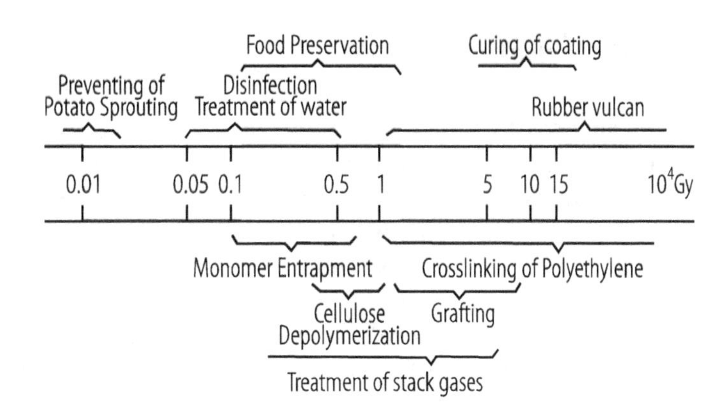

Fig. 11.19 Doses for electron irradiation [35]

Very high doses in the range 500–1000 kGy are needed to degrade polytetrafluoroethylene, which can be ground into fine particles or powder to be used as an additive to grease, engine oil, printing inks, coatings and thermoplastics.

Polypropylene and other polymers can be degraded by irradiation in air. This effect increases the melt flow and decreases the melt viscosity, which improves extrusion processes. By blending irradiated polymer with unirradiated material the desirable mechanical properties can be obtained. The doses for these processes range from 15 to 80 kGy.

An overview of applied doses is shown in Fig. 11.19.

The treatment of stack gases can be performed with electron beam radiation according to the reaction:

$$SO_2 + NO_x + 3/2 \cdot H_2O + O_2 + NH_3 \xrightarrow{\text{electron irradiation}} (NH_4)_2SO_4 + NH_4NO_3$$

The flue gas concentration of SO_2 and NO_x are reduced by adding oxygen and ammonia to a large extent depending on the applied energy dose. High-energy electron beams (few MeV) can penetrate in air through a thin foil and they are therefore applicable for medical applications. Low dose beams will be used to treat tumors by direct radiation or by production of high energetic X rays. With special structured filters a three dimensional irradiation is possible which reduces the damage of the healthy tissue.

High energetic electron beams (5–10 MeV) with high dose rates are used for sterilization of medical products and food e.g. spice. These kinds of machines produce a huge radiation level and must be installed in a locked safety radiation area. Typically, an endless treadmill is used to transport the items to the irradiation area. The electron beam is scanned in air over a stripe with a width of some cms.

The majority of electron accelerators for chemical processes are electron linacs in some cases as superconducting installations. Some special designed accelerators are the superconducting "Helios" compact electron synchrotron (Oxford Instruments) and the "Rhodotron" (Ion Beam Applications, Louvin-la-Neuve, Belgium).

11.2.3 *Ion Accelerators*

The structure of condensed matter materials can be modified by adding additional elements in order to achieve the special behavior required for a specific application. For this purpose, the application of accelerators is ideal, and enables high selectivity to be achieved in both the species of atoms to be implanted as well as their kinetic energy which allows control of the desired composition, including depth and thickness of the layers.

The application of accelerated ion beams in industry has two main directions, one for the analysis of materials and the other for the modification and production of materials. A new topic is the use of ion beams in medicine to treat cancer by irradiation with protons or carbon ions. In contrast to electron beams the energy transfer of heavy particles is concentrated at low kinetic energies (Bragg peak) and allows a three dimensional control of the desired cell damage.

11.2.3.1 Materials Modifications

The implantation of atoms into solid structures allows tailored changes in the properties of materials to be achieved. Targets are metals, semiconductors and insulators.

Tribological applications like the change of wear and friction are one of the domains in that field. Particularly the modification of surface hardness, ductility and lubrication effects are of interest. The implantation of nitrogen into metals and also alloys producing hardening of the surfaces has improved dramatically the efficiency of cutting tools.

Magnetic properties of materials have been modified by ion implantation [39]. The ion implantation into ceramics has started another industrial application [40].

In semiconducting materials, ion implantation is one of the most important processes for producing integrated electronic circuits. Mainly the basic semiconductors consisting of pure silicon and germanium are doped by elements of the third and fifth group to produce suitable acceptors and donors for transistor functions. Whereas for logic devices low energy implantation down to a few eV is important, the production of high power devices requires ion energies up to several MeV.

The profiles of implanted species are superior to those achieved by previously used diffusion processes. Ion implantation, however, produces radiation damage in the sample material which needs an annealing process in every case to produce functional electronic devices. Three dimensional ICs can be produced by a sequence of implantation and annealing if the necessary insulating layers can be provided. In many cases these layers are also produced by ion implantation.

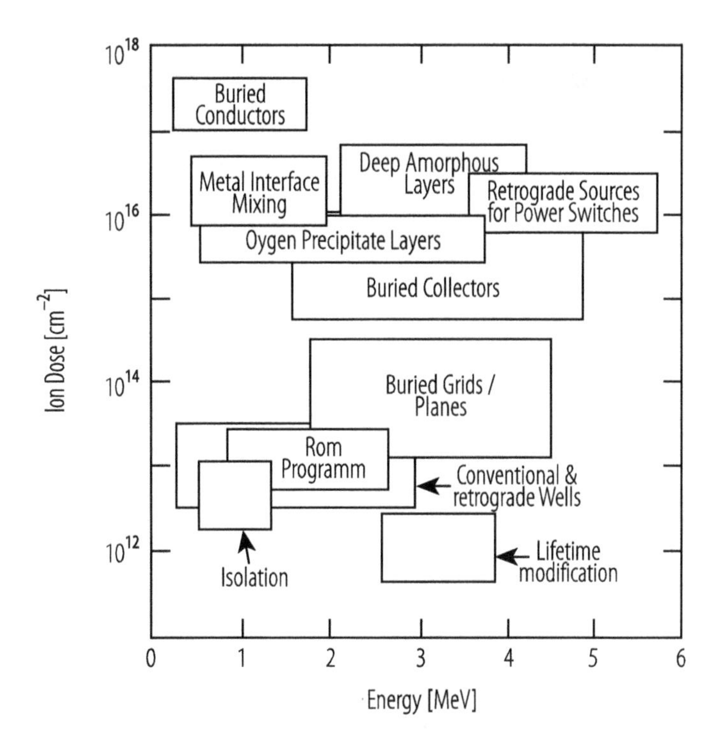

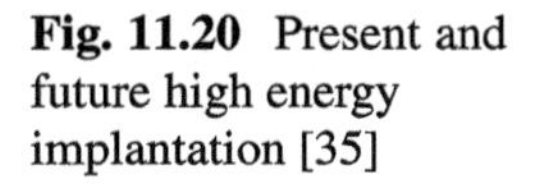

Fig. 11.20 Present and future high energy implantation [35]

The implantation can be used not only for elemental semiconductors, but also for compound semiconductors as GaAs, InP or superlattices. Figure 11.20 shows diagrammatically the whole field as dose dependent on ion energy [34]. To increase the performance of high power devices high energy ion beam irradiation is also used to reduce the charge carrier life time. In contracts to the common used life-time killing with electron beams, the damage area of ions is concentrated at the Bragg peak. This allows the reduction of the carrier concentration without a changing of the main electrical specification of the device.

Accelerators used for that work are van de Graaff accelerators, dynamitrons and in increasing number also RFQs. In all these types of accelerators the controllability of the ion energy is one of the major advantages.

11.2.3.2 Analysis of Materials

The analysis of materials uses many methods of the original nuclear research work, originally developed mainly for basic research. In many laboratories the RBS (Rutherford Back Scattering) is applied. The principle and one schematic spectrum are shown in Fig. 11.21. Ions impinge with energy E_0 on the sample, which is composed of the species A_A and A_B. After scattering they leave the sample with energy E_1 and are than energy analyzed. A typical spectrum is shown in the lower part of Fig. 11.21. According to the kinematics for each element an upper energy is measured which indicate the elements on the surface of the sample. With

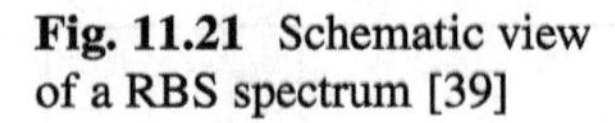
Fig. 11.21 Schematic view of a RBS spectrum [39]

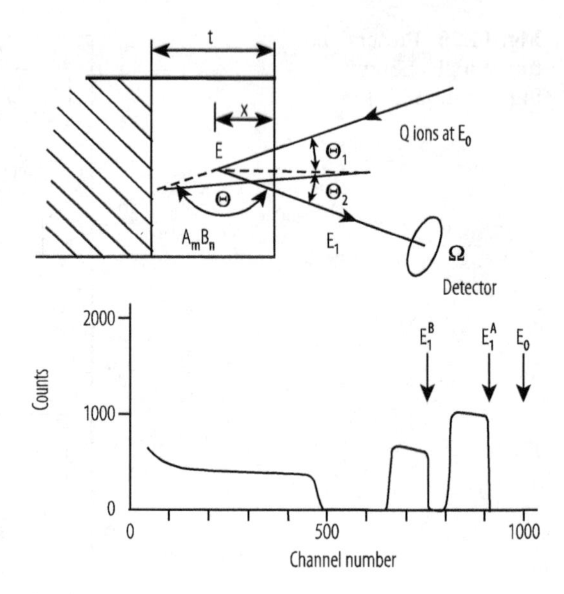

increasing penetration the energy loss of the incoming and outgoing particles has to be considered. Important quantities are the sensitivity, the mass resolution and also the depth resolution. They all depend on the available parameters such as the mass and energy of the projectile, and the energy resolution of the detector as well as on the cross section of the scattering process. In most cases ^{4}He with energies 0.5 MeV $< E <$ 2.5 MeV are used as projectiles. The Coulomb repulsion is sufficient for the analysis as Rutherford scattering. At energies $E >$ 2.5 MeV Rutherford scattering is applicable for higher masses, however, if the energy of the α-particles is high enough Non-Rutherford scattering and the influence of nuclear reactions has to be taken into account. In these cases the cross sections have to be determined experimentally.

The sensitivity for low target masses is limited due to the underlying spectrum from the heavy substrate. The mass resolution is optimal for lower target masses, poor for heavier species and can also be limited by intrinsic detector resolution. For increasing beam energy the mass resolution improves linearly.

The depth profiling is dependent on the energy loss factor which incorporates the energy loss of the incoming and outgoing particles. Small energy loss factors result in large profiling depth, but limited resolution. The resolution is limited at high energy by the energy loss factor and at lower energies by an increasing energy straggling.

At high energies with broad resonances in the excitation functions the limitations are similar to the conventional backscattering.

Many other ions are suitable for RBS like ^{12}C, ^{16}O, ^{19}F or ^{35}Cl with energies >5 MeV. The beam energy has to be so chosen, that Rutherford cross sections can be applied. In some cases screening corrections are required particularly for heavy target species. Increased beam energy cancels potential increase in sensitivity arising from higher Z_1. It is less sensitive for lighter species than conventional backscattering due to increase in cross section. Mass separation improves with increasing beam energy. An increasing energy loss factor compared to light ions leads to shallower profiling depths and superior potential depth resolution. All target species heavier than the beam ion may be analyzed simultaneously.

The most frequently applied detector is the surface barrier detector. In some installations time-of-flight detectors are used.

A further well established method is ERD (Elastic Recoil Detection) which is also called Forward Recoil Spectrometry (FRES). In this process heavy projectiles are used to measure the content of hydrogen. For that method an accelerator with sufficiently high energy is required.

A general review of the methods of backscattering analysis is given e.g. by J.A. Leavitt et al. in [41].

Many nuclear reactions are used for the analysis of materials. In particular, those reactions which exhibit resonances in their excitation functions. By varying the energy of the projectiles to excite these resonances and exploit their enhanced cross sections, materials samples can be scanned with higher sensitivity.

A very special method of analysis is the charged particle activation analysis. In Table 11.2 [42] for a few particles, their energy and the reactions used are listed.

Table 11.2 Reactions for activation analysis [39]

Incident ion	Energy [MeV]	Reaction	Sampling depth
p	10–30	(p,n)	100 μm to few mm
		(p,2n)	
		(p,pn)	
		(p,α)	
d	3–20	(d,n)	10 μm to 2 mm
		(d,2n)	
		(d,p)	
		(d,α)	
t	3–15	(t,n)	10 μm to 100 μm
		(t,p)	
		(t,d)	
^{3}He	3–20	(^{3}He,n)	few μm to 100 μm
		(^{3}He,2n)	
		(^{3}He,p)	
		(^{3}He,α)	
α	15–45	(α,n)	10 μm to some 100 μm
		(α,2n)	
		(α,3n)	
		(α,p)	
		(α,pn)	
		(α,αn)	

11.2.4 Accelerator Mass Spectroscopy

Many methods and also facilities which were developed for solving basic scientific questions have also found their way into the applications. One of these methods is the Accelerator Mass Spectrometry (AMS). It was developed for the measuring of very small ratios of radioisotopes to stable isotope concentrations by detecting atoms, rather than by detecting their radioactive decay. The samples which contain the isotopes under investigation are incorporated in the ion source either in gaseous form or as solid material in a sputter source. One characteristic isotope is ^{14}C, which has a half life of 5760 years, and is used for age determination in many organic substances. Trees, e.g., incorporate also ^{14}C as long as they live. From the analysis of the amount of remaining ^{14}C the age of a piece of wood can be determined.

Tracing the radioactivity in organic substances can also determine e.g. metabolism or the paths of drugs and medicine. By measuring the concentration of the parent ions in beams, rather than their detecting radioactive decay products, AMS can determine values within the low 10^{-16} range for the ratio of $^{14}C/^{12}C$. The accuracy is of the order of 0.3%. The measurements need much less time than detecting the radioactive decay products, an important factor in industrial application.

For industrial applications particularly compact instruments have been developed [42, 43].

As example—in Fig. 11.22 the installation of VERA (Vienna Environmental Research Accelerator) [42]—as a quite universal installation is shown. In this installation the measurement of negative as well as positive ions is possible. Particularly for the detection of ^{14}C the measurement of negative ions is important. The mass difference of ^{14}C and ^{14}N is so small that a magnetic separation of both is impossible, but since nitrogen forms no negatively charged ions ^{14}C appears as a very pure line in the spectrum.

Therefore the detection of ^{14}C allows the analysis of pharmaceuticals which is one of the dominant industrial applications of AMS.

11.2.5 Conclusion

The present experience has shown that accelerators with energies below 100 MeV fulfill the needs of industrial applications because electron and ion beams can steer fabrication processes in much smaller areas compared to conventional methods. Particularly the well-defined energies and doses are essential for tailoring layered structures in some cases also with isolating layers.

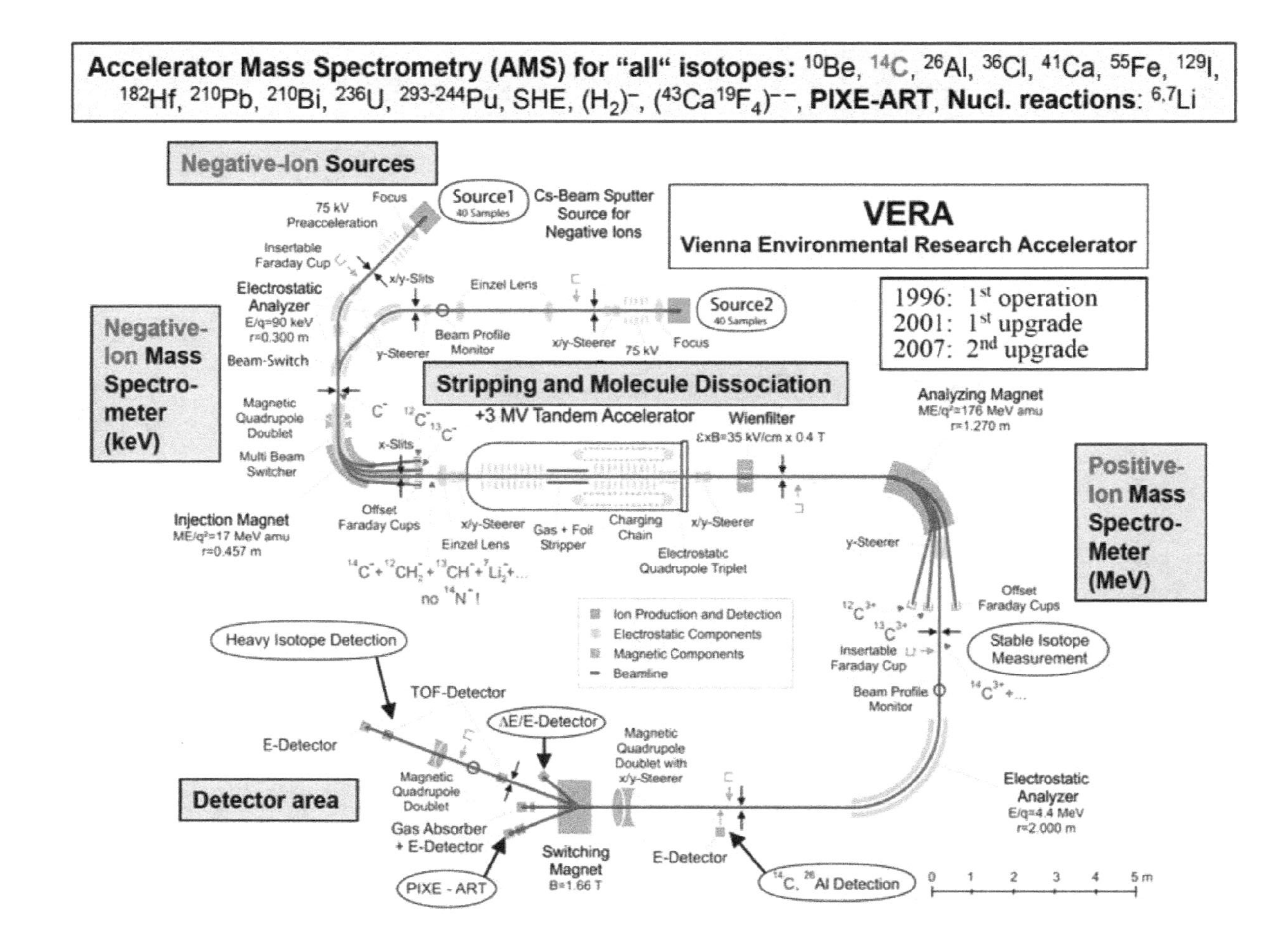

Fig. 11.22 Outlay of the VERA installation [42] (courtesy W. Kutschera)

11.2.6 Accelerator Suppliers

National Electrostatics Corp., Middleton, Wi, USA
High Voltage Engineering Europa B.V., Amersfoort, NL
Ion Beam Applications S.A., Louvain-la-Neuve, B

11.3 Accelerators in Medicine: Applications of Accelerators and Storage Rings

U. Amaldi · G. Magrin

Electron and hadron accelerators are widely employed in the production of radionuclides—used in diagnostic and brachytherapy—and in *teletherapy* i.e. in the formation of radiation beams directed from outside the body towards tumours and malformations. In this section these two topics are discussed in a historical perspective by describing both, the status of the art and the challenges that accelerator developers have to face today to meet the highest standards in nuclear diagnostic and cancer therapy.

11.3.1 Accelerators and Radiopharmaceuticals

11.3.1.1 History

The years that followed the invention of cyclotron by Ernest Lawrence were very prolific in defining what would be later known as nuclear medicine. In 1932, Lawrence, Stand and Sloan were able to produce, with their new 27-in. cyclotron, a proton beam of 4.8 MeV. In 1934 two important discoveries influenced the future use of that accelerator, alpha-induced radioactivity by Frédéric Joliot and Iréne Curie [44] and the neutron-induced radioactivity by Enrico Fermi [45]. Those discoveries convinced Lawrence to fully employ its accelerated proton, deuteron, neutron, and alpha particle beams for the production of artificial isotopes.

The medical exploitation of the newly radionuclides was clear in Ernest Lawrence's mind when in 1935 he called his brother John, a medical doctor from Yale school of Medicine, to join him in Berkeley to study the use of the new radioisotopes. Although the mainstream activity was nuclear research, a number of cyclotron-produced radionuclides was used for health applications in studies of physiology in animals and humans. The medical applications of radioisotope—radiotracing, endotherapy, and diagnostic—were defined at that time.

Radionuclides have the same chemical and physiological characteristics of the stable elements, so, when a radio-labeled organic compound is supplied to the organism, it follows the normal metabolism. From the mid 1930s, radionuclides were supplied to animals and humans and traced, with Geiger counters, in the internal organs following the physiologic uptake. Today the same concept is used in sophisticated diagnostics procedures as *Single Photon Emission Computer Tomography* (SPECT)—also called 'scintigraphy'—and *Positron Emission Tomography* (PET). Since the pioneering years, the challenge was to find the right radioisotope that, as part of a molecule, would enter in the physiological processes and would be preferably taken by a specific organ. The right isotopes must be chosen for their half-life and decay type, both compatible with the physiology of the processes investigated, the detection procedure, and the radiation protection of the patient.

Radio Phosphorus-32 (beta-emitter with half-life of 14.3 days), made from the 27-in. cyclotron was used as a tracer to study the absorption and metabolism of phosphorus so that the cellular regeneration activity could be followed over several weeks [46]. In 1936 John Lawrence gave for the first time Phosphorus-32 to a leukemia patient beginning the therapeutic use of artificial radionuclides [47].

In 1936 radioactive Sodium-24 (half-life of 15 h), obtained at the Berkeley cyclotron bombarding ordinary sodium with deuterium, was one of the first artificial radioisotopes applied in physiology to study transport and uptake in animal and humans and to determine the speed of absorption on the circulatory system. Radioactive sodium was first given to treat leukemia patients by Hamilton in 1937 [48].

In those years radionuclides were also used to study the transfer of traceable elements and compounds across the cell membrane. Today cytopathology and molecular biology are using radiotracers to study cellular diseases and physiology.

Larger cyclotrons were produced in the late 1930s at Berkeley, Harvard, and Leningrad. Berkley's 60-in. accelerator built in 1939 was used by Ernest Lawrence to produce radioactive gasses (nitrogen, argon krypton, and xenon) that were used to study the decompression sickness suffered by the aviators of the Second World War. The ^{81m}Kr gas is still used after 80 years in some studies of lung ventilation and functionality.

At the time of the first experiments in humans, the effects of X-rays on healthy biological tissues were known and avoiding overdose of ionizing radiation was a concern. It is important to mention that in 1936 John Lawrence, comparing identical doses of X-ray and neutrons [49], recognized different biological effects giving one of the earliest contribution to the knowledge of radiation qualities in correlation with internal and external radiotherapy and with radiation protection.

The isotope 131 of Iodine was selected by Joseph Gilbert Hamilton because its half-live of 8 days was considered to be the optimal time to treat the thyroid and avoid side effects to the patients. The radionuclide was isolated from a sample of tellurium bombarded with deuterons and neutrons from the 37-in. cyclotron. From 1941 Iodine-131 was used in diagnostic of thyroid functionality and, from 1946, in cancer therapy. In the 1950s Iodine-131, ingested as a liquid solution, became a common treatment. Commercially producer in nuclear reactors, it dominated for 40 years the market of radiopharmaceuticals for cancer treatment and it is still widely used today.

György Hevesy gave a decisive impulse to the studies with radiotracers of human metabolism and physiological processes. He carried on a program, supported by private funding, to introduce Sodium-24, Phosphorus-32, Cobalt-60, Technetium-99, and Iodine-131 into medical practice. In 1941 the first cyclotron fully dedicated to production of radionuclides for medical purposes was built at the Washington University of St Louis and used for the production of radionuclides for diagnostic and internal radiotherapy.

In the years that followed, the production of neutron-generated radioisotopes was moved to more efficient nuclear reactors. Nevertheless a number of cyclotron facilities, fully dedicated to medical applications, were built with the purpose of continuing the research of new products. In the 1950s, the use of Thallium-201 as tracer for cardiac flow revitalized the use of medical cyclotrons. Thallium-201 is produced via Lead-201 generator obtained in cyclotron from protons or deuterons beams.

In the late 1960s, a turning point that drastically transformed the medical application of cyclotrons was the idea of developing the radiochemistry labs inside the cyclotron facility, possibly close to the clinics where they were the patients were injected. Organic molecules containing short-living positron-emitting isotopes could be immediately transformed in pharmaceutical products and distributed in the facilities of the region to be used within few hours. The most important product was the Flourodeoxygloucose, FDG, made for the first time in 1976 with Fluorine-18, a positron emitter with a half-live of 110 min [50]. FDG was developed by a scientific collaboration of the Brookhaven National Laboratory, the US National Institute of Health and the University of Pennsylvania with the purpose of studying the brain metabolism.

Since then, FDG radiochemistry has evolved to fast and efficient production processes and today it is used in 75% of PET medical imaging [51]. The applications include diagnoses of brain diseases like epilepsy and dementia, examination of heart functionality, and detection of evolution of many different tumours.

New radiopharmaceuticals labeled with short half-lives positron emitters (in particular Fluorine-18 and Carbon-11) are still studied and developed today. Although Oxygen-15 is not commercially used for PET due to the fast decay time of 2 min, the Company Ion Beam Applications (IBA), built a 3 MeV deuterium cyclotron dedicated to the production of Oxygen-15 labeled gasses [52].

11.3.1.2 Accelerator for Radioisotope Production

Today the number of cyclotrons used for medical purposes, officially registered by the last IAEA survey published in 2006, is 262 [53]. The real number is certainly higher and it is increasing continuously by approximately 50 units per year, mainly because of new small PET accelerators. A survey made in 2010 combining inputs of production of the four major manufacturers already estimated in 671 the number of cyclotrons in operation [54].

The technology of cyclotrons has advanced in parallel to the development of PET. In the 1970s The Cyclotron Company (TCC) [55] studied the acceleration on negatively charged hydrogen (H^-). The extraction of the beam is in this way simplified since it can be obtained with a thin stripping foil conveniently positioned on the beam trajectory. In the foil all electrons are stripped leaving the protons, positively charged, deviate to the opposite direction under the effect of the magnetic field. The result is a clean extraction with efficiency very close to 100%. The elimination of beam loss (of the order of 25% with normal extraction) drastically reduces the activation of the cyclotron components simplifying the maintenance, the regular inspection, and the decommissioning. A remarkable characteristic of these cyclotrons is the possibility of obtaining simultaneous multiple beams making partial extraction with additional more foils.

In recent years several companies entered into the production of cyclotrons, machines that are stable, reliable, and can run continuously for days with limited supervision and maintenance requirements. The accelerators design has been adapted to the radiopharmaceutical request that is mainly concentrate on PET products. The cyclotrons dedicated to production of Fluorine-18, Carbon-11, Nitrogen-13, and Oxygen-15, have abandoned the versatile characteristics of the traditional machines and are made to accelerate only a specific ion (typically H^-), run at low energy (20 MeV and below), and at relative low current (approximately 100 μA).

Commercial cyclotrons characteristics are chosen taking into account the half-life of the radionuclides they produced. For the optimal logistics the cyclotrons should be homogeneously distributed in the territory with a large number of small cyclotrons, 10 MeV and 20 MeV, to fulfill the requests of FDG and other radiopharmaceuticals made on ^{18}F, ^{11}C, ^{13}N, ^{15}O, and a sparse distribution of cyclotrons with wider range of energy, 30 MeV or 70 MeV, able to produce all PET and SPECT radioisotopes used in research and in less common diagnostic (Fig. 11.23). Some of the higher-energy accelerators have the possibility of changing the

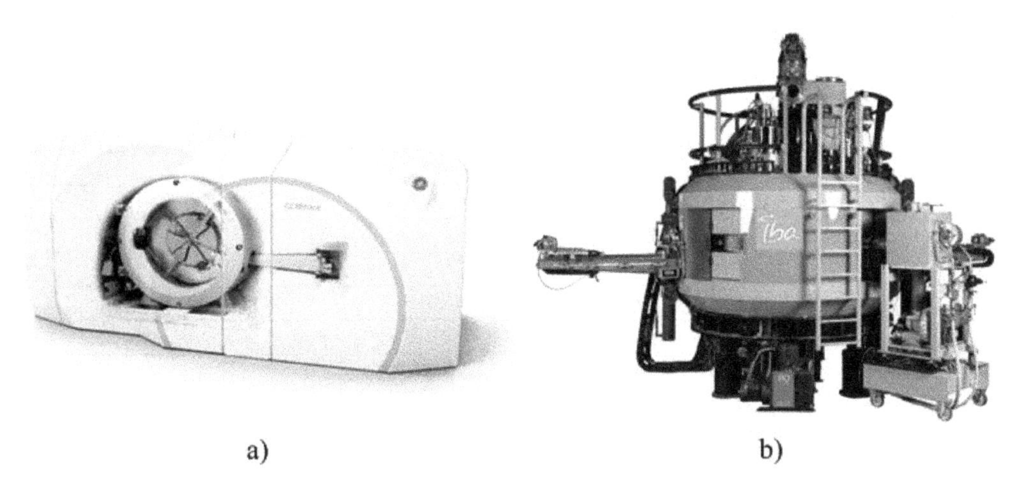

Fig. 11.23 Examples of two commercial cyclotrons. (**a**) The self-shielded 7.8 MeV cyclotron GENtrace produce by General Electric (courtesy of GE Healthcare) for common PET radioisotopes. (**b**) The 30 MeV cyclotron Cyclone 30 produced by IBA for a variety of PET and SPECT radioisotopes (courtesy of IBA SA)

energy moving the stripping foil along the cyclotron radius to adapt to the different nuclear reactions.

The commercial cyclotrons available today are presented here in two lists, Table 11.3 describes the cyclotrons in the range between 10 and 20 MeV, which correspond to the vast majority of machines today in operation for PET applications; Table 11.4 describes the cyclotrons with energies of 22 MeV and above for a more general use.

Besides cyclotrons, other particle accelerators have been considered for medical applications. Studies for low energy, high current accelerators have been financed in the 1980s, by the Star Wars program of the US Department of Defense. RFQ accelerators have been designed to obtain beam intensities of hundreds of milliampere at energies below 10 MeV. In the framework of the same program a 3.7 MeV, 750 μA Tandem Cascade Accelerator was designed, realized, and put into operation by

Table 11.3 Characteristics of commercial cyclotrons for range energies 10–20 MeV (from [53, 56, 57])

Company	Model	Description
Advanced Cyclotron Systems, Inc.	TR 14	11–14 MeV H^-, 100 μA
	TR 19	14–19 MeV H^-, 300 μA
Advanced Biomarkers Technology	BG-75	7.5 MeV H^+, 5 μA
Best Cyclotron System, Inc.	15	15 MeV H^-, 400 μA
China Inst. Atomic Energy	CYCCIAE14	14 MeV H^-, 400 μA
D.V. Efremov Institute	CC-18/9	18 MeV H^-, 9 MeV D^-, 100 μA
EuroMeV	Isotrace	12 MeV H^-, 100 μA
General Electric Healthcare	MiniTrace	9.6 MeV H^-, 50 μA
	PETrace	16.5 MeV H^-, 8.6 MeV D^-, 100 μA
Ion Beam Applications	Cyclone 3	3.8 MeV D^+, 60 μA
	Cyclone 10/5	10 MeV H^-, 150 μA, 5 MeV D^-,
	Cyclone 11	11 MeV H^+, 120 μA
	Cyclone 18/9	18 MeV H^-, 9 MeV D^-, 150 μA
Japan Steel Works	BC168	16 MeV H^+, 8 MeV D^+, 50 μA
	BC1710	17 MeV H^+, 10 MeV D^+, 60 μA
	BC2010N	20 MeV H^-, 10 MeV D^-, 60 μA
KIRAMS	Kirams-13	13 MeV H^+, 100 μA
Oxford Instrument Co.	OSCAR 12	12 MeV H^-, 60 μA
Scanditronix Medical AB	MC17	17.2 MeV H^+, 8.3 MeV D^+, 60 μA 12 MeV $3He^{++}$, 16.5 $4He^{++}$, 60 μA
Siemens	Eclipse	11 MeV H^-, 2 × 60 μA
Sumitomo Heavy Industries	HM 7	7.5 MeV H^-, 3.8 MeV D^-
	HM 10	9.6 MeV H^-, 4.8 MeV D^-
	HM 12	12 MeV H^-, 6 MeV D^-, 60 μA
	HM 18	18 MeV H^-, 10 MeV D^-, 90 μA

Table 11.4 Characteristics of commercial cyclotrons for energies of 22 MeV and above (from [53, 56, 57])

Company	Model	Description
Advanced Cyclotron Systems, Inc.	TR24	24 MeV H^-, 300 μA
	TR30/15	30 MeV H^-,1000 μA/15 MeV D^-, 160 μA
Best Cyclotron System, Inc.	35p	15–35 MeV H^-, 1000 μA
	70p	70 MeV H^-, 700 μA
China Inst. Atomic Energy	CYCCIAE70	70 MeV H^-, 750 μA
Ion Beam Applications	Cyclone 30	30 MeV H^-, 1500 μA/15 MeV D^-
	Cyclone 70	30–70 MeV H^-, 2 × 350 μA 35 MeV D^-, 17.5 MeV H_2^{++}, 70 MeV He^{++}, 50 μA
	Cyclone 235	240 MeV H^-
Japan Steel Works	BC2211	22 MeV H^+, 11 MeV D^+, 60 μA
	BC3015	30 MeV H^+, 15 MeV D^+, 60 μA
KIRAMS	Kirams-30	15–30 MeV H^-, 500 μA
Scanditronix Medical AB	MC30	30 MeV H^+, 15 MeV D^+, 60 μA
	MC32NI	15–32 MeV H^-; 8–16 MeV D^-, 11–23 MeV $3He^{++}$, 15–31 $4He^{++}$, 60 μA
	MC40	10–40 MeV H^+, 5–20 MeV D^+, 13–53 MeV $3He^{++}$, 10–40 $4He^{++}$, 60 μA
	MC50	18–52 MeV H^+, 9–25 MeV D^+, 24–67 MeV $3He^{++}$, 18–50 $4He^{++}$, 60 μA
	MC60	50 MeV H^+, 60 μA
	K130	6–90 MeV H^-, 10–65 MeV D^-, 16–173 MeV $3He^{++}$, 20–130 MeV $4He^{++}$, 60 μA
Sumitomo Heavy Industries	AVF series	30, 40, 50, 70, 80, 90 MeV H^+, 60 μA
	Ring Cyclotron 400	400 MeV H^+ (K = 400), 60 μA
	Ring Cyclotron 540	240 MeV H^+ (K = 540), 60 μA
	C235	240 MeV H^-, 60 μA

Science Research Laboratory in Massachusetts for the production of Nitrogen-13, Oxygen-15, and Fluorine-18 for PET [58]. Today, AccSys Technologie proposes a linear proton accelerator for PET isotope productions. This RFQ accelerates protons to 3 MeV with a current of 150 μA. The energy can be upgraded to 10.5 MeV coupling the RFQ to a Drift Tube Linac (DTL). The system is suitable for supplying PET radioisotopes to a single diagnostic centre.

11.3.1.3 The Radionuclides Used in Nuclear Medicine

In the world the request of radionuclides for imaging is constantly increasing. Multimodality scanning systems have been favorably accepted by the medical doctors. The leading companies report that 35% of the SPECT scanners produced are sold combined with Computed Tomography (CT). In the same way the request of PET/CT and PET/MRI systems that combine metabolic and morphological imaging is stably increasing and so is the demand of positron emitter radionuclides. In the years to come many factors will contribute to define the role of accelerators in the commercial production of medical isotopes. Unforeseen events as the shortage of Molibdenum-99 described at the end of this section can weaken a situation that was considered stable. The primary role of reactors, to which it is ascribed approximately 80% of the medical radioisotope production, can be diminished by downsides as the rigidity of a centralized production, the risk of incident in nuclear power plant, and the problems connected to radioactive waste. These drawbacks are strongly reduced with radioisotope production based on accelerators and this can play in their favor.

Today PET imaging is certainly the largest medical use of the cyclotron radioisotopes. IAEA reports in its 2006 survey that 75% of the cyclotrons are dedicated to FDG [53]. Concerning SPECT, some radioisotopes for (gamma- and gamma/beta-emitters) are produced only with cyclotrons (Iodine-123, Indium-111, Gallium-67, Cobalt-57) some others, as Copper-67 and Rhenium-186, are produced both with cyclotrons or reactors.

Immuno-positron emission tomography is an imaging technique that, using radio-labeled monoclonal antibodies, allows tracking and quantifying their distribution in the body, and can be applied to imaging of human malignancies. Zirconium-89 has a primary role in immunoPET being used in most of the diagnoses based on radioactive antibodies [59]. Its favorable characteristics are a half-life of 3.3 days which allows a manageable production and distribution, and the favorable decay mode where the unavoidable gamma rays have energies well distinguishable from PET photons. Finally Zirconium-89 production, via proton-neutron reaction in Yttrium-89, is optimal at energies of 14 MeV, widely available in cyclotron facilities [see Table 11.3].

Radiopharmaceuticals for therapy represent only 5% of the total production [60], nevertheless the research in the field of cancer targeted radionuclide treatments is active and the role of cyclotrons is becoming more and more important. The isotopes that better conform to the therapy are those that decay transferring locally all their energy. The favorable decay products are alpha particles, beta particles and Auger electrons. The optimal radionuclide is selected based on essential characteristics, which are (i) the possibility to associate it to molecules or, in case of radio-immunotherapy, to monoclonal antibody, with affinity to tumour cells, (ii) the right decay time to allow the kinetics and avoid overdose, (iii) the availability of the product in reliable quantities, and (iv) the affordable cost. The research is trying to balance these demanding needs and the number of radionuclides under examination is getting large.

Alpha-particles emitters are Astatine-211, produced with the reaction ^{209}Bi($^{4}{}_{2}$He,2n) by 28 MeV alpha particle beams, and Bismuth-212, obtained from a

Actinium-225 generator produced with the reaction ^{226}Ra(p,2n) by 22 MeV protons. Bromine-77, an Auger-electron emitter produced with either alphas (27 MeV) or protons (in various possible reactions starting at the energy of 13 MeV), is a good candidate for its simple association with physiological molecules. Among the beta emitters, Renium-186 is favorably regarded since, it belongs to the same chemical family of Technetium and therefore it can follow similar the chemical process well establishes for ^{99m}Tc.

Brachytherapy with radioactive seeds is an established way of endoradiotherapy based almost exclusively on reactor-produced radionuclides. One exception is Palladium-103, an Auger-electron emitter produced in reasonable quantities by 18 MeV cyclotrons.

Interest is rising in endoradiotherapy based on pre-therapeutic PET dosimetry. Two radioisotopes of the same element can be used for complementary proposes. First the pre-therapeutic positron emitter is injected to trace the physiological uptake of the element and then the therapeutic isotope is administered to the patient according to the dose estimation. Isotope pairs of Copper and Scandium (β^+ emitted from ^{64}Cu or ^{44}Sc used for PET imaging, β^- emitted from ^{67}Cu or ^{47}Sc used for therapy) are considered for these procedures [61].

It is important to underline the steering role played by the medical community in determining the future of diagnostic, with SPECT and PET, and of endoradiotherapy, which is today only a marginal segment. Medical doctors—who today use preferably one diagnostic modality, SPECT, and one radionuclide, Technetium-99m—with their future orientations will influence the development on new tools than could complement or substitutes the existing one. A challenge is also the complexity of transforming an effective radioisotope to an approved pharmacological product. The most important factors to consider are the availability and cost of the raw material, the access to accelerators or reactors with appropriate energy and fluxes, the existence of fast radiochemistry, and the logistic to delivery the radioisotopes on time before they decay.

To address the needs of the research of new medical applications, and also to partially take over the aging accelerators, some dedicated facilities have been put in place worldwide. Among others (i) ARRONAX in Nantes, financed by French regional and National authorities and the European Union, that hosts a cyclotron for H^- (up to 70 MeV, 350 μA protons), H_2^+, D^-, and He^{++}; (ii) LANSCE in Los Alamos, USA, that produces from 2005 medical radioisotopes from a high-current 100-MeV proton linear accelerator; (iii) the PEFP center in South Korea, a facility based on a 20 mA proton linear accelerator (an RFQ followed by two Drift Tube Linacs to energies of 20 MeV and 100 MeV) for medical and industrial applications.

A motivation for studying new applications of accelerators came from the worldwide shortage of Molybdenum-99, the generator of Technetium-99m. The production dropped to about 50% of the market needs for 16 months between 2009 and 2010. Commercial Molybdenum is a fission product of highly enriched uranium target produced almost exclusively in five reactors that are old and need continuous maintenance. Each year 30 millions diagnostic studies, 80% of all medical scans that use radioisotopes, are made with Technetium-99m (6 h half-life) from a Molibdenum-99 generator (66 h of half-life).

The shortage of Molybdenum-99 stimulated several initiatives for finding alternative solutions and the use accelerator was considered for the production of either Molybdenum-99 or directly Technetium-99m. If, from one side, a single accelerator facility is unable to compete in production with large reactors, on the other side it allows a better distributed production with advantages for the logistics, the possibility of producing directly technetium-99m for locally distributed end users, and undisputable environmental benefits. The accelerators considered for different productions processes are electrostatic accelerators, electron linacs, and cyclotrons.

Two Canadian initiatives were established to produce Technetium-99m trough the reactions ^{100}Mo(p,2n)^{99m}Tc [62]. The projects use existing cyclotrons available for positron emission tomography (PET) from General Electric (130 μA, 16.5 MeV) and Advance Cyclotron Systems, Inc. (300 μA, 19 MeV) and the productions should start in 2017. The clinical trial which compared reactor- and cyclotron-produced Technetium-99m was concluded in 2017 receiving the approval from Health Canada.

Some project plan to substitute reactor-generated neutrons with accelerator-generated neutrons. To reach the capabilities of a reactor production, the neutron rate should exceed 10^{14} neutrons per second. The concept developed by the company Shine is based on the fusion of deuteron and tritium for producing neutrons (and helium). The electrostatic accelerated deuterons (300 keV, 60 mA) are directed toward a tritium target. The resulting neutrons ($5 \cdot 10^{13}$ neutrons per second) cross a natural uranium target for neutron multiplication and irradiate a low enriched uranium target producing, among other fission products, Molybdenum-99. The first facility is in construction phase.

The company NorthStar is developing a facility for the production of Molybdemum-99 from photo-nuclear reactions ^{100}Mo(γ,n)^{99}Mo using electron linacs. Two accelerators (40 MeV, 3 mA each) direct the electron beams from opposite sides to a Molybdenum-100 target where the bremsstrahlung photons are created. The production is expected to start in 2017 [63].

Another project linked to TRIUMF is the development of a high-current electron linac (50 MeV, 100 mA) for the generation of photons that, directed to a target of Uranium-238 produce Molybdenum-100 from photo-fission.

Electron linacs for the generation of neutrons are foreseen in different project from Shine (35 MeV, 0.6 mA), and Niowave (superconductive 40 MeV, 2.5 mA). The generated neutrons irradiate a target containing a solution of depleted uranium.

Several other ways of using accelerators for production of Molibdemun-100 or Technetium-99m have been proposed and are still under study and development [64].

11.3.2 Accelerators and Cancer Therapy

11.3.2.1 History

Conventional Therapy

The roots of brachytherapy (discussed in section 11.3.1.3) and teletherapy date back to the discoveries of X-rays and radium made by Roengten and the Curies in the years 1895–1897. X-ray tubes and gamma-ray sources were soon employed in medical or technological uses. Teletherapy with X-ray photons was performed to cure superficial tumours few years after the discovery.

The first electron linac was built in 1947 at Stanford by Bill Hansen and his group for research purposes [65] and was powered by a klystron produced by Varian. Soon after, this new tool superseded all other electron/photon sources. Also in 1947, in England, Fry and collaborators build a 40-cm linac that accelerated electrons from 45 to 538 keV [66].

In the years that followed, those two groups and others, in particular the group of John Slater at MIT, studied the parameters of the irises to improve the power efficiency of the structure and to adapt the traveling wave parameters of wavelength and phase velocity. Around 1950 megavoltage' tubes were built and, complementing the 'cobalt bombs' entered clinical practice. The positive surprise was that, due to the longer ranges of the electrons—put in motion by the gammas of the beam mainly through the Compton effect—these 'high-energy' radiations had a much better sparing of the skin.

In those years a process was initiated to adapt the complex machines developed for research purposes to the clinical environment and to the treatment needs. After a short season in which the X rays of energy larger than 5 MeV were produced with medical 'betatrons', electron linacs, running at the by-now standard 3 GHz frequency, became the instrument of choice. Varian in 1960 produced for the first time a linac that, mounted on a gantry, rotated around the patient and in 1968 the first medical machine based on standing-wave acceleration.

From the side of the traveling-wave accelerators, further studies have improved the efficiency and the start up time of the radiation so that today both designs are still used in medical linac: the two major manufacturers, Varian and Elekta (see Table 11.5) produce accelerators the first based on standing-wave and the second on travelling-wave. The average accelerating fields are in the range 10–15 MV/m.

In 2017 about 12,000 [67] linacs are installed in hospitals all over the world and in the developed countries there is one linac every 200,000–250,000 inhabitants. It is interesting to remark that all the electron linacs used to treat patients are close, as far as the overall length is concerned, to the 27 km circumference of LHC.

About 50% of all tumour patients are irradiated as exclusive treatment or combined with other modalities so that radiotherapy is used every year to treat about 20,000 patients on a Western population of 10 million inhabitants. On average, 40% of the patients who survive 5 years without symptoms have been irradiated.

Table 11.5 A selection of commercial electron linacs for radiotherapy

Company	Complete system	Acceleration characteristics	Maximum photon energies	Tumour conformation
AccuRay	Cyberknife®	Standing-wave, X-band	6 MV	Computer-controlled robotic arm
Elekta AB	Versa™	Traveling-wave, diode injection, S-band	18 MV	80 multileaf collimator
Siemens Medical Solutions	Artiste™	Standing-wave, triode injection, S-band	18 MV	82 multileaf collimator
AccuRay	TomoTherapy®	Standing-wave, S-band	6 MV	64 multileaf collimator
Varian Medical	HyperArc™	Standing-wave, triode injection, S-band	20 MV	120 multileaf collimator

This enormous development has been possible because of the advancements made in computer-assisted treatment systems and in imaging technologies, such as CT imaging, PET scans, MRIs. The most modern irradiation techniques are the *Intensity Modulated Radiation Therapy* (IMRT), which uses 6–12 no coplanar and non-uniform X-ray fields, and the *Image Guided Radiation Therapy* (IGRT), a technique capable of following tumour target which moves, mainly because of patient respiration.

Varian is the worldwide market leader with more than 50% of the share of electron linac for radiotherapy. One of its last development, HyperArc, is an irradiation system in which the uniform prescribed dose is delivered in a single revolution of the gantry. The times are drastically reduced thanks to the computer assisted continuous variation of the beam intensity and the use of a multileaf collimator that adapts the irradiation fields to the requirement of the treatment planning.

Cyberknife produced by Accuray (USA) is an original development in which a robotic arm substitutes the gantry to support the accelerator. To reduce dimension and weight, a X-band, 6 MV linac is chosen. The resulting pencil beam can penetrate and aim to the tumour with the optimal direction to spare the organs at risk.

In helical TomoTherapy a narrow x-ray beam constantly irradiates the tumour while rotating around the patient and, at same time, produces an image of the patient organs. The combination with the movement of the couch creates the helical irradiation which is modulated in intensity and collimated by a multileaf system to conform to the requirements of the panning. In 2011 the company TomoTherapy has been bought by Accuray.

The so-called MRI-guided radio therapy, i.e. the simultaneous combination of radiation therapy treatment and MRI scanning, has been performed for the first time by the company ViewRay. The system based on a 6 MeV linac, employs a MRI with 0.35 T magnets. A similar system which combines a 1.5 Tesla MRI and a

7 MeV linac, was developed and built by a collaboration between Elekta, Philips, and several clinics from Europe and North America [68]. The advantages of MRI scanning are the capability of identifying movements of soft tissue and organs, which are not detectable with other imaging media, and the absence of ionizing radiation exposure. Today the average frame rate during dynamic acquisitions is of the order of 10–20 frames per seconds with spatial resolution of the order of 2 mm.

The major companies producing electron linacs for photon therapy and the characteristics of some of the most advanced models are listed in Table 11.5.

It must be mentioned that, because of the increase risk due to the non-negligible production of photoneutrons, the use of the 16–20 MV photons is limited to the cases for which the size of the patient and the depth of the tumour require it. In normal conditions 6–8 MV energies are employed.

Neutron Therapy

At variance with the case of conventional radiotherapy, the use of atomic nuclei to treat cancer has benefited of a long series of accelerator developments, which is still continuing today. This is the reason for which we focus, in this second Part, on hadron accelerators.

'Hadron therapy' ('hadronthérapie' in French, 'hadronentherapie' in German, 'adroterapia' in Italian, 'hadroterapia' in Spanish) is a collective word which covers all forms of radiation therapy which use beams of particles made of quarks: neutrons, protons, pions, and ions of helium, carbon, neon, silicon, and argon have been used for treating cancer. Furthermore, lithium and boron ions, as well as antiprotons have been investigated as potential candidates. 'Hadron therapy', 'particle therapy', 'ion-beam therapy', 'heavy ion therapy' and 'light ion therapy' are other terms often used to indicate the same procedure, and every performing facility has its favorite term.

The first hadrons used for treatment purposes are the neutrons. The time progression in implementing neutron therapy was so fast that is astonishing today. The neutron was discovered by James Chadwick in 1932. Soon after, Ernest Lawrence and his brother John were experimenting with the effects of fast neutrons on biological systems. Following a paper by Gordon Locher [69], who in 1936 underlined the therapeutic potentialities of both, fast and slow neutrons, at the end of September 1938, the first patient was treated at the Berkeley 37-in. cyclotron. The first study on 24 patients, which used single fractions, was considered a success and led to the construction of the dedicated 60-in. Crocker Medical Cyclotron. Here Robert Stone and his collaborators (Fig. 11.24) treated patients with fractionated doses using fast neutrons. The tuning of the dose to control the tumour and prevent, at the same time, the complication of the normal tissue was still developing. The dose given to healthy tissues was too high, so that in 1948 the program was discontinued [70].

Neutron therapy was revived in 1965 by Mary Catterall at Hammersmith Hospital in London and later various centres were built where fast neutrons were used for many years. In particular in the 1970s radiation oncologists of Chicago worked with Robert Rathbun (Bob) Wilson, Fermilab Director, to build the 'Neutron Therapy

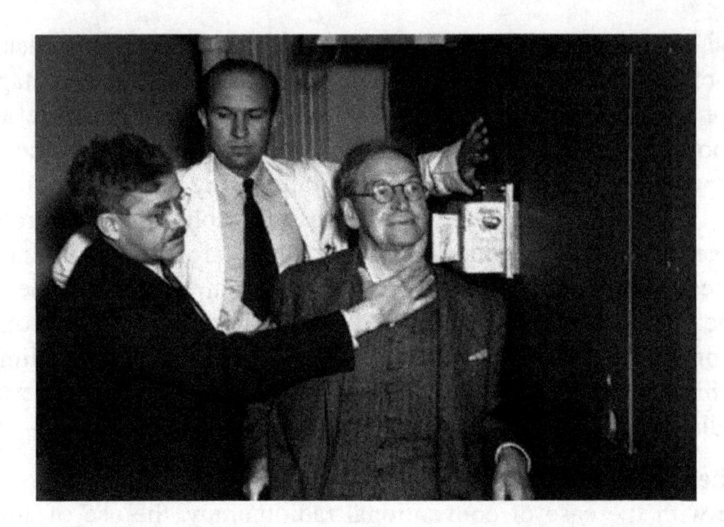

Fig. 11.24 Robert Stone is watched by John Lawrence while aligning a patient in the neutron beam produced by the 60-in. cyclotron

Facility' at Fermilab based on the injector linac [71]. At Michigan University, Henry Blosser built for the Harper Hospital a proton cyclotron rotating around the patient.

The worldwide effort for high-current 40–60 MeV proton cyclotrons was large. However, at present, this technique is rarely used. The reason is the poor depth-dose distribution of fast neutron and the fact that, all along the path, the biological effects, due mainly to highly ionizing protons put in motion by the neutrons, are difficult to determine and tissue-dependent. As discussed in the following, carbon ion beams are a much better solution to deliver doses being highly ionizing particles, i.e. particles which have energy losses (Linear Energy Transfers—or LET—in the medical parlance) much larger than the electrons put in motion by X rays.

Thermal and epithermal neutron are also used in Boron Neutron Capture Therapy (BNCT). This cancer treatment is based in two subsequent procedures, first the injection of the boron carrier, conceived to be absorbed preferably by the tumour, and second the irradiation of the patient with low energy neutrons. Alpha particles resulting from the reaction $^{10}B(n,{}^{4}{}_{2}He)^{7}Li$ release all their energy close to the point where they are originated affecting the tumour cells.

Proton Therapy

The use of protons in teletherapy was proposed in 1946 by Bob Wilson [72], who was asked by Lawrence to measure, at the Berkeley Cyclotron, proton depth-dose profiles. Describing the significant increase in dose at the end of particle range, the so-called Bragg peak, which had been observed 40 years before in the tracks of alpha particles by W. Bragg, Wilson recognized in his paper the advantages in treating tumours. The Bragg peak—which can be 'spread' with modulator wheels—can be used to concentrate the dose sparing healthy tissues better than with X-rays. It is interesting to remark that in his paper Wilson discussed mainly protons but mentions also alpha particles and carbon ions.

In 1954, the first patient was treated at Berkeley with protons, followed by helium treatment in 1957 and by neon ions in 1975 [73]. In these treatments—as in most of the following facilities—the beam was distributed over the target volume using 'passive' shaping systems, like scatterers, compensators, and collimators that were adapted from the conventional photon therapy. The first treatments consisted of irradiation to stop of the pituitary gland from producing hormones that stimulated some cancer cells to grow. Between 1954 and 1957 at Berkeley, under the leadership of Cornelius Tobias, 30 pituitary glands and pituitary tumours were treated with protons [74].

In 1957 the first tumour was irradiated with protons at the Uppsala cyclotron by Börje Larsson [75], but the facility that made the largest impact on the development of proton therapy is certainly the 160 MeV Harvard cyclotron, which was commissioned in 1949 (Fig. 11.25).

The Harvard staff got interested in using protons for medical treatments only after proton therapy was started at both, LBL and Uppsala. In 1961, Raymond Kjellberg, a young neurosurgeon at Massachusetts General Hospital in Boston, was the first to use the Harvard beam to treat a malignant brain tumour.

The results obtained for eye melanoma—by Ian Constable and Evangelos Gragoudas of the Massachusetts Eye and Ear Hospital—and for chordomas and chondosarcomas of the base of the skull [76]—by Herman Suit, Michael Goitein and colleagues of the Radiation Medicine Dept of Massachusetts General Hospital—convinced many radiation oncologists of the superiority of protons to X rays in particularly for tumours that are close to organs at risk. In 2002, when the cyclotron was definitely stopped, more than 9000 patients had been irradiated and the bases were put in place for the following development of the field. The competences developed in Boston were soon transferred to the new hospital-based facility of the Massachusetts General Hospital, now called "Francis H. Burr Proton Therapy

Fig. 11.25 Picture of the just completed Harvard cyclotron. Many years later Norman Ramsey was awarded the Nobel prize, together with Wolfang Paul

Table 11.6 The pioneers of proton therapy

Facility	Country	Years in operation
Lawrence Berkeley Laboratory	USA	1954–1957
Uppsala	Sweden	1957–1976
Harvard Cyclotron Laboratory	USA	1961–2002
Dubna	Russia	1967–1996
Moscow	Russia	1969–
St. Petersburg	Russia	1975–
Chiba	Japan	1979–2002
Tsukuba	Japan	1983–2000
Paul Scherrer Institute	Switzerland	1984–

These facilities, that today are out of operation except for the centres Moscow, St. Petersburg, and PSI, treated together 10,200 patients

Center", which is equipped with the first commercial accelerator for therapy—built by the market leader, IBA.

Soon after the start-up of the Harvard facility other nuclear physics laboratories in USSR and Japan assigned horizontal beams of their research facilities to proton therapy. As shown in Table 11.6 in 1984 the Paul Scherrer Institute, Switzerland, did the same. This facility has been fundamental to the progress of proton therapy with its pioneering realizations of the gantry to change the angle of the beam penetration and the active scanning [77], the technique that, avoiding passive elements, moves the pencil beam in three dimensions painting the tumour with Bragg spots (a short description of active scanning is provided in the following paragraph on dose distribution).

Not by chance the first hospital-based centre was built at the Loma Linda University (California), where the determination of James Slater lead to the agreement with Fermilab in 1986, to design and built the 7-m-diameter synchrotron which accelerates protons to 250 MeV.

A smooth conversion from a physics laboratory to a hospital facility took place in Japan. The University of Tsukuba started proton clinical studies in 1983 using a synchrotron constructed for physics studies at the High Energy Accelerator Research Organization (KEK). A total of 700 patients were treated at this facility from 1983 to 2000. In 2000, a new in-house facility, called *Proton Medical Research Center* (PMRC), was constructed adjacent to the University Hospital. Built by Hitachi, it is equipped with a 250 MeV synchrotron and two rotating gantries [78].

Light Ion Therapy

Ions heavier than protons, such as helium and argon, came into use at Berkeley in 1957 and 1975, respectively. At the old 184-in. cyclotron 2800 patients received treatments to the pituitary gland with helium beams: the lateral spread and range straggling being much smaller than in the proton case described before.

About 20 years later, argon beams were tried at the Bevalac in order to increase the effectiveness against hypoxic and otherwise radioresistant tumours, i.e. tumours that need deposited doses 2–3 times higher if they are to be controlled with either photons or protons (radiation oncologists speak of 'Oxygen effect'). But problems arose owing to non-tolerable side effects in the normal tissues. After the irradiations of some 20 patients, Cornelius Tobias and collaborators decided to use lighter ions, first silicon for 2 patients and then neon, for 433 patients. Only towards the end of the program it was found that the neon charge ($Z = 10$) is too large and undesirable effects were produced in the traversed and downstream healthy tissues [79]. The Bevalac stopped operation in 1993 and, at the conclusion of this first phase of hadrontherapy, almost 3600 patients have been treated worldwide with pions, helium ions or other ions.

The larger effects that ions have with respect to the ones produced by cobalt-60 gammas of identical dose is quantified by introducing the 'RBE' (*Relative Biological Effectiveness*). RBE depends upon the cell type, the radiation used (particle, energy, dose) and the chosen effect. For a given cell type and effect RBE varies with the LET of the ionizing particles. Only at the beginning of the 1990s carbon ions ($Z = 6$) were chosen as the optimal ion type. Indeed light ions produce a radiation field which is qualitatively different from the ones due to either photons or protons and succeed in controlling also radioresistant tumours.

The carbon choice was made in Japan by Yasuo Hirao and collaborators of the National Institute of Radiological Science, NIRS [80], who proposed and built HIMAC (Heavy Ion Medical Accelerator in Chiba) in the Chiba Prefecture (Fig. 11.26).

In 1994, under the leadership of Hirohiko Tsujii, the facility treated the first patient with a carbon-ion beam of energy smaller than 400 MeV/u, corresponding to a maximum range of 27 cm in water. By the end of 2015 about 10,500 patients have

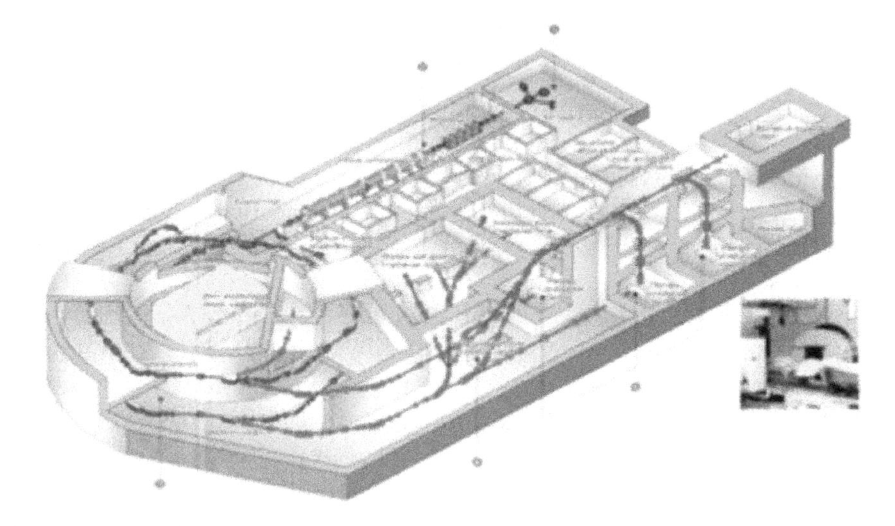

Fig. 11.26 HIMAC features two large synchrotrons, injected by a Alvarez linac, and three treatment rooms for a total of two horizontal and two vertical beams

been treated and it has been shown that many common tumours otherwise difficult to cure (e.g. lung and liver) can be controlled [81, 82].

In 1987 in Europe an important initiative was launched to create a full-fledged European light-ion therapy centre. The characteristics of the needed hadron beams were defined in a series of expert meetings. EULIMA, the European Light Ion Medical Accelerator project, financed by the European Commission, was led by Pierre Mandrillon and involved many European laboratories and centres. The European therapy synchrotron was never built but instead national-based projects in Germany and Italy were pushed through giving to the radiation oncologists facilities similar to the HIMAC at Chiba and the Hyogo Ion Beam Medical Center.

In 1993 Gerhard Kraft and colleagues obtained the approval for the construction of a carbon ion facility at GSI (Darmstadt), later called the "pilot project". Hadron therapy started in 1997 and, at the time of the closure in 2009, 440 patients [83] were treated with carbon-ion beams. The medical and technical competences were both used to build the Heidelberg Ion-Beam Therapy Center (HIT), which started operation in 2009 [84] and transferred to Siemens Medical, which built the HIT treatment rooms. The main new features of the GSI pilot project where: (i) the active 'raster' scanning system [85]; (ii) the sophisticated models and codes that take into account the biological effects of each irradiated sub-volume in the treatment planning system; (iii) the in-beam PET system which determined 'on-line' the location and shape of the irradiated volume by detecting the β^+ radioactivity produced by the incident carbon ions, mainly ^{11}C and ^{15}O [86–88].

11.3.2.2 The Bases of Cancer Radiation Therapy

The absorbed dose due to a conventional beam of photons has a roughly exponential absorption in matter after a slight initial increase. For instance, the highest dose for beams having a maximum energy of 8 MeV, is reached at a depth of about 2–3 cm of soft tissue and the dose is about one third of the maximum at a depth of 25 cm. Because of this non-optimal dose distribution, the unavoidable dose given to the healthy tissues represents the limiting factor to obtain the best local control of the pathology in conventional radiation therapy. In this connection, it has to be remarked that even a small increase of the maximum dose can be highly beneficial in terms of tumour control.

The radiation is given to the patient balancing two opposing goals, deliver a sufficient dose to control the tumour and spare the normal tissue from dose levels that could cause complications. To increase the dose to the tumour and not to the healthy surrounding organs it is essential to 'conform' the dose to the target. In order to selectively irradiate deep-seated tumours with X-rays, radiation oncologists use multiple beams from several directions (portals), usually pointing to the geometrical centre of the target. This is achieved by using a mechanical structure containing the linac which rotates around a horizontal axis passing through the isocentre ('isocentric gantry'). The already mentioned IMRT makes use of up to 6–12 X-

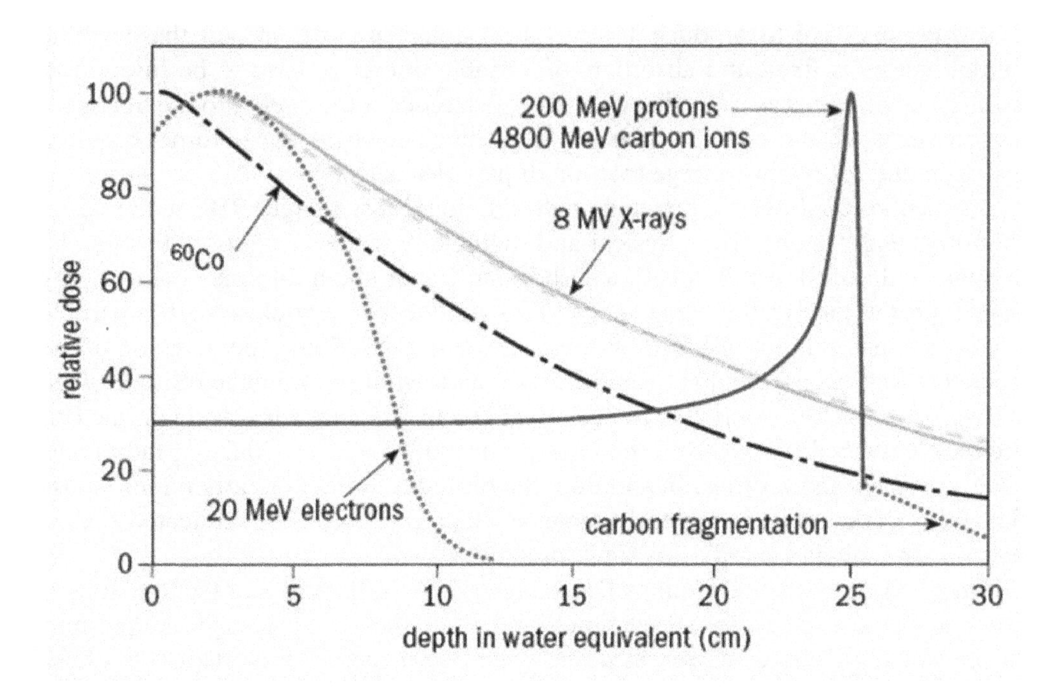

Fig. 11.27 Comparison of depth dependence of the deposited dose for each radiation type, with the narrow Bragg peak at the end. For monoenergetic beams, the carbon peak is three times narrower than the proton beam, a feature which is not represented in this schematic drawing

ray beams; the beams may be non-coplanar and their intensity is varied across the irradiation field by means of computer-controlled multi-leaf collimators [89].

The depth-dose curves of proton and light ion beams are completely different from those of X-rays, because these charged particles give the highest dose near the end of their range in the Bragg peak, just before coming to rest. Figure 11.27 illustrates how hadrons are more suitable to treat deep-seated tumours as they can be used to target profound regions preserving the more superficial healthy tissue.

In order to reach depths of more than 25 cm in soft tissues—necessary to treat *deep-seated tumours*—proton and carbon ion beams must have an initial energy not lower than 200 MeV and 4500 MeV (i.e. 375 MeV/u), respectively. In practice, the standard maximal energies are 230 MeV for protons and 400 MeV/u for carbon ions, which correspond to a 32 cm and 27 of water range, respectively. The minimal energies, used for shallow tumours, are 60 MeV and 100 MeV/u respectively.

For this reason sizeable particle accelerators are needed in hadron therapy. These machines are not demanding in terms of output current since 2 nA and 0.2 nA are sufficient for treating patients with protons and carbon ions, respectively, when active spreading systems are used. Higher currents are needed for 'passive' delivery systems in which the beam is reduced in energy with a variable absorber, spread by multiple scattering with two 'scatterers' and transversally shaped with collimators.

Cyclotrons have to produce 10–20 times larger currents. In fact the machine output energy is fixed and absorbers of variable thickness have to be inserted in a 15–20 m long Energy Selection System which reduces the energy to the treatment requirements. In the process of beam degrading, down to the minimal required energies, and re-shaping a large fraction of particles is lost.

As mentioned above, carbon and other light ions have a larger RBE with respect to X-rays and protons. The physical and radiobiological arguments of this can be summarized as follows. In a cell, a carbon ion leaves about 24 times more energy than a proton having the same range. This produces—very close to the particle track—a dense column of ionization, especially near the Bragg-peak region of the track, causing many 'Double Strand Breaks' and 'Multiple Damaged Sites', when crossing the DNA contained in the cell nucleus. In this way, the effects on the cell are *qualitatively* different from the ones produced by sparsely ionizing radiations, such as X-rays and protons. In addition, the biological effect of carbon ions is less dependent of the oxygenation of the target and therefore their use is indicated against hypoxic and otherwise radio-resistant tumours.

Due to the much more complex DNA damage, the RBE values of the light ions at the Bragg peak can be about three times larger than the one for X-rays and protons. In the slowing down of an ion in tissue this effect becomes important when LET becomes larger than about 20 keV/μm. For carbon ions this happens in the last 5 cm of their range in water.

11.3.2.3 Cyclotrons and Synchrotrons in Hadron Therapy

Proton Therapy

The first facilities that treated patients with protons and ions were based on existing accelerators built for fundamental research in nuclear and particle physics: LBL at Berkeley (USA); GWI in Uppsala (Sweden); JINR in Dubna (Russia); PMRC-1 in Tsukuba (Japan); UCL, in Louvain La Neuve (Belgium); MPRI-1 in Indiana (USA) 60 MeV protons at Chiba (Japan). Moreover pions where used at TRIUMPH and PSI (SIN at that time) in the periods 1979–1994 and 1980–1993, respectively. The clinical results have shown that pions are not superior to protons and light ions neither to obtain conformal dose volumes nor in the treatment of radioresistant tumours.

The proton therapy facilities running in the world are listed in Table 11.7. The data are extracted from PTCOG online database [90] and the data on the number of patients are updated to December 2015.

Table 11.7 includes three centres producing 60–70 MeV proton beams which are exclusively used for the treatment of eye tumours and malformations. The most recent facilities in the table are hospital-based, in the sense that they feature an accelerator built specifically for medical purposes and have more treatment rooms so that several hundred of patients can be treated every year. The companies which have built the accelerators and the high-tech parts of these centres are: Optivus (USA),

Table 11.7 Proton therapy facilities in operation [90]

Centre	Country	Acc.[a]	Max. energy [MeV]	Beam direct.[b] (del. method)[c]	Start date	Total patients
ITEP, Moscow	Russia	S	250	H(P)	1969	4368
St. Petersburg	Russia	S	100	H(P)	1975	1386
CPT, PSI, Villigen	Switzerland	C	250	2G(A),H(P)	1984	5458
Clatterbridge	England	C	62	H(P)	1989	2813
J. Slater PTC, Loma Linda	USA, CA	S	250	3G,H(P)	1990	18,362
CPO, Orsay	France	C	230	G,H(P)	1991	7560
CAL/IMPT, Nice	France	SC	230	G,H(P)	1991	5478
NRF—iThemba Labs	South Africa	C	200	H(P)	1993	524
UCSF-CNL, San Francisco	USA, CA	C	60	H(P)	1994	1839
TRIUMF, Vancouver	Canada	C	72	H(P)	1995	185
HZB, Berlin	Germany	C	250	H(P)	1998	2750
NCC, Kashiwa	Japan	C	235	2G(P,A)	1998	1560
JINR 2, Dubna	Russia	C	200	H(P)	1999	1122
PMRC 2, Tsukuba	Japan	S	250	2G(P,A)	2001	4502
MGH Francis H. Burr PTC, Boston	USA, MA	C	235	G,H(P,A)	2001	8358
INFN-LNS, Catania	Italy	C	60	H(P)	2002	350
Shizuoka Cancer Center	Japan	S	235	3G,H(P)	2003	1873
WPTC, Wanjie, Zi-Bo	China	C	230	2G,H(P)	2004	1078
UFHPTI, Jacksonville	USA, FL	C	230	3G,H(P,A)	2006	6107
MD Anderson Cancer Center, Houston	USA, TX	S	250	3G,H(P,A)	2006	6631
KNCC, IIsan	South Korea	C	230	2G,H(P)	2007	1781
STPTC, Koriyama-City	Japan	S	235	2G,H(A)	2008	2797
RPTC, Munich	Germany	C	250	4G(A),H(P)	2009	2725
ProCure PTC, Oklahoma City	USA, OK	C	230	G,H,OB(P)	2009	2079
Chicago Proton Center, Warrenville	USA, IL	C	230	G(A),H(P),OB(P)	2010	2316

(continued)

Table 11.7 (continued)

Centre	Country	Acc.[a]	Max. energy [MeV]	Beam direct.[b] (del. method)[c]	Start date	Total patients
Roberts PTC, UPenn, Philadelphia	USA, PA	C	230	4G(P,A),H(P)	2010	3376
HUPTI, Hampton	USA, VA	C	230	4G(P),H(P)	2010	1399
MPTRC, Ibusuki	Japan	S	250	3G(A,P)	2011	1654
Fukui Prefectural Hospital PTC, Fukui City	Japan	S	235	2G(A,P),H(P)	2011	646
IFJ PAN, Krakow	Poland	C	230	G(A),H	2011	128
PTC Czech r.s.o., Prague	Czech Republic	C	230	3G(A),H	2012	780
ProCure Proton Therapy Center, Somerset	USA, NJ	C	230	4G(A,P)	2012	1892
WPE, Essen	Germany	C	230	4G(A),H	2013	366
Nagoya PTC, Nagoya City, Aichi	Japan	S	250	2G(A),H	2013	1095
S. Lee Kling PTC, Barnes Jewish Hospital, St. Louis	USA, MO	SC	250	G(P)	2013	270
SCCA ProCure Proton Therapy Center, Seattle	USA, WA	C	230	4G(A,P)	2013	844
UPTD, Dresden	Germany	C	230	G(A)	2014	106
APSS, Trento	Italy	C	230	2G(A),H	2014	92
Hokkaido Univ. Hospital PBTC, Hokkaido	Japan	S	220	G(P)	2014	
Aizawa Hospital PTC, Nagano	Japan	C	235	G(P)	2014	1
Scripps Proton Therapy Center, San Diego	USA, CA	C	250	3G(A),2H	2014	400
Willis Knighton PTCC, Shreveport	USA, LA	C	230	G(A)	2014	151
Provision Center for Proton Therapy Knoxville	USA, TN	C	230	3G(A)	2014	856
Samsung PTC, Seoul	South Korea	C	230	2G(P)	2015	4

(continued)

Table 11.7 (continued)

Centre	Country	Acc.[a]	Max. energy [MeV]	Beam direct.[b] (del. method)[c]	Start date	Total patients
The Skandion Clinic, Uppsala	Sweden	C	230	2G(A)	2015	1431
Chang Gung Memorial Hospital, Taipei	Taiwan	C	230	4G(A,P)	2015	
Ackerman Cancer Center, Jacksonville	USA, FL	SC	250	G(P)	2015	140
Mayo Clinic PBTC, Rochester	USA, MN	S	220	4G(A)	2015	186
Laurie Proton Center of Robert Wood Johnson Univ. Hospital, New Brunswick	USA, NJ	SC	250	G(P)	2015	50
St. Jude Red Frog Events PTC, Memphis	USA, TN	S	220	2G(A),H	2015	1
Texas Center for Proton Therapy, Irving	USA, TX	C	230	2G(A),H	2015	1
Tsuyama Chuo Hospital, Okayama	Japan	S	235	G(P)	2016	
Mayo Clinic Proton Therapy Center, Phoenix	USA, AZ	S	220	4G(A)	2016	
Orlando Health PTC, Orlando	USA, FL	SC	250	G(P)	2016	
Maryland PTC, Baltimore	USA, MD	C	250	4G(A),H(A)	2016	
UH Sideman CC, Cleveland	USA, OH	SC	250	G(P)	2016	
Cincinnati Children's PTC, Cincinnati	USA, OH	C	250	3G(A)	2016	
Beaumont Health Proton Therapy Center, Detroit	USA, MI	C	230	G(A)	2017	

[a]Cyclotron (C), Synchrotron (S), Synchrocyclotron (SC)
[b]Gantry (G), horizontal (H), oblique (OB)
[c]Active pencil-beam scanning (A), Passive beam spread (P)

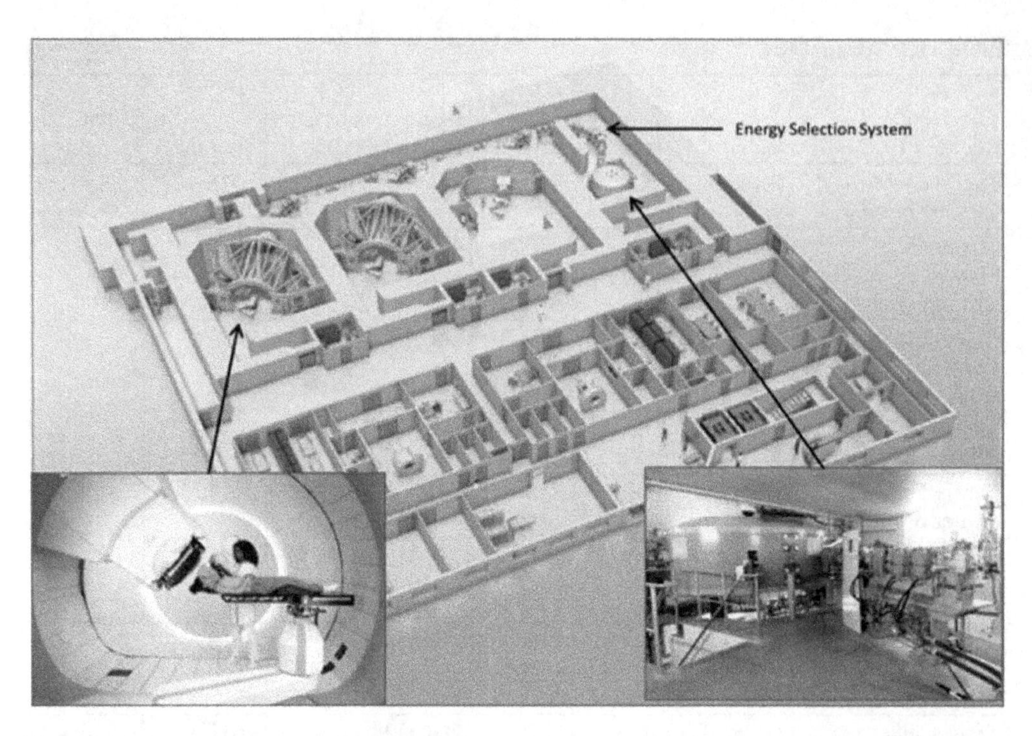

Fig. 11.28 An Ion Beam Applications (IBA) proton therapy centre featuring two gantries and a fix-beam room (insert at the bottom left). The magnetic channel, called ESS (Energy Selection System), is needed to reduce the 230 MeV proton energy produced by the cyclotron (insert at the bottom right) to lower energies, down to 70 MeV (with courtesy of IBA SA)

IBA (Belgium), Varian/Accel (USA/Germany), Hitachi (Japan), and Mitsubishi (Japan).

A centre by IBA, which is the market leader of protontherapy, is shown in Fig. 11.28.

The PTCOG is the organization, which associates experts from the technical and the clinical sectors in the field of hadron therapy. In its website [90] statistics on patients and the information about new proton and carbon-ion centres are updated with data from the companies and the clinics. By summing the numbers of all the centres in operation and out of operation one obtains that the number of patients treated with protons until the end of 2015 exceeded 131,000.

The most recent figures show that 39 new proton therapy centres are in construction and should start operation within 2020 while 24 more centres are planned. The geographical areas of major expansions are the USA with nine new centres under construction, China with seven, UK with six, and Japan with four. Proton therapy will be performed for the first time in Danemark, the Netherland, and Slovak Republic, and, outside Europe, in India, Saudi Arabia, Singapore, and Taiwan. Centres are planned to be built for the first time in South America (at the Instituto de Oncologia Angel Ruffo Hospital, Boenos Aires, Argentina), in Australia (at the Australian Bragg Centre for Proton Therapy and Research in Adelaide), and in North Africa (at the Children's Cancer Hospital Foundation, Cairo, Egypt).

Carbon-Ion Therapy

As far as carbon ions are concerned, the situation is summarized in Table 11.8. It is worth remarking that all these centres have the possibility of treating patient with both, carbon-ion and proton beams, so that it the name 'dual centres' are more appropriate than 'carbon-ion centres'. Several centres opted to use in clinic exclusively carbon ions in particular in Japan where the number of proton therapy facilities is large and well distributed in the territory.

NIRS promoting role has continued after the first experience of HIMAC and its activity was instrumental in the development and consolidation of carbon-ion therapy in Japan. Starting in 2004, NIRS was involved on the design of a new compact carbon-ion synchrotron. The new machine, that was employed for the first time in Gumna, is adapted to the medical use accelerating carbon ions up to 400 MeV/u and has a circumference of 63 m, more than two times smaller than HIMAC accelerator which was conceived for silicon ions up to 800 MeV/u [91]. Mitsubishi is commercializing the accelerator offering a complete turn-key solution as implemented in Saga-HIMAT. The centre Kanagawa i-ROCK is the result of the partnership of NIRS and Toshiba.

Table 11.8 Carbon-ion and dual facilities in operation

Centre	Country	Acc.[a]	Max. energy [MeV/u]	Beam direction[b] (scanning method)[c]	Treatment start	Total patients
HIMAC, Chiba	Japan	C	800/u	H,V,G (A,P)	1994	10,769
HIBMC, Hyogo	Japan	p, C	320/u	H,V (P)	2002	2366
IMP-CAS, Lanzhou	China	C	400/u	H (P)	2006	213
HIT, Heidelberg	Germany	p, C	430/u	2H,G (A)	2009	2086
GHMC, Gunma	Japan	C	400/u	3H (P)	2010	1909
CNAO, Pavia	Italy	p, C	480/u	3H,V (A)	2012	591
Saga-HIMAT, Tosu	Japan	C	400/u	3H,V,OB (P)	2013	1136
SPHIC, Shanghai	China	p, C	430/u	3H (A)	2014	1498
MIT, Marburg	Germany	p, C	430/u	3H,OB (A)	2015	0
Kanagawa i-ROCK, Yokohama	Japan	C	430/u	4H,2V (P)	2015	0
MedAustron, Wiener Neustadt	Austria	p, C	430/u	2H,V (A)	2016	67

The data are extracted from PTCOG online database [90] and the statistics on patients are compiled basing on update of December 2015 and form direct information. All accelerators are synchrotrons

[a]Accelerated protons (p), accelerated carbon ions (C)

[b]Horizontal (H), vertical (V), oblique (OB), gantry (G)

[c]Active pencil-beam scanning (A), Passive beam spread (P)

Not linked to NIRS developments is the synchrotron-based facility of HIBMC in Hyogo, which was developed by Mitsubishi and is in operation since 2001. This was the first clinical centre featuring dual modality, protons with energy between 70 MeV and 230 MeV and carbon ions between 70 MeV/u and 320 MeV/u.

Today the five carbon-ion therapy centres in operation in Japan treat, combined, more than 2000 patients each year, with a total at the end of 2016 of 18,000 treatments [92].

Built by Siemens Medical, the centre of HIT in Heidelberg applies the competences developed by GSI for its pilot project including the active scanning capabilities. This is the first dual centre featuring a gantry. Its construction has been a real challenge since it weighs 600 tons and consumes, at maximum energy, about 400 kW. The dual centre of Magburg was initiated as the second centre of Siemens Medical, and, when the company withdrew from hadron therapy it was bought by Marburger Ionenstrahl-Therapiezentrum which recommissioned and started treating patients in 2015.

In 1996 the *Proton Ion Medical Machine Study* (PIMMS) [93] was initiated at CERN with the participation of TERA Foundation (Italy), MedAustron (Austria), and Oncology 2000 (Czech Republic) to study a proton and carbon-ion accelerator for the therapy. The project of the high-tech part of the *Centro Nazionale di Adroterapia Oncologica* (CNAO)—shown in Fig. 11.29—is due to the TERA Foundation [94] which in 2002 obtained the financing of the centre by the Italian Health Minister. The synchrotron, features a unique 'betatron core' to obtain a very uniform extracted beam. The centre, built in Pavia, treated the first patient with protons in 2011, a year later, treated the first patient with carbon ions. The MedAustron centre built in Wiener Neustadt, uses the synchrotron design of CNAO and treated the first patient with proton beams in December 2016.

A new carbon-ion therapy facility is under construction in China (HITFil, Lanzhou). A dual center will be operational starting in 2018 in South Korea (KIRAMS, Busan). By the year 2020 two more carbon-ion centres will be operational in Japan, in the cities of Osaka and Yamagata. All these centers are based on synchrotrons.

Dose Distribution Systems

In order to maximize the outcomes of the therapy, the optimal hadron accelerators systems should deliver a biological equivalent dose defined with an uncertainty lower than 5% to tumour volumes that varies from few cubic milliliters up to 2 L preserving as much as possible the healthy surrounding tissue. The desirable dose rate is 2 grays/min/L, where 1 Gy = 1 J/kg.

Till the end of the last century, in all facilities, except PSI, the conformation of the dose to the tumour was realized with 'passive' dose delivery systems, using individually machined collimators, boluses, and passive absorbers to spread longitudinally the Bragg peak. With this technique, which is still widely used, the same spread out Bragg peak applies to the whole transversal section of the tumour, so that large volumes of healthy tissue, that surrounds the tumour, are exposed to the same dose as the tumour tissues.

Fig. 11.29 Above. CNAO has been built in Pavia by the CNAO Foundation in collaboration with INFN, the Italian National Institute for Nuclear Physics. To reduce the overall dimensions, the 7 MeV/u linac (built by GSI) is placed inside the 25 m diameter synchrotron. Below, MedAustron built in Wiener Neustadt (© Kästenbauer/Ettl.): the same synchrotron structure of CNAO hosts sources and the linac in a separate room, not visible in the picture

Modern system have been develop to minimize the over dosage of the healthy tissue. The spot scanning technique developed for protons by PSI and the raster scanning technique developed for carbon ions by GSI operate dividing the whole region to be treated in sub-volumes of the order of few cubic millimeters (named *voxel* from *volume cells*), directing beams with reduced transversal section (*pencil* beams) toward them, and delivering a dose that is varying from one position to the next. To further minimize the dose to healthy tissue the radiation is directed to the patients from several directions (typically three). These techniques, which have various practical realizations, are collectively called 'active dose spreading systems'.

The accelerating systems that realize such treatments require the highest beam stability, the possibility of changing promptly and frequently beam intensity and energy eider acting on the acceleration or using passive Energy Selection Systems with rapidly moving absorbers.

A recent development that resulted from the common effort of NIRS and HIMAC is the implementation of fast active spot scanning system. The fast beam scanning speed of 100 mm s^{-1} allows a high repetition rate of the scanning spots so that a 'volumetric multipainting' of the tumour target is achievable. One-liter tumour volume at a rate of 100 Hz can be repainted by the pencil beam several times in few minutes reducing the statistical uncertainty in the delivered dose by a factor $\sqrt{n}$ (where n is the number of re-paintings) and minimizing any local accidental under-dosage or over-dosage. This method also improves the synchronization of the irradiation with the breathing movements. A second development concerns the superconductive gantry. The structure, studied for carbon-ion beams up to 430 MeV/u, has a radius of 5.5 m and a wait of approximately 200 tons, comparable with normal-conductive proton therapy gantries and sensibly lighter than the 600 tons gantry built in Heidelberg.

The ultimate goal is the treatment of moving organs, a challenge that is more critical in hadron therapy than in conventional therapy because of the dose is concentrated in the Bragg peak. In this case the accelerating system and beam delivering system require a feedback that, based on on-line movement detection, adapt in few milliseconds the beam characteristics to the three dimensional movements of the target.

11.3.2.4 Present and Future Challenges

There is still space for improving the dose delivery, for instance, the active dose delivery systems for protons and carbon ions have been used for a small part of the patients treated with protons or ions. In spite of the fact that most new centres are featuring 'active' scanning systems and some of the existing centres are upgrading to them, the implementation in the clinical practice has been for many years quite slow.

As far as the accelerators and transfer lines are concerned, two main lines of research have to be intensively pursued:

- systems which actively scanned are able to follow the tumours which are subject to movements,
- accelerator and delivery systems which are lighter, smaller, more efficient, and less power consuming than the present cyclotrons

Fast Cycling Accelerators to Follow Moving Organs

In conventional radiotherapy various techniques have been introduced to determine in real time the position of an irradiated tumour which moves, for instance, because of the respiration cycle and to deliver the dose following these movement in IGRT. In this case computer controlled multileaf collimators are used. In hadron therapy the tumour position can be detected with the same methods and a better follow-up can be obtained by rapidly moving the Bragg spot so to compensate for *three-dimensional* movements.

In the transverse plane the active scanning of tumours with hadron pencil beams is obtained by adjusting *two* perpendicular magnetic fields located many meters upstream of the patient. The time needed to move the beam transversally, following the indication of a precise feedback system, is of the order of milliseconds.

Due to the respiration cycle, the target organs can also move longitudinally, i.e. in the direction of the beam. A good example of the problems concerning the movements is a rib which enters the irradiation field of a lung tumour. In cyclotrons and synchrotrons the time needed to change the energy, and thus the depth of the Bragg peak, is of the order of 1 s because of the delay necessary either to move energy absorbers (in cyclotrons) or to change the energy (in synchrotrons). An accelerator with an energy cycle of the order of a few milliseconds would allow a longitudinal follow-up of the tumour similar to the transversal one.

In synchrotrons, the beam is delivered to the treatment rooms only for fraction of the machine time. Ramping up and down the magnets after the extraction of the beam and waiting for the magnetic field to stabilize are time-consuming operations. Few seconds are needed in this procedure and to reduce this dead time the centre of HIT studied a system in which probes for the measurements of the field in the magnet were inserted on the feed-back loop of the regulation of the magnetic fields in the elements of synchrotron and beam lines. The result was an overall increase in efficiency of 24% and a reduction of the time between two successive extractions to approximately 1 s. If such time intervals are still not compatible with movement compensation, nevertheless the improved efficiency has important impacts on the facility, since a larger number of patients can be treated and the cost per treatment decreases [95].

NIRS and HIMAC optimized the efficiency of their extraction system allowing to extract of multiple energies, during a single machine cycle [96]. It is important to remind that, in a treatment based on active scanning, the first layer irradiated is the deepest, and then step by step all other layers. The number of particles accelerated to certain energy in general exceeds the total number needed for irradiating the corresponding layer at the planned dose. The beam not used is dumped and this corresponds also to a waste of time. At NIRS, a system was studied to avoid beam dumping using the synchrotron to decelerate and use the remaining beam to the next more superficial layer. The process takes approximately 100 ms and, since the procedure can be repeated several times within the same beam cycle, the optimization of the irradiation time is consistent. Similar techniques are feasible to compensate tumour movements.

Recently two new fast cycling accelerators—particularly suited for the movement compensation and repainting of moving organs—have been considered: high frequency linacs and Fixed Field Alterenating Gradient (FFAG) accelerators.

Since 1993 the linac approach is pursued by the TERA Foundation with the choice of using the same frequency (3 GHz) employed in the construction of electron medical linacs [97]. Since the acceleration of very low velocity protons (and carbon ions) is problematic, it was proposed to use as injector a commercial 30 MeV cyclotron. The cyclotron-linac complex has been dubbed 'cyclinac'. Between 1998 and 2003 a TERA, CERN, INFN collaboration built and tested a 3 GHz 1-m long side-coupled standing-wave structure which accelerated protons from 62 to 74 MeV [98] and could stand gradients as large as 27 MV/m, corresponding to surface

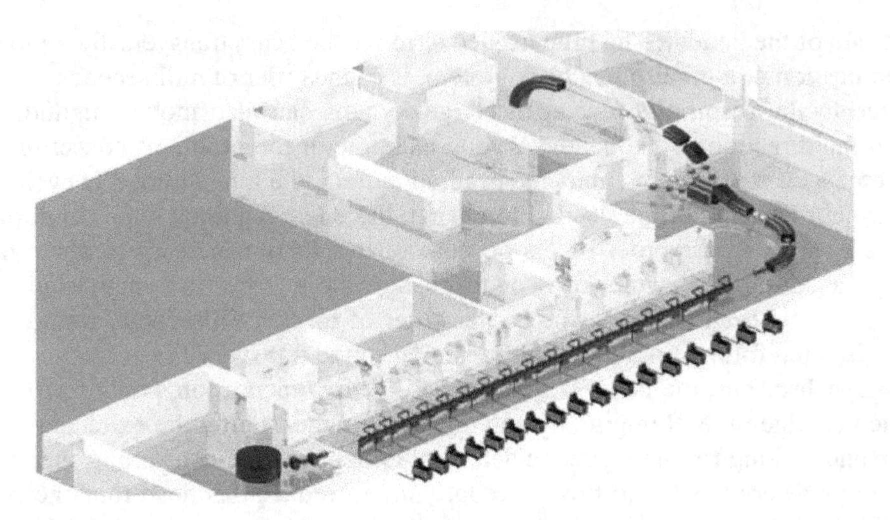

Fig. 11.30 This design of CABOTO (CArbon BOoster for the Therapy in Oncology) features a 150 MeV/u superconducting isochronous cyclotron followed by a 5.7 GHz standing-wave linac powered by seventeen 12 MW klystrons (courtesy of TERA Foundation)

electric fields larger than 150 MV/m. In parallel an *all-linac* solution was studied by L. Picardi and collaborators [99] which has been adopted for the TOP-IMPLART a proton-therapy facility under development at the Laboratories of Enea, in Frascati, Italy. The 150 MeV linac will be installed in the IFO Hospital, in Rome.

In the last years various cyclinacs have been designed: for proton therapy a 15-m long linac could be used to accelerate the protons between 30 MeV and a maximum of 230 MeV. A second cyclinac design is based on a superconducting cyclotron that accelerates carbon ions C^{+6} (and $H_2{}^+$ molecules) to 150 MeV/u (Fig. 11.30) with a 300–400 Hz repetition rate [100]. The linac is subdivided in modules, which interspaced with permanent magnetic quadruples are so short that the beam energy can be continuously varied in a couple of milliseconds between the cyclotron energy and the maximum by varying the output power of the klystrons.

Fixed Field Alternating Gradient (FFAG) accelerators have been proposed as fast cycling accelerators for hadron therapy since the magnetic fields of the ring are constant in time—as in a cyclotron—and the frequency of the accelerating electric field can vary up to 500–1000 Hz. The many bending magnets are arranged in triplets: the central magnet bends the circulating beam inwards while the two external ones bend it outwards. During acceleration the orbits maintain the same oscillating shape but increase in average radius so to remain inside the wide vacuum chamber and find stronger and stronger deflecting fields. In the recent 'non scaling' FFAGs the orbits of increasing radius do not maintain the same shape so that the full excursion requires bending magnets which are radially smaller than in classical 'scaling' FFAGs. It has to be noted that in every case beam extraction at different energies needs fast kickers [101].

A 150 MeV FFAG has been built in Japan for high-current applications [102] and low-current uses of non-scaling FFAGs in proton therapy have been proposed [103, 104].

In the case of carbon ions, due to the high magnetic rigidity of the beam, a solution with three concentric non-scaling FFAG machines has been studied [105]. In this case there is the added complication of the many injection and extraction systems which are needed in multi-ring facilities. This and other arguments are discussed in a recently published comparison between linacs and FFAGs for hadron therapy [106].

Compact Accelerators and 'Single Room' Facilities

An important line of development concerns what are now known as 'single room' proton facilities. The rational can be best appreciated by browsing Table 11.9 which has been constructed by using the results of the epidemiological studies performed in Austria, France, Italy and Germany in the framework of the EU funded network ENLIGHT [107]. They can be summarized by saying that in the medium-long term about 12% (3%) of the patients treated with high-energy photons would be better cured with fewer secondary effects if they could be irradiated with proton (carbon ion) beams. The table presents the number of treatment rooms needed for a population of 10 million people living in a developed country. The estimated numbers of rooms turn out to be in the easy to remember proportions 1:8:8^2. Since a typical hadron therapy centre has 3–4 rooms, the above figures tell that a proton (carbon ion) centre would be needed every about 5 (40) million people.

For proton therapy this shows the way to a flexible and patient-friendly solution: instead of a multi-room centre with its large building, one should develop single-room proton accelerator/gantry systems, constructed on a relatively small area (approximately 500 square metres) attached to existing hospital buildings homogeneously distributed in the territory. It is worth remarking that the overall cost of a full proton (dual) centre with three treatment rooms is of the order of 100 M€ (200 M€). Single-room proton therapy facilities, which cost approximately 25 M€ [108] are strategic for the future expansion of proton therapy in existing clinical facilities where cost and size is the most important factors.

Single room facilities are offered by Mevion, IBA, Varian, Sumitomo and other companies. Two lines of developments have to be considered: compact accelerators

Table 11.9 Estimate of the number of X ray and hadron treatment rooms

Radiation treatment	Patients per year in 10^7 inhabitants	Av. number of sessions per patient	Sessions/d in 1 room (d = 12 h)	Patients/y in 1 room (y = 230 d)	Rooms per 10 million people	Relative ratio
Photons	20, 000	30	48	370	54	8^2
Protons (12%)	2400	24	36	345	7.0	8
C ions (3%)	600	12	36	690	0.87	1

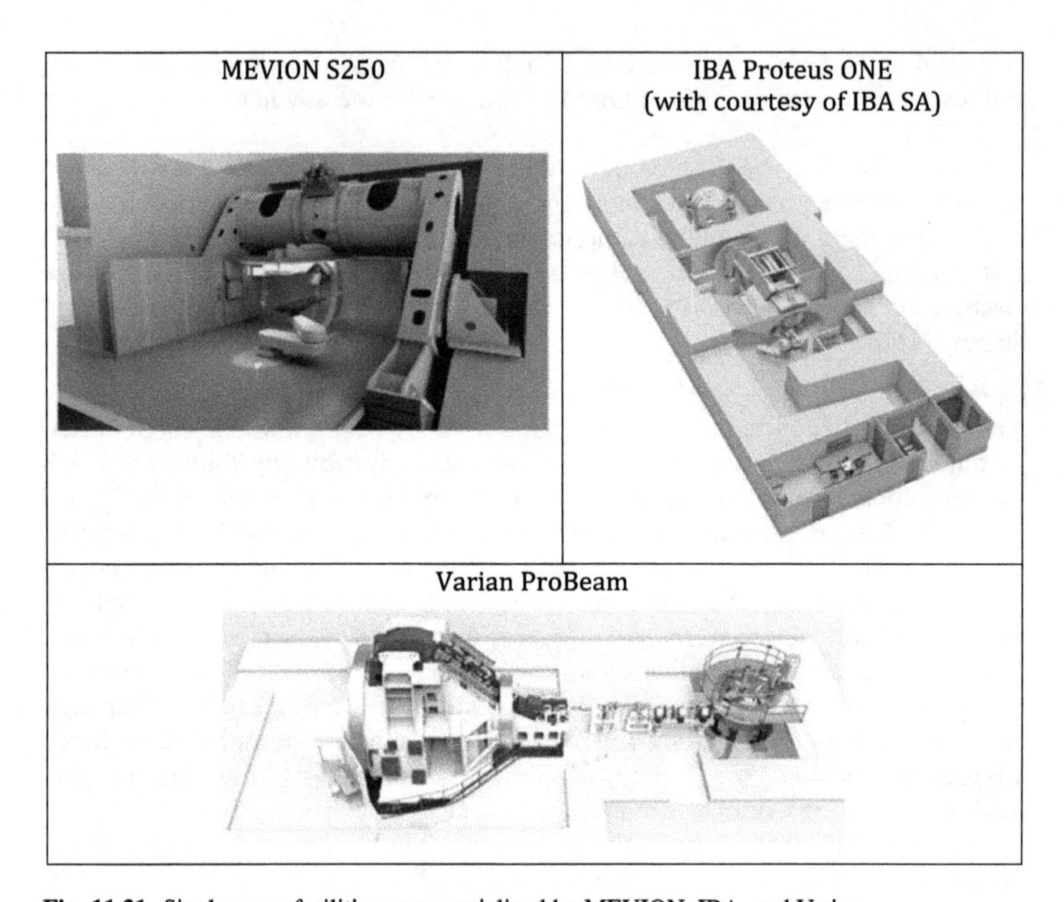

Fig. 11.31 Single-room facilities commercialized by MEVION, IBA, and Varian

followed by short magnetic channels and accelerators which rotate around the patient.

A very high field 15-tons niobium-tin superconducting 250 MeV synchrocyclotron was developed by Mevion Medical Systems Inc. in collaboration with the Massachusetts Institute of Technology (MIT) and the Massachusetts Medical Hospital. Mevion S250 is the firstly constructed system based on the concept of single-room facility which treated the first patient in December 2013. The accelerator is mounted directly on the 190° isocentric gantry (Fig. 11.31a) with a superconductive coils producing a 8.7 T central magnetic field. The first version of the system which featured a passive beam delivery system has been updated to a 'pencil beam' system for active scanning.

The Proteus ONE system of IBA consists of a compact 230 MeV superconducting synchrocyclotron, named S2C2. The energy selection system is mounted on the 220° isocentric gantry (Fig. 11.31b). In September 2016 the Lacassagne Centre in Nice treated the first patient with S2C2 accelerator.

The single-room system ProBeam of Varian system combines a 250 MeV cyclotron—the first commercial superconductive accelerator developed for proton therapy used at PSI since 2007—and the 360° gantry (Fig. 11.31c).

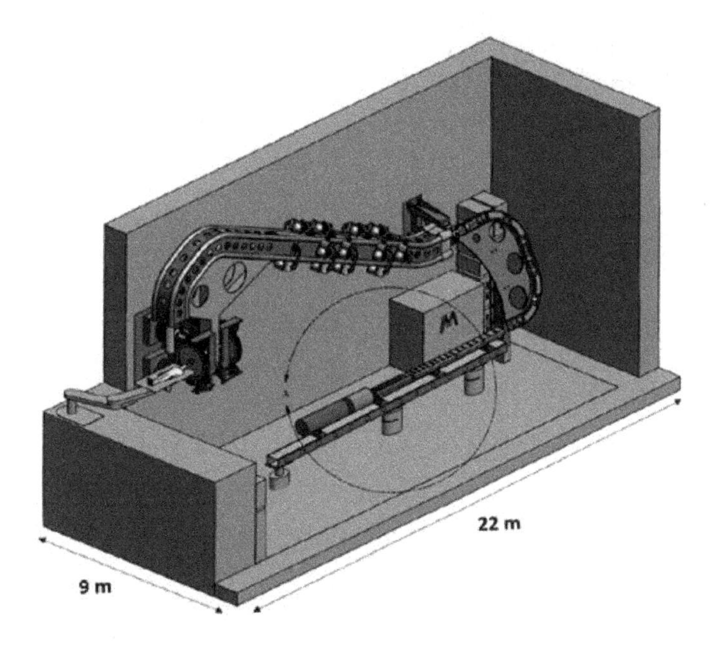

Fig. 11.32 TULIP (Turning LInac for Proton therapy) all-linac solution (courtesy of Mohammad Vaziri—TERA Foundation)

A proposed system is the high-frequency proton linac rotating around the patient—according to a scheme named TULIP and patented by TERA [109] which is based on the same 5.7 GHz linac designed for CABOTO (Fig. 11.32).

An original development concerns laser-driven ion beams and is based on protons accelerated by illuminating a thin target with powerful (10^{18}–10^{20} W/cm^2) and short (30–50 fs) laser pulses. When the laser pulse hits the target—with thickness of the few micrometers—electrons are violently accelerated producing an electric field of the order of teravolt per meter which draws behind the protons for the surface of the target itself. The phenomenon has been studied experimentally reaching proton energies of more than 10 MeV with promising results [110]. The energy spectrum is continuous and computations show that, using proper combination of target shapes and laser power, a 3% energy spread can be obtained [111]. The advantages of a laser-driven system are that the ion beams can be generated very close to the patient and that optical elements can substitute the cumbersome and heavy beamline elements including the gantries. The most relevant challenge is probably the short pulse duration which is linked to the unknown biological effects of the high dose rates, the impossibility of controlling the beam intensity during the irradiation, and the potential risk of overdose. An important issue is also linked to very low repetition rate of few hertz which would extend the irradiation to unacceptable time. The implementation process is complex and still several years are foreseen to create, around the laser-driven 200 MeV proton beam, the whole infrastructures for a therapy facility [112].

Acknowledgment
The authors are grateful to the International Atomic Energy Agency and in particular to Joao Alberto Osso and Amirreza Jalilian for the discussions and for providing

statistics and data on worldwide radioisotope production, to Eiichi Takada and Hikaru Souda for the information about the recent developments on carbon-ion therapy accelerators in Japan, and to Saverio Braccini for the precious contributions and comments he provided on the text.

11.4 Spallation Sources

M. Lindroos · S. Molloy · G. Rees · M. Seidel

11.4.1 Introduction

Spallation is a nuclear process in which neutrons at different energies are emitted in several stages following the bombardment of heavy nuclei with highly energetic particles. In this chapter we limit ourselves to the description of the accelerator that drives spallation sources. For a more detailed discussion of the spallation process itself, the target, the moderators and the physics at such facilities see (for example) [113]. However, it is worth noting that there are other ways to produce neutrons with accelerators, for example photo-fission induced by an intense electron beam. The spallation process is the most practical and feasible way of producing neutrons for a reasonable effort (or simply cost) of the neutron source cooling system, see Table 11.10. Research reactors also require fissile material handling, potentially a major constraint for both handling and licensing.

Spallation sources come in at least three types: short pulse sources (a few μs), long pulse sources (a few ms) and continuous sources. In general, synchrotrons or accumulator (compressor) rings provide short neutron pulses, linear accelerators provide long neutron pulses, and cyclotrons provide continuous beams of neutrons.

Pulsed neutron sources are more efficient in their use of neutrons than continuous sources, in a majority of all applications, according to a survey performed for the Autrans meeting in 1996 [115]. The time-of-flight of the neutrons—readily measured from a pulsed source—allows us to determine the neutron speed and energy. Additional tools are needed to select or determine the neutron speed in a

Table 11.10 The number of fast neutrons produced per joule of heat energy where the energy in joule is taken as heat produced over energy consumed [114]

Fission reactors	~10^9	in ~50 litre volume
Spallation	~10^{10}	in ~1 litre volume
Fusion	~2×10^{10}	in huge volume
Photo neutrons	~10^9	in ~0.01 litre volume
Nuclear reaction (p, Be):	~10^8	in ~0.001 litre volume
Laser induced fusion	~10^4	in ~10^{-9} litre volume

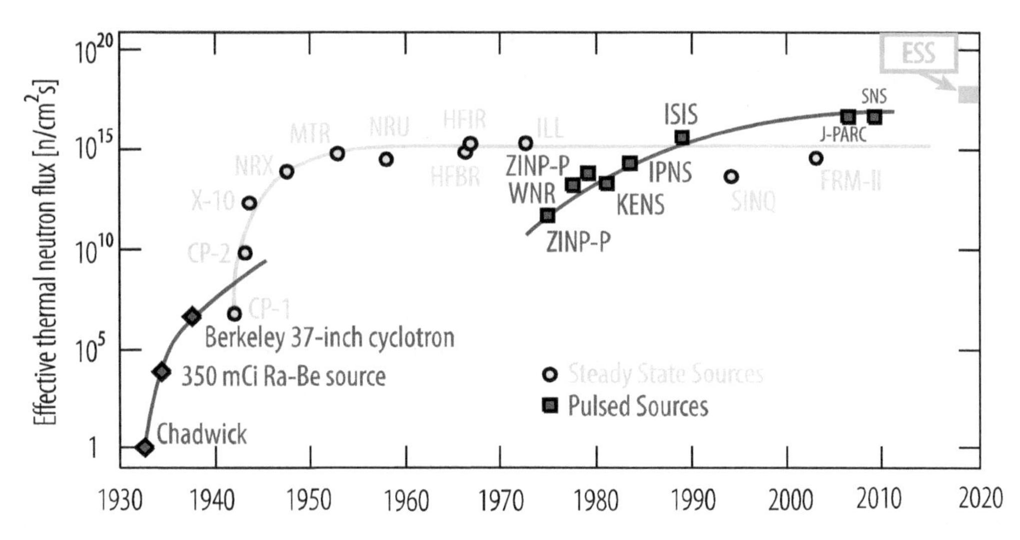

Fig. 11.33 The instant neutron flux at different research facilities plotted as a function of time

standard research reactor source. For example, a pulsed structure can be achieved by chopping a continuous beam with a shutter. Consequently many of the neutrons are then lost. Figure 11.33 plots the instantaneous neutron flux at different facilities as a function of the year of their completion. It is not possible to extract the average neutron flux from the figure without detailed knowledge of the time structure of the different facilities. However, the future European Spallation Source (ESS) will be the first spallation source with a time average neutron flux as high as that of the most intense research reactors.

The spallation cross section for protons on heavy nuclei increases as a function of proton energy up to several tens of GeV [116]. Nonetheless it is generally agreed that a kinetic proton energy between 1–3 GeV is optimal for practical target and moderator designs, and in order to keep the shielding requirements reasonable.

The first spallation source, ZING-P, based on a synchrotron, was built and operated at the Argonne National Laboratory. It provided a short pulse well suited for Time-Of-Flight (TOF) techniques at the "instruments" (experiments). Spallation sources at Argonne were superseded by the Intense Pulsed Neutron Source (INPS), by the KEK Neutron Science Center (KENS) in Japan, and by the ISIS neutron facility at Rutherford in UK. These synchrotron based spallation facilities were recently exceeded in power by J-PARC, at Tokai in Japan.

The first linear accelerator used to provide protons for a spallation source – the Los Alamos Meson Physics Facility (LAMPF) in the USA – later became the Los Alamos Neutron Science Center (LANCSE). The highest power Spallation Neutron Source (SNS) in Oak Ridge combines a full energy linear accelerator with an accumulator ring to provide very high intensity short pulses of neutrons to the instruments. The European Spallation (ESS) source will provide even higher intensities, but is developing instruments able to use longer linac pulses directly for spallation, avoiding the need for a costly and performance-limiting accumulator ring

[117]. Synchrotrons and accumulators, alike, have a intensity limits due to space charge effects that introduce tune spreads and cause beam instabilities, resulting in significant beam losses.

11.4.2 The Linear Accelerator

For many spallation-based sources, a good choice of technology for the acceleration of the proton beam is a linear accelerator (linac). These very complex machines are conceptually quite simple—a beam of H+ ions or protons is generated by a source and passed through a series of accelerating structures on its way to the spallation target. Since the spallation process means that the time structure of the neutron pulses will match that of the proton beam, the requirements of the experiments directly dictate the time structure of the proton current, and thus the high-level design of the linac.

11.4.2.1 High-Level Machine Design

Typically the specifications given by the experiments are the following.

- The desired repetition rate of the facility, usually less than 120 Hz.
- The duty factor or the length of the neutron pulses.
- The beam power on target; either average or maximum.

Given these parameters, the accelerator designers then begin sketching out the specifications of the design.

Kinetic Energy and Current
The maximum beam power on target constrains the maximum energy and current of the proton beam, and engineering capabilities will place additional limitations on these quantities. For example, the spallation target may have an energy threshold beyond which it will no longer survive.

As mentioned previously, the optimal beam energy for a spallation target system is in the range 1–3 GeV, and so an accelerator designer is likely to pick an energy in this range. From this, the required beam current is easily derived.

After this, these parameters will be optimised in an iterative way. Engineering limitations will first be reached in either the power transmission capabilities of the couplers on the accelerating structures, or the acceleration gradients required within the cavities. Once these have been determined, the designer can iteratively tweak the beam energy and current in order to balance the performance requirements of the cavities.

RF Frequency
Due to unavoidable losses in the generation and transmission of the RF, the installed capability will need to be more than a factor of two larger than the beam power. The

high power nature of these machines means that one of the largest cost drivers will be the power sources for the accelerating RF. In order to save cost it is advisable to take advantage of already-designed components, and thus RF frequencies previously adopted by other facilities.

Typically this results in acceleration RF whose frequency is an integer multiple of 176.105 MHz or 201.5 MHz.

Since it is possible to achieve higher acceleration gradients with higher frequencies, the self-force (i.e. space charge) of the beam at the lower energy regions of the machine mean that the strong focusing that results from high frequencies can be problematic. It can therefore be optimal to use multiple acceleration frequencies, with that in later sections of the linac being a multiple of two or four of the low energy acceleration.

11.4.2.2 Linac Layout

A linac is laid out as a series of consecutive stages of acceleration. The following sections each provide some detail on the most common technologies.

Proton Source

The ion source for a long pulse spallation source delivers protons rather than H- ions, since there is no need to inject into an accumulator ring using charge exchange injection (see section on synchrotrons). Proton sources can be implemented using various techniques. For example, compact Electron Cyclotron Resonance (ECR) sources such as VIS [118] and SILHI [119] are well suited to the task, delivering continuous proton beams of up to 100 mA. The diverging beam leaving the source has to be transported and matched to RFQ, usually using Einzel lenses or solenoids in the Low Energy Beam Transport (LEBT). The LEBT has to be designed carefully because it defines the beam quality throughout the rest of the linac [120].

Radio Frequency Quadrupole

Modern proton sources typically output a continuous beam of protons, however it is not possible to directly accelerate such a CW current. The structure immediately following the proton source must be one that compresses the current into a series of bunches at the fundamental frequency of the accelerating RF. In the majority of linacs a Radio Frequency Quadrupole (RFQ) performs this job.

An RFQ is a resonant cavity loaded with four vanes or rods in such a way that the beam experiences a quadrupolar electrical field that strongly confines the size of the beam. These vanes or rods are modulated in such a way that the beam also experiences longitudinal forces. Initially these forces are tuned so that the continuous current will begin to coalesce into a series of bunches at the frequency of the accelerating RF.

As the beam proceeds further through the RFQ, the modulation of the vanes/rods will be adapted so that there will be an acceleration force in addition to the focusing terms. This will be increased throughout the length of the RFQ in a way that preserves the quality of the beam, while keeping the structure to a reasonable length.

For a given frequency the four-vane structure is the most power efficient, for low current applications 4-rod structures may be cheaper to build. Currents of up to 100 mA can be handled with a single RFQ in the frequency range of 80–400 MHz, while higher currents require beam from two parallel RFQs to be combined in a funnelling section that operates at half the frequency of the main linac.

Drift Tube Linac

Once the beam has reached an energy of approximately 3 MeV, the efficiency of the acceleration provided by an RFQ will drop to a level where it would be optimal to use an alternative method of acceleration. In high-power proton linacs, a Drift Tube Linac (DTL) is often chosen to succeed the RFQ due to its high efficiency, and well-understood design principles.

A DTL is a large RF cavity through which the beam passes, absorbing power from the accelerating field. The time taken for the beam to traverse this structure is normally many tens of RF periods, and so metallic drift tubes are added to shield the beam from the decelerating phase of the RF. The drift tubes must then be designed to have the correct length and location for the beam to see the correct amplitude and phase of the accelerating fields.

The DTL accelerates the bunched beam from the RFQ to energies between 50 and 100 MeV [121]. The accelerating efficiency of DTLs drops at energies above 50 MeV and therefore a change of structure is required. The problem is that the physics & engineering effort required to build a more efficient structure costs more than any potential operational savings while increasing the risks. The normal conducting alternatives are Separated DTLs (JPARC), Side Coupled DTLs (Los Alamos and SNS), and Cavity coupled DTLs (LINAC4). Transverse focusing is achieved by permanent or electromagnetic quadrupoles housed inside the DTL drift tubes and/or between RF cavities.

Intermediate Energy Acceleration

In the energy range between approximately 100 MeV and 500 MeV, there tends to be a split in the technologies chosen for beam acceleration.

For machines with a duty cycle of more than a few percent, the ohmic losses due to the finite resistance of the metallic cavities is a significant driver of the operational costs of the accelerator. For this reason, these machines tend to choose superconducting structures in order to increase the fraction of the RF power that accelerates the beam. Despite the running cost of the cryogenic facility, superconducting facilities with significant duty cycles will cost substantially less to operate than room-temperature machines. An emerging alternative to higher energy DTL-variants is the superconducting spoke cavity technology. Spoke resonators have a large transverse and longitudinal acceptance and are mechanically very stiff, reducing their sensitivity to microphonics and to Lorenz force detuning compared to elliptical resonators in this energy range.

For machines whose duty cycle is less than this cut-off, the long time taken for filling the superconducting cavities with RF before beam arrival becomes significant, thereby degrading the efficiency. For these machines, the higher RF power costs are easily outweighed by the removal of the large cryogenic facility.

These linacs tend to favour the alternatives mentioned in the previous section—separated DTL's, Side-Coupled DTL's, and Cavity-Coupled DTL's.

Superconducting Structures

In the high-energy part of the linac, superconducting technology tends to dominate, however care must be taken to set the energy at which the linac transitions to superconducting cavities at a point that properly balances costs and benefits.

Due to several decades of R&D by the ILC community, elliptical cavities are a very good choice for these structures. The complex manufacturing processes are very well understood, and the design is very robust to mechanical disturbances.

High-energy efficiency requires that each RF structure is designed to have an accelerating mode with a very high Quality Factor (Q). Consequently, each Higher Order Mode (HOM) also tends to have long damping time, with a significant risk that its amplitude will still be at a high level when the subsequent pulses arrives. Thus, it is necessary to consider a scheme to damp HOMs. In general two solutions are available to the accelerator designer.

One solution is to mount one or more HOM couplers in locations where it is expected that the more destructive parasitic fields will have large amplitudes. These ports couple HOM power into an external load, dramatically shortening the decay times of the modes. It is common for accelerating structures to join multiple resonant cells into a single cavity structure. In this case, it is possible (as shown in Fig. 11.34) that the amplitude of a particular HOM may be negligible in the region of the extraction coupler, while remaining high in the body of the cavity. This is an inherent weakness of the multiple-cell structure, and care should be taken in the design of the cavity to ensure that such "trapped modes" do not couple well to the beam.

An alternative HOM damping solution is to include low Q, lossy, material around the beam pipe between one cavity and the next, in order to induce losses in any field that extends into this volume. This is implemented in a region where the amplitude and losses of the fundamental accelerating field harmonic are negligible. Nonetheless, a reduced fundamental Q is an unavoidable consequence of these designs.

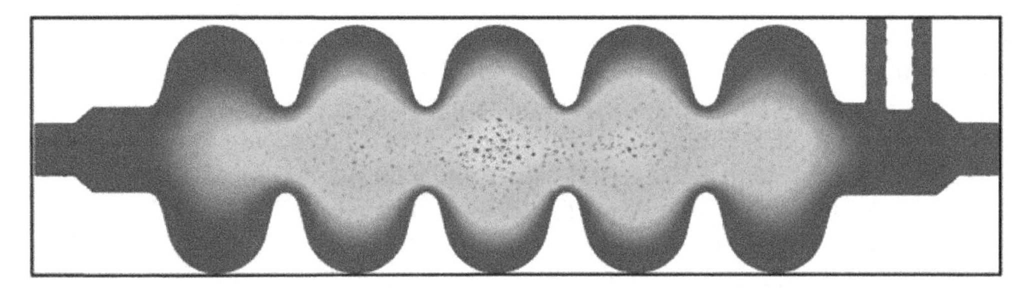

Fig. 11.34 Field magnitude for a particular dipole HOM oscillation within a five-cell accelerating cavity

HOM couplers typically have an inductive and a capacitive component, such that there is a very sharp rejection at the frequency of the accelerating mode. Although the addition of this filter successfully limits the impact on the accelerating efficiency of the cavity, the requirement for a non-zero capacitance between the central conductor and the cavity wall strongly increases the risk of a resonant cascade of electrons being field-emitted from the metallic surfaces. This would severely limit the successful operation of the cavity.

It has been shown [122] possible to instrument the signals excited on HOM couplers, to act as a diagnostic device measuring the 4D transverse location of the bunch (with position and angular resolutions of 4 μm and 140 μrad in FLASH), as well as to measure the phase of the bunch with respect to the accelerating RF (with a resolution of 0.08° at 1.3 GHz in FLASH). A determination of the cavity-cavity transverse alignment using the excited HOMs has also been demonstrated. The ability to instrument every accelerating cavity in this way is a major advantage to the HOM coupling scheme.

11.4.3 Rapid Cycling Accelerators for Short Pulse Spallation Neutron Sources

Accelerator schemes proposed for short pulse spallation neutron sources include:

1. Charge exchange injection from a pulsed H^- linac to a fast cycling synchrotron
2. Charge exchange injection from a pulsed H^- linac to a pulsed compressor ring
3. Direct proton injection from a pulsed proton linac to a fast cycling synchrotron
4. Direct proton injection from a pulsed proton linac to a pulsed compressor ring
5. Direct proton injection from a pulsed proton linac to a fast cycling FFAG ring

11.4.3.1 Charge Exchange Injection from a Pulsed H^- Linac to a Fast Cycling Synchrotron

A rapid cycling proton synchrotron (RCS) was the basis of a first, short pulse source for neutron scattering research. It was proposed by J Carpenter in 1972, and developed at ANL as the IPNS. The 450 MeV, 6.4 kW 30 Hz RCS source continued successfully for 26 years [123] and other sources soon followed. In Japan, KENS was developed at KEK, with a spallation target fed from their 20 Hz, 500 MeV, 3.5 kW proton RCS. ANL then studied higher power sources, but next came ISIS at RAL, (1979–1984) with a 70 MeV H^- linac and a 50 Hz, 800 MeV, 160 kW, RCS, [124], exceeding the beam power of the IPNS by a factor of 25.

There followed a Central European Initiative for a spallation source in 1993–1994, involving detailed feasibility studies [125] at both Vienna and CERN. Austron proposed to use a 0.13 GeV, H^- linac with a 1.6 GeV, 213 m circumference, proton

RCS to deliver neutron target beam powers of 205 (410) kW, at 25 (50) Hz. Despite the extent of the study, the project was not approved.

Now, Japan's Proton Accelerator Research Complex, J-PARC [126] has a 3 GeV RCS for a spallation neutron source as a part of a larger facility. The initial phase had a 25 Hz, 0.18 GeV, H^- linac with a 25 Hz, 3 GeV RCS, for a 0.5 MW beam power, but the H^- linac energy has been raised to 0.4 GeV to double the beam power. The 348.3 m circumference RCS has three superperiods, and uses novel, MA (magnetic alloy) loaded cavities for its harmonic number 2, RF system.

A RCS based, spallation source, the CSNS, is being built in China. It has a 81 MeV, H^- injector linac and a 25 Hz, 120 kW, 1.6 GeV, proton synchrotron [127]. Later linac energies of 134 and 230 MeV will allow beam powers of 240 and 500 kW, respectively. The 41.5 m long H^- linac has a 324 MHz, RFQ and DTL. An initial hybrid lattice of doublet straights and FBDB arc cells has been replaced by an all triplet lattice.

ISIS experience showed the importance of reliable 50 Hz, ion sources, long-life H^- stripping foils, RF shields for ceramic vacuum chambers, glass bonding for ceramic sections, low loss fast extraction, and quick release for flanges and water fittings. H^- charge exchange injection is a key issue and includes painting smooth beam distributions, reducing proton foil traverses (to limit foil temperatures), collecting foil stripped electrons, and locating a momentum and a betatron collimation system, allowing hands-on and "active" maintenance, for improved reliability and availability. Anti-correlated is preferred to correlated painting, as the latter produces some large amplitude protons in both transverse planes, which may be lost via coupling. Every fifth, RCS pulse is now diverted, at 10 Hz, to a new target station, which has become a major development.

A sub-tunnel within a main ring tunnel may reduce air activation near the collectors. Fractional loss design levels are set at $\approx 10^{-3}$ at collimators and 10^{-4} elsewhere in the ring, with the latter leading to a beam loss of <1 W/m. A RCS needs rectangular vessels to avoid vertical loss of any un-trapped beam, which spirals to protective momentum collectors, set in a high, normalized dispersion region. Rectangular vessels also aid transverse halo collection. Betatron collectors may be in three adjacent straights, with three long secondary units placed at 17°, 90° and 163°, transverse, betatron phase shifts after a primary scatterer.

Beam density limitations due to Liouville's Theorem are circumvented in charge exchange injection. Two designs are available for low-loss H^- injection systems for RCS rings. In one, the merging of the H^- ions and the circulating protons occurs at a ring dipole and, in the other, at the centre of an orbit chicane of four fixed and eight variable dipoles, in a very low dispersion region (SNS scheme) [128]. The former does not need an injection chicane, or an injection septum unit, or horizontal painting magnets, but it is not as flexible as the latter for longitudinal painting.

Both injection schemes aim to scrape the H^- beam (~$\pm 10\sigma$) in the input line at $\pm 5\sigma$ and set the beam centre at 5σ (~5 mm) from the adjacent, stripping foil edges. Input steering errors (of one polarity) result in some H^- ions missing the foil. Partially stripped H^o atoms appear in a range of quantum states [129] and, together with unstripped H^- ions, diverge from the protons in the region's bend field, and

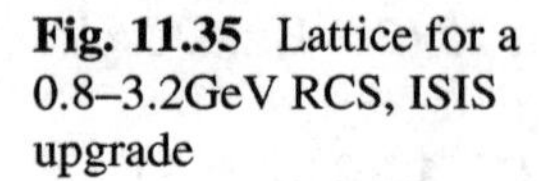
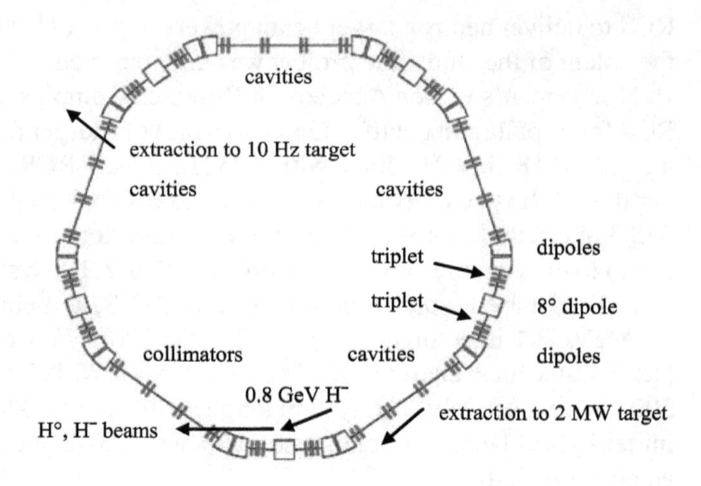

Fig. 11.35 Lattice for a 0.8–3.2GeV RCS, ISIS upgrade

traverse a second stripper and septum extraction unit(s) to exit the ring. Atoms with low principal quantum number, n, remain as H^o up to the second foil, while atoms of high n strip rapidly and are accepted as protons. Graphs of H^o lifetime versus field show "gaps" between n = 4 and 5 and between n = 5 and 6 states. A "gap" may be chosen by optimum choice of the region's field.

An upgrade study for ISIS [130] has considered a 0.8–3.2 GeV RCS, fed either from the ISIS ring for a 1 MW source at 50 Hz, or from a new 800 MeV H^- linac for a 2 (or 5) MW source at 30 (or 50) Hz. The injection is in an arc dipole as shown schematically in Fig. 11.35. The ring has been designed with five superperiods, each with arcs of triplets and straight sections of doublets (one pair of which is back to back). The lattice has been designed to minimize peak stored energy in the bend magnets and provide sufficient straight section space for economic cavity and RF system designs. In comparison with an all triplet lattice, it has 20 quadrupoles less, 20 m more of straight sections and ~67% less, total dipole stored energy.

11.4.3.2 Charge Exchange Injection from a Pulsed H^- Linac to a Pulsed Compressor Ring

Charge exchange injection from a pulsed H^- linac to a pulsed compressor ring was considered for a short pulse, 5 MW, European Spallation Source (ESS) shortly after the success of ISIS [131]. The study was overtaken in the USA, where it was decided to build a similar source, the SNS, at Oak Ridge. Injection was similar to that proposed for the RCS rings. SNS parameters were set at a 1.4 MW beam power at 60 Hz and 1 GeV, with a later upgrade to 3 MW at 1.3 GeV. Operating experience at the SNS [132] soon identified problem areas for future high power short pulse sources. These included the beam losses in the H^- linac following intra-beam scattering, and during H^- charge exchange injection in the compressor ring. Use of a H^- linac injector for a compressor ring or RCS now needs to be re-evaluated and other schemes also considered.

11.4.3.3 Direct Proton Injection from a Pulsed Proton Linac to a Fast Cycling Synchrotron

In a new scheme, a proton linac replaces the H^- linac and a ring with only one type of lattice cell is assumed, as a H^- injection insertion isn't needed. A more stable, higher current, lower emittance linac beam is realised and intra-beam and injection stripper losses are avoided. The RCS, compressor or FFAG ring is fed directly from the proton linac, using a two plane, multi-turn injection scheme, as proposed for a heavy ion fusion research project in 1998 [133]. Direct proton injection is similar for all three rings. A tilted, corner, electrostatic septum unit is set at a long drift section centre point. The septum wires have an effective 1 mm thickness and are at a 40–60° angle, θ, to the horizontal. Control of closed orbits at the septum is via vertical and horizontal bump magnets. Correlated painting is used, with closed orbits reduced during injection to reach the design emittances.

Basic parameters are input beam and ring emittances, angle θ, ring betatron tunes, Twiss parameters, space charge levels, and positions and directions, at the septum output, of the incoming beam centre and the ring's closed orbit. The ring lattice parameters may be kept constant during the injection or varied, to minimise injection beam losses. Evolution of beam distributions under the non-linear space charge forces, during and after injection, was simulated in the study [133]. Emittances were found to continue to grow after injection and evolve towards a stationary state so, to avoid loss, the late beam-septum separations were increased. A maximum number of turns injected without loss, for F = 10–20, and $(\varepsilon h\ \varepsilon v)_{ring}$ and $(\varepsilon h\ \varepsilon v)_{inj}$ the respective products of ring and injected emittances, was found to be approximately: $N_{inj} = (1/F)\ (\varepsilon h\ \varepsilon v)_{ring}/(\varepsilon h\ \varepsilon v)_{inj}$.

Ring circumference and straight sections must be large enough to achieve the required injected beam. RAL studies [134] have compared combined function, DFD and FDF triplet, and dFDFd and fDFDf, pumplet lattice cells. Magnets d and D provide vertical focusing (horizontal defocusing) and f and F give opposite focusing. Doublet cells are avoided as beam waists aren't at an injection straight centre. RCS pumplet cells are similar to those proposed for FFAGs [135], but their fields are linear and their magnets all have positive bends. The bending radii are thus larger and magnetic fields much reduced. For equal bending and corrector lengths, triplet and pumplet cells have the same length. Pumplets are preferred as the Twiss parameters are lower and hence the magnet apertures smaller. Ring designs use only one type of pumplet lattice cell (dFDFd or fDFDf), and a low dispersion in the injection straight section is advantageous.

An ISIS upgrade considers replacing the present ring by a large emittance, 0.6–1.8 GeV pumplet RCS. A fDFDf cell is preferred to a dFDFd as it has lower $\beta(h)$ values and dispersion. The 12 cells have a 26.0 m mean and a 12.5 m bend radius, and a 0.686 T maximum on-axis magnetic field. Assumed is a 70% chopped, 0.1 A linac proton beam, of 3 (π) mm mr, full un-normalised (y, x) emittances, F = 20, two-plane painting, a 450 (π) mm mr, ring vertical acceptance, and 150 and 300 (π) mm mr, respective, full un-normalised (y, x) ring emittances. These allow 75 10^{+12} protons to be injected in 172 μs over 250 turns, at a peak space charge tune shift $\approx$0.23 and a 1.08 MW beam power. ISIS second harmonic cavities may be used at a h = 2, frequency range of 2.91–3.45 MHz.

11.4.3.4 Direct Proton Injection from a Pulsed Proton Linac to a Pulsed Compressor Ring

A compressor ring and a RCS may be compared for the case of direct proton injection and equal beam powers. The compressor ring requires a higher energy, longer, more costly proton linac injector, a little larger ring circumference, slightly smaller ring acceptances, and a simpler, lower power radio frequency system. A detailed cost estimate for both linacs and rings is needed in order to determine the choice.

11.4.3.5 Direct Proton Injection from a Pulsed Proton Linac to a Fast Cycling FFAG Ring

FFAG accelerators have non-linear fields and focusing, defined by $B = Bo\,(1 + x/r_o)^K$, with x and r the distances from the ring centre and $K = \rho\, r\, (B'/B\, \rho) = \rho\, r\, Kv$. They may be non-scaling or scaling (NSFFAG or SCFFAG) [134, 135], with the former more common for higher beam energies. Two types have been considered for a 26.0 m radius ISIS upgrade. One, proposed by S. Machida [136], has a novel type of DF spiral lattice cell and construction of a model magnet cell is under study.

The second, by G.H. Rees [134], has 12 [O f(+) o D(–) o F(+) o D(–) o f(+) O] pumplet cells, where the non-linear magnets, d & D, give vertical focusing (horizontal defocusing), the f & F provide opposite polarity focusing, straight sections are O and o and bend directions are (+) or (–), with (–) the reverse bends.

Ring closed orbits needed for direct, multi-turn, proton injection into the FFAG are influenced by the non-linear main fields. Injection orbit bumps reduce the ring periodicity below that set by the number of lattice cells, and the design must ensure the beam dynamic aperture is not reduced during the injection.

FFAG-RCS differences include the former's lower chromaticities and beam-size dependent tune shifts and the latter's lower injection dispersion. Errors in FFAG's non-linear magnets are harder to correct than those in RCS's linear magnets. A RCS has the advantage of no reverse bends so, in the case of equal outer radii, room temperature magnets, it may have a higher energy and beam power than a FFAG. It may also include trim quadrupoles, sextupoles and octupoles to adjust the respective tunes, chromaticities and tune spreads, and so aid in obtaining beam stability.

11.4.4 High Intensity Cyclotrons

Cyclotrons have a long history in accelerator physics and are used for a wide range of medical, industrial and research applications [137]. First cyclotrons were designed and built by Lawrence and Livingston [138] back in 1931. The cyclotron is a resonant accelerator concept with several properties that make it well suited for the acceleration of hadron beams with high average intensity.

As a circular accelerator the cyclotron repetitively uses the same accelerating resonators and radio frequency (RF) sources, thereby utilizing these expensive components in an effective and economical way. Cyclotrons allow continuous injection, acceleration and extraction of a beam. Neither RF frequency nor magnetic bending field must be cycled. The continuous wave (CW) operation results in low bunch charges, which leads to moderate space charge effects in comparison to pulsed accelerator concepts. Basic components of a separated sector cyclotron, suited for high intensity operation, are the bending magnets in sector shape, RF resonators for beam acceleration and the injection/extraction elements. The Fig. 11.36 shows the layout of the PSI Ring cyclotron [139, 140] as an example of a high intensity, separated sector cyclotron.

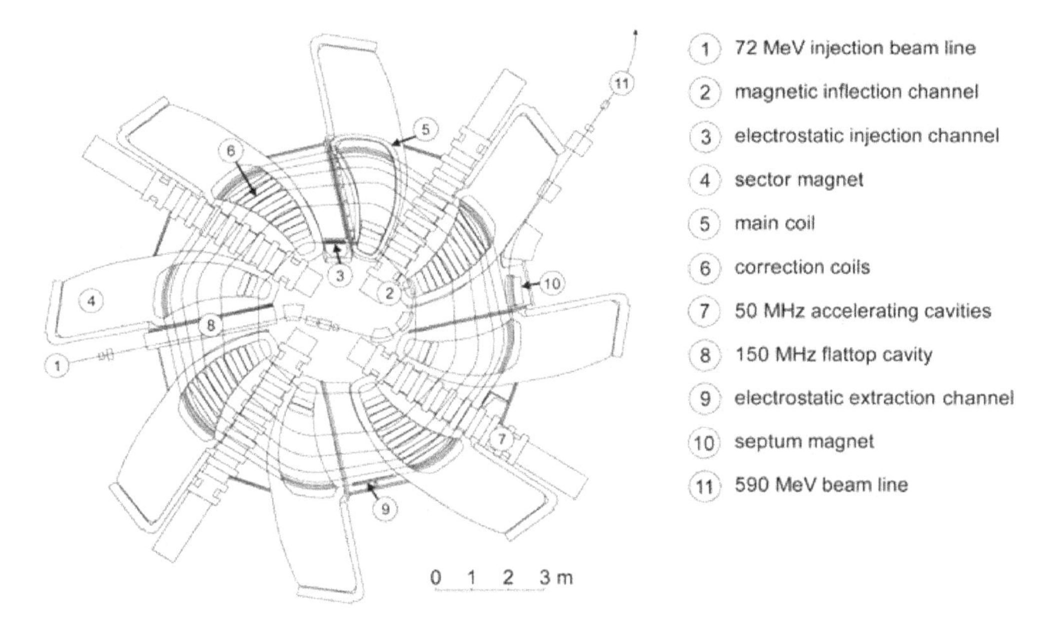

Fig. 11.36 The PSI Ring cyclotron generates a 1.4 MW proton beam at 590 MeV kinetic energy

During the course of acceleration the revolution time of the beam must be kept constant in order to ensure synchronicity with the accelerating RF voltage. At relativistic energies this is achieved by raising the average magnetic bending field in proportion to the relativistic γ factor towards larger orbit radii. This positive slope of the bending field as a function of radius would result in a loss of vertical focusing if the field was uniform azimuthally. Sufficient focusing is achieved due to the azimuthally varying field strength in sector magnets and intermittent gaps (Thomas focusing [141]), and by the introduction of spiral magnet shapes that enhance the edge focusing effect of the bending magnets.

In practice the strength of vertical focusing is typically just sufficient for operation, but weak in comparison with alternate gradient focusing. In the PSI Ring cyclotron the number of vertical betatron oscillations per turn is slightly smaller than 1 which reflects the weak focusing properties of the cyclotron magnet lattice.

The key for the acceleration of a high intensity beam in a cyclotron is a clean extraction with small losses. In the PSI cyclotron the extraction is realized by deflecting the beam with an electrostatic element whose 50 μm tungsten electrode has to be placed between last and second last turn. Particles in the beam halo may hit this electrode, are scattered out of the beam and are lost in the extraction beamline. According to practical experience losses of the order of 100 W can be accepted with respect to this major loss mechanism in the PSI facility. The highest activation levels in the extraction beamline typically amount to 10 mSv/h. All efforts to optimize the accelerator towards higher beam powers are focused on lowering the particle density in the beam tails and maximizing the turn separation at the extraction point. One dominating effect generating beam tails is the longitudinal space charge force which increases the correlated energy spread inside the bunch. The bending fields transform the energy spread finally into transverse beam tails that may interact with the extraction electrode. As Joho has shown [142], the strength of this effect scales quadratically with the time the beam needs to be accelerated, i.e. with the number of turns. In addition turn separation at the extraction of the cyclotron scales inversely with the number of turns. Thus in total the losses caused by longitudinal space charge scale with the third power of the turn number. Because of this strong scaling it is very beneficial to realize the highest possible accelerating voltages. For a cyclotron designed purposely for high intensity beams the achievable radius increment per turn is an important design criterion. Under the condition of constant revolution frequency the turn separation can be expressed in the following way:

$$\Delta R = \frac{R}{\gamma\left(\gamma^2 - 1\right)} \frac{U_t}{m_0 c^2} \,. \tag{11.89}$$

Here R is the orbit radius and U_t the energy gain per turn. Thus the ideal high intensity cyclotron exhibits a large diameter, provides a large energy gain per turn and accelerates to moderate energies $\leq$1 GeV, since for large γ the turn separation decreases very fast.

An alternative way to realize a clean extraction is charge stripping with a thin foil. In this case incompletely stripped ions are accelerated in the cyclotron, for example H^- or ${H_2}^+$. At the extraction foil the ions change their charge and are separated from the circulating beam by the action of the bending field. This scheme is employed in the TRIUMF cyclotron to extract proton beams up to 520 MeV [143]. A disadvantage of this method is the non-negligible rate of stripping that occurs in the bending field, causing losses and activation. The ${H_2}^+$ molecule has a higher binding energy in the ground state, resulting in significantly lower dissociation probability. However, disadvantages are the lower charge to mass ratio requiring stronger fields, and the complex extraction path across the centre of the cyclotron, resulting from a reduction of the bending radius by a factor 2 after stripping.

Transverse space charge forces present a limitation for the maximum feasible beam intensity. Above a certain charge density the repelling space charge forces exceed the relatively weak vertical focusing forces in a cyclotron. With the PSI concept the realization of cyclotrons for beam powers up to 10 MW seems feasible.

Classical cyclotron magnets represent rather heavy and complicated devices since they need to cover a wide radius variation of the beam orbit. To ensure the condition of constant revolution time the radial field shape must be precisely controlled. Radially distributed trim coil circuits are used for fine tuning on the basis of beam phase measurements. Modern cyclotrons like the RIKEN SRC employ superconducting magnets [144, 145] to obtain higher bending strength. In view of turn separation and efficient extraction it is generally not desirable to design very compact high intensity cyclotrons with the help of superconducting magnets. However, with a large radius the gained space can be used to maximize the installed RF voltage.

For the production of MW-class beams the energy efficiency of the accelerator systems is important. Cyclotrons with continuous RF generated by tetrode amplifiers and utilizing copper resonators can reach a relatively high efficiency. A comparison of different accelerator concepts suited for producing high intensity proton beams is presented in [146]. At the PSI facility a fraction of 18% of the total grid power is converted to beam power.

In Table 11.11 selected properties of three cyclotrons are listed.

In summary cyclotrons present an effective and relatively compact solution to generate high intensity proton beams at energies below 1 GeV and beam powers of a few MW.

Acknowledgements
Mohammad Eshraqi, ESS
Steve Peggs, ESS and BNL
Romuald Duperrier, CEA
Jim Stowall, Los Alamos
John Galambos, SNS
Ference Mezei, ESS

Table 11.11 Parameters of existing cyclotrons [147]

Cyclotron	K [MeV]	$N_{mag.}$	Harmonic number	R_{inj} [m]	R_{extr} [m]	Extraction method	Overall transmission	P_{max} [kW]
TRIUMF	520	6 (sect.)	5	0.25	3.8..7.9	H^- stripping	0.70	110
PSI-Ring	592	8	6	2.1	4.5	Electrostatic channel	0.9997	1400
RIKEN s.c. Ring	2600	6	6	3.6	5.4	Electrostatic channel	0.63	1 (^{86}Kr)

11.5 Heavy Ion Accelerators for Nuclear Physics

N. Angert · O. Boine-Frankenheim

11.5.1 Accelerator Facilities for Heavy Ion Nuclear Physics: Background and Aims

Nuclear spectroscopy, reaction studies, and nuclear astrophysics using beams of accelerated heavy ions have been research fields in nuclear physics since the middle of last century. An especially interesting topic is the search for new super-heavy elements (SHE) using fusion reactions between medium-mass ions, with energies of 4–5 MeV/u (around the Coulomb barrier), and very heavy target nuclei. Such studies have been conducted for many years now at LBL, Berkeley; GSI, Darmstadt; JINR, Dubna; RIKEN, Saitama, and other laboratories [148]. Using the cold fusion technique to produce super-heavy compound nuclei, the element chart has been expanded up to element $Z = 118$ in recent years using actinide target nuclei [149]. Today, a growing research community is conducting experiments in nuclear spectroscopy, mass and life time measurements as well as in the nuclear chemistry of super-heavy elements [150].

Since the mid-1980s the availability of medium-energy ion beams up to 100 MeV/u in a new generation of normal conducting sector cyclotron [151] and superconducting cyclotron facilities [152] has provided access to radioactive isotopes in unexplored regions of the nuclear chart, far from stable isotopes. Fragmentation and fission of heavy projectile nuclei in thin light element targets were used to produce fast exotic nuclei using the in-flight method. Light exotic nuclei produced in this way such as ^{11}Li have exhibited the new phenomenon of neutron halos [153, 154]. In other exotic nuclei produced using this method, changes have been detected in the neutron shell structure for large N/Z-ratios shedding new light on nucleon-nucleon interaction. Precision studies of masses, life times, and new decay channels have revealed interesting phenomena, extended the understanding of nuclear forces, and raised new questions. In addition, experiments with unstable nuclei have contributed to a better understanding of an increasing number of observations in astronomy.

A new era in nuclear astrophysics has opened up with the rare-ion beam facilities dedicated to the measurement of nuclear reactions involving post-accelerated short-lived nuclides of particular relevance to astrophysics. Whereas at large facilities such as RHIC and LHC the aim is to simulate the early state of the universe in the laboratory, exotic beam facilities will increase our understanding of the development of stars and the creation of heavy elements. It is estimated that more than 8000 isotopes may remain bound, with only about a third of them having already been identified. Nuclear behaviour is expected to change significantly in the yet unexplored regions. Figure 11.37 shows, as an example, the landmark results

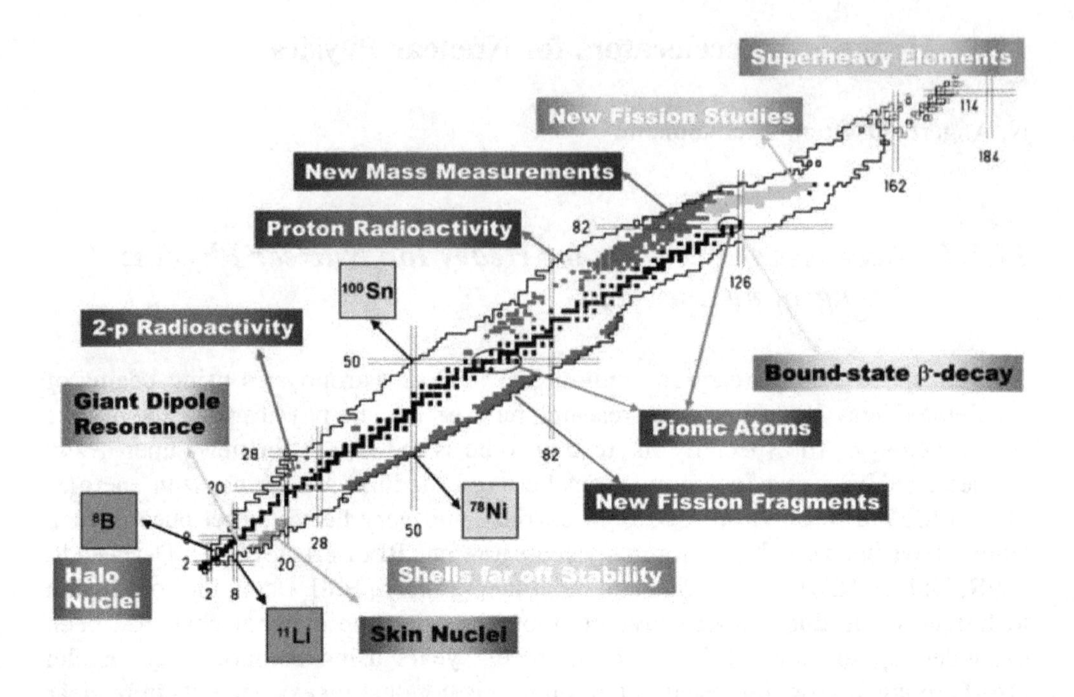

Fig. 11.37 Landmark results from experiments with in-flight produced super heavy elements (SHE), with exotic beams at the Synchrotron-FRagment-Separator facility (SIS-FRS), and at the Experimental Storage Ring (ESR) at GSI Darmstadt.

from experiments with in-flight produced super-heavy elements, exotic beams at the Synchrotron-FRagment-Separator facility (SIS-FRS) and at the Experimental Storage Ring (ESR) at GSI Darmstadt [155].

Another method of producing isotopes far off stability and which can deliver higher phase-space density exotic ion beams is the ISotope On-Line technique (ISOL) [156]. This method is to some degree complementary to the in-flight method. ISOL uses intense proton or light-ion beams in the energy range from 20 to 1400 MeV to generate radioactive nuclei using spallation, fission, or fragmentation reactions induced by the projectile in thick targets of heavy elements. Unstable isotopes produced in this way have thermal kinetic energy. After effusing out of the hot target into an ion source they are singly ionized and separated in a high-resolution magnetic separator. Then they are either trapped for precision mass measurements or nuclear spectroscopy in a Penning trap, for example, or further ionised in a charge breeder and post-accelerated to low energies in the keV/u to MeV/u range. The principles of the in-flight and the ISOL methods are shown in the diagrams in Fig. 11.38.

To summarize: rare isotopes either stopped ones or those with very low energy (0–100 keV) are used for precision measurements of masses, moments, and symmetries. Re-accelerated isotopes (0.2–20 MeV/u) are used for detailed nuclear structure studies, high-spin studies, and measurements of astrophysical reaction

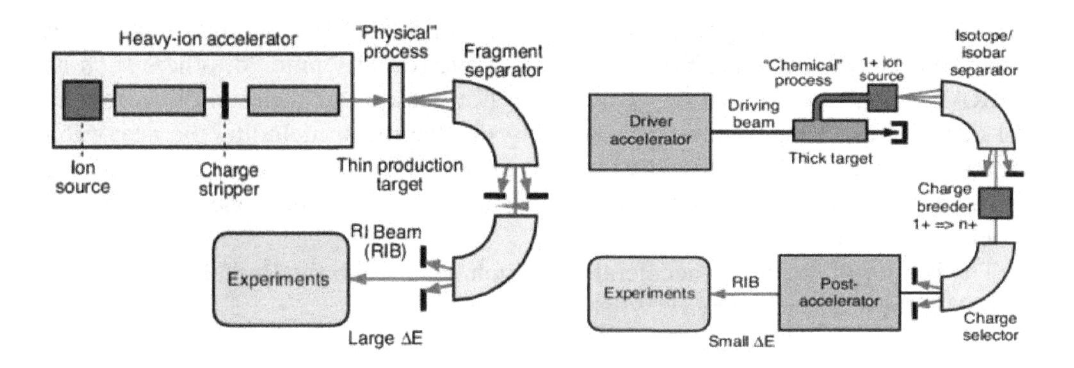

Fig. 11.38 Diagram of the in-flight production method for radioactive isotopes (left) and of the ISotope On-Line (ISOL) production method (right).

rates. In-flight produced fast isotopes (>100 MeV/u) make possible the farthest reach from stability to the limits of existence and to the shortest life times. Generally speaking, with the expansion of nuclear physics research to areas far off stability in the nuclear chart, the atomic nucleus has become a laboratory for studies of fundamental interactions and symmetries at the interface of particle, nuclear, atomic, and astrophysics. This program requires the availability of both stable beam and radioactive beam facilities, along with the development of new experimental techniques and instrumentation.

Also, new facilities dedicated to delivering high-intensity heavy ion beams are needed to further extend the possibilities for synthesis and experiments involving super-heavy elements. In addition, smaller accelerator facilities are needed to develop and test new instruments, and to educate the next generation of scientists.

11.5.2 Accelerators

11.5.2.1 Introduction

In the last two decades exciting challenges in physics described above have stimulated many laboratories to begin research with radioactive isotope beams (RIB) in existing or expanded accelerator facilities. Progress in ion source and superconducting accelerator technology has facilitated this development.

At present there are a number of small and large in-flight and ISOL facilities. The larger ones are NSCL at Michigan State University (MSU); ISOLDE at CERN; ISAC (I and II) at TRIUMF, Vancouver; SPIRAL1 at GANIL, Caen; RIKEN RIBF, Saitama; and SIS/ESR at GSI, Darmstadt. Large facilities planned for the near future are GSI-FAIR at GSI and FRIB at MSU [157].

SPES in Legnaro and SPIRAL2 at GANIL are facilities on the way to EURISOL, which is still in the planning stage. The quest for and the study of super-heavy elements has been an ongoing effort at JINR, GSI, and RIKEN, in particular, but other laboratories such as JYFL, Jyväskylä, have recently entered; others such as SPIRAL2 plan to do so. In this article it is not possible to mention the complete list of facilities. An overview of nuclear physics facilities, including the heavy ion accelerators that exist worldwide, is given in the IUPAP Report 41 [157]. Indeed, there has been something of a renaissance in low- and medium-energy heavy ion accelerators in recent years.

The history of heavy ion accelerators, which began with the Berkeley cyclotrons in the 1950s, is described by E. Wilson in Chap. 1.

11.5.2.2 Special Issues of Heavy Ion Accelerators and Storage Rings

Some specific issues must be considered when designing accelerators and storage rings for heavy ions. These issues are not relevant for proton or light ion machines. Ion sources are needed which can generate highly charged, intense ion beams of accelerator-compatible beam quality for the full spectrum of elements, with their varying physical and chemical properties. During the acceleration of partially ionized, highly charged heavy ions, charge-changing processes occur, whether by accident or design. Knowledge of the equilibrium ion beam charge after it passes targets (strippers), and of the yield in the desired charge state after stripping, is important when designing and optimising an accelerator layout. The high specific energy loss and irradiation effects are much more important issues for heavy ions than for light ones. Knowledge of charge-changing cross sections is important when estimating beam transmission in synchrotrons and beam lifetimes in storage rings for given vacuum conditions, as well as recombination processes during electron cooling. In synchrotrons and storage rings, heavy-ion-induced gas desorption processes are crucial with respect to intensity limits and lifetimes of ion beams. The reference atom for the design of the large in-flight driver accelerators to be described later is ^{238}U. Therefore, the explanation of aspects specific to heavy ions will mainly be confined to that element.

11.5.2.2.1 Ion Sources

The first ion sources used for multiply charged ions in accelerators were been Penning ion sources. In the 1950s, the cold-cathode Penning type was developed at Berkeley for ion beam production [158]. Later, the hot-cathode Penning ion source, which provided higher ion-beam currents for medium-charged ions, was the favourite choice both for linear and cyclotron heavy ion accelerators up to the mid-1980s [159]. It was now possible to deliver beams of up to several 100 μA for U^{10+} [160]. However, the source lifetime was limited by erosion of the discharge electrodes. In addition, in view of the fact that beams from metallic elements were produced by using the sputtering technique, the lifetime of sputter electrodes was a limiting factor, too, along with rather high material consumption. This was a serious drawback when beams from expensive isotope-enriched materials were needed in super-heavy element research.

For the production of short-pulsed high-charge-state beams of light ions Laser Ion Sources (LIS) were used, for purposes such as injection into the Synchrophasotron in Dubna beginning in the mid-1970s; now they are used for injection into the Nuclotron [161]. Pulsed Electron Beam Ion Sources (EBIS) were used there for medium-mass ion beams up to krypton [162]. Both low duty cycle ion source types (LIS and superconducting EBIS) are now in use at the Nucletron based collider facility NICA for high charge state light and heavy ions, respectively [163]. A powerful superconducting EBIS is delivering intense beam pulses of highly charged heavy ions up to uranium for the RHIC injector at BNL [164, 165]. EBIS-type devices are also used as charge breeders at various ISOL facilities [166].

The most influential development on the field of heavy ion sources has been the Electron Cyclotron Resonance Ion Source (ECRIS), which has replaced the Penning type ion sources at most heavy ion accelerators over the years and has substantially increased the capabilities of cyclotron-, linac-, and synchrotron-facilities for research on the field of nuclear physics over the last decades. The ECRIS concept evolved from plasma devices for fusion research and has been investigated as a source for highly charged ions at several laboratories, such as Grenoble, Jülich, Karlsruhe, Louvin-la-Neuve, Marburg, and Oak Ridge since the 1970s. In the ECR ion source, a plasma-confining magnetic configuration is generated by the superposition of an axial solenoid mirror and a radial hexapole field. Energetic electrons are generated for the ionization process using electron cyclotron heating by GHz microwaves in magnet field regions which meet the electron-cyclotron-resonance condition for the plasma electrons.

R. Geller's group in Grenoble developed this device by introducing hexapole magnets to improve ion confinement to a compact efficient generator for highly charged ions (Minimafios) [167, 168]. An ECRIS was used for the first time at an accelerator for ion injection in CYCLONE at Louvain-la-Neuve begin of the 1980s [169]. The first ECRIS used at a large-scale accelerator was installed at the CERN facility. An ECRIS, operated with a 10 GHz klystron, delivered 40 μA oxygen 6+ in pulsed mode for injection into the CERN PSB-PS-SPS complex in the mid-1980s [167]. Other accelerator laboratories followed CERN's lead. Figure 11.39 shows LBL's superconducting Versatile ECR ion source for NUclear Science (VENUS), as an example of the third generation of ECR ion sources which are used at various laboratories today [170]. These third generation ECR sources, such as SECRAL and SECRALII at IMP (Lanzhou) [171], SuSI at NSCL/MSU (East Lansing) [172], the RIKEN-SCECR at Nishina-Center (Wako) [173], and the KBSI-SCECR for RAON (Deajeon) [174], are essential for the capabilities of existing and next generation heavy ion accelerator facilities for Nuclear Physics.

Third-generation ECRIS devices with superconducting magnets and operated with microwave generators for electron heating up to 28 GHz are the standard now. Developments in the various laboratories and operating experience over many years have led to steady improvements of performance. These ion sources can deliver now beams of several mA of oxygen 6+. Record beam currents of more

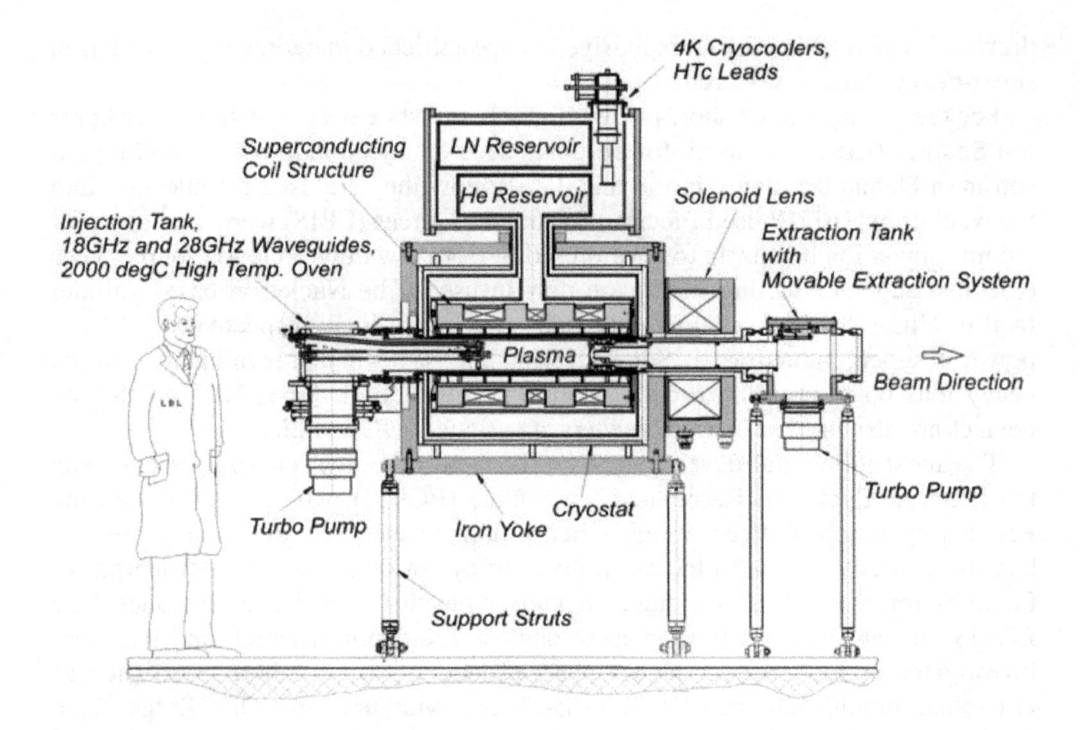

Fig. 11.39 Mechanical layout of the LBL VENUS ion source and cryogenic system.

than 400 μA have been reached for U^{31+} with VENUS [175], 680 μA for Bi^{31+} and 10 μA for Bi^{50+}, respectively, with SECRAL [176]. In long term accelerator operation typically about half of these record beam currents are used in order to have stable operating conditions over days or weeks. Figure 11.40 shows a charge-state spectrum for uranium achieved with VENUS, when operated in the high-current mode for medium charge states.

For cyclotrons and superconducting linacs, the ECRIS is the ideal choice. It can be operated at a duty cycle of 100%. It has no consumable electrodes; therefore, it can run for weeks when operated with gaseous elements. For metal-ion production, furnaces have been developed at many laboratories for a variety of elements up to uranium (e.g. [177–180]) [175]. Material consumption is at least an order of magnitude lower than for Penning ion sources, in the range of 10 mg/h or below, an important issue if beams from enriched isotope materials are needed. For injection into synchrotrons the pulsed afterglow mode can be used which provides higher peak currents of very highly charged ions than the cw operation [181, 182]. A survey on the progress, challenges, and experiences in intense highly charged ion beam production and long term operation with the third generation ECRISs is given in [176].

The impact of ECRIS on the ion-accelerator field is best illustrated by the following example. For the Berkeley 88-in. cyclotron (K = 140) it was possible to

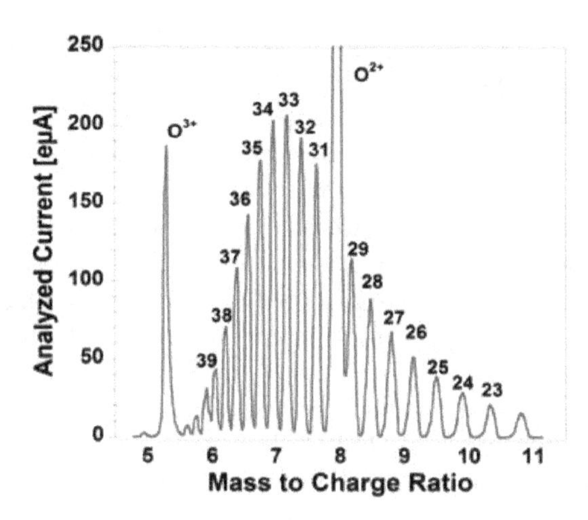

Fig. 11.40 Spectrum of uranium charge states from the LBL VENUS in the high-current mode.

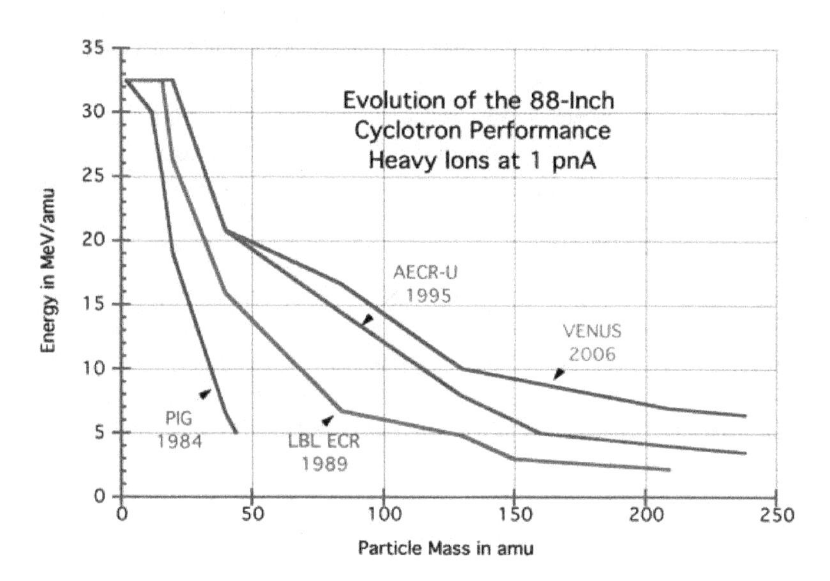

Fig. 11.41 The evolution of beam energies vs. element mass at the 88-in. (K = 140) cyclotron at LBL from the mid-1980s, using the best Penning ion sources, up to the present using the third-generation superconducting ECRIS of the VENUS type. [Courtesy of LBL Berkeley, C.M. Lyneis]

expand the mass range, for which energies of more than 5 MeV/u can be provided, from argon ($A = 40$) in the 1980s using Penning ion sources to medium-mass ions (xenon) using the normal conducting AECR-U ion source, and finally to uranium ($A = 238$) using the superconducting VENUS (Fig. 11.41) [183]. There are many cyclotrons worldwide with $K \geq 100$, which can take advantage of the high charge-states delivered by the ECRIS to reach beam energies in order to perform nuclear physics experiments with heavy ions.

ECRIS can also be used for charge breeding of ISOL-produced or stopped in-flight-produced rare isotopes before post- or re-acceleration [184, 185].

Based on the experience and progress with third generation ECRIS in increasing beam intensities of highly charged ions in the last decade, proposals to build next generation ECRIS for microwave frequencies beyond 28 GHz have been discussed since some years. The 1st fourth generation ECRIS is under construction now at IMP, Lanzhou, to be operated at 45 GHz, with correspondingly higher magnetic fields, based on Nb3Sn- instead NbTi- superconducting technology [186]. Expectations are to reach with that fourth generation two to three times higher beam currents in the future.

With respect to the formation and transport of beams from ECR ion sources, one has to bear in mind that the extraction system is placed in a region with a superposition of stray fields from the axial coil and the radial hexapole field, affecting beam formation and beam density distribution. In addition, beams with different ion charge states have different emittances due to the magnet-field structure, the production and confinement processes inside the source. The ion-current density is not uniform within the beam [187–189]. In third generation ECRIS, with higher magnetic fields and microwave frequencies, beam emittance and low energy beam transport is an issue [176].

In contrast to cw-operated cyclotrons and superconducting linacs, synchrotrons require injection at high peak currents within a brief period. Those peak beam currents of highly charged ions can be delivered from laser ion sources, or by an EBIS, or by an ECRIS along with an accumulator ring such as LEIR, which uses fast extraction to provide the requested peak intensities for the LHC [190]. For the FAIR synchrotrons, high-current short-pulse ion sources are used, which can deliver many mA (e.g. 15 mA of U^{4+}) of ions in a low charge state [191, 192]. After pre-acceleration and intermediate stripping in the injector linac (Unilac) the charge state is reached (U^{28+}) which is needed in the booster synchrotron SIS18 (section "Facility for Antiproton and Ion Research (FAIR) at GSI").

11.5.2.2.2 Charge-Changing Processes

The design of the first stage of a heavy ion accelerator depends on the charge state delivered from the ion source for the heaviest ion species to be accelerated (highest A/q-ratio). In accelerator facilities for medium to relativistic energies (some 10–1000 MeV/u), usually one or two strippers are needed to increase the charge state of the ions for an efficient acceleration scheme. In a stripper target, the initially charge-homogeneous beam splits up into several charge components. One gets a charge state distribution centred on the equilibrium charge state. To optimize the overall accelerator layout, it is important to know the ion and energy dependence of the equilibrium charge states and the yield of the charge states as well.

Pioneering theoretical work on these processes was done by Bohr [193]. Bohr assumed that a fast heavy ion passing a gas stripper retains the electrons which have orbital velocities which are greater than the ion velocity (Bohr's Criterion). This physically reasonable criterion has been proven to be a good first-order approximation for estimating the charge states which can be reached. Semi-empirical approximation formulas were developed later to describe the energy and target dependence of the equilibrium charge and the width of charge-state distributions. For the low energy range, that means for energies from about 1 to some 10 MeV/u, the equilibrium charge state for heavy ions can be approximately described by a simple parameterized formula of the type

$$1 - (\overline{q}/Z_P) = C\ \exp(-\delta\beta/\alpha)\,, \tag{11.90}$$

with $\overline{q}$ = average charge state, Z_P = nuclear charge of the ion, $\beta = v/c$, $\alpha = 1/137$, $\delta = Z_P^{-\gamma}$, see among others [194, 195]. For the stripping data gathered for example in the operation of the Unilac (GSI) over the years, this formula is used with $C = C^* + 140/Z_P^2$, $C^* = 1.0285$. Averaging the experimental data, γ-values of 0.56 for foil and 0.65 for nitrogen-gas-jet data have been calculated [196]. Foil strippers deliver higher charge-states, because the higher collision frequency in solid materials leads to higher electron loss probability [195], but also suffer from degradation due to beam intensity. For the range of $0.2 \leq \overline{q}/Z_P \geq 0.8$, which means if the ion has a sufficient number of electrons with comparable binding energy, the charge state distribution is roughly Gaussian in shape. The width of the Gaussian curve is proportional to $Z_P^{1/2}$ (see e.g. [195]).

Simple exponential relations for the average charge begin to deviate from measured data, and charge state distributions are no longer Gaussian-like in shape, when the electron configuration of the ion approaches inner shells with large steps in ionization energies. This was discovered in the case of bromine ions by C.D. Moak et al. in 1967 when approaching the L-shell by the stripping process in the energy range between 1 and 2 MeV/u [197]. Therefore, after estimations with simple semi-empirical formulas, experimental data should be taken into account whenever available.

For beams of energies in the range from 10 MeV/u to 80 MeV/u, the computer program ETACHA was developed by Rozet et al. to calculate charge-exchange cross-sections, charge-state evolutions in targets, and equilibrium charge-state distributions for highly charged ions, taking into account the electronic structure of inner shells [198]. Experimental results in this energy range can be found in [199], for example.

An overview of theoretical and experimental results for Xe, Au, and U projectiles impinging with kinetic energies from 80 to 1000 MeV/u on solid and gaseous targets ranging from Be to U is given in [200]. The calculations are compared to data from experiments carried out at the BEVALAC (LBL) and at the heavy ion synchrotron SIS (GSI) and associated facilities. Measured equilibrium charge state distributions for uranium in the energy range from 1 to 1000 MeV/u are shown in Fig. 11.42 [200].

For high intensity beams, gas or liquid strippers have advantages compared to foil strippers concerning their durability. Gas strippers lead to much lower equilibrium

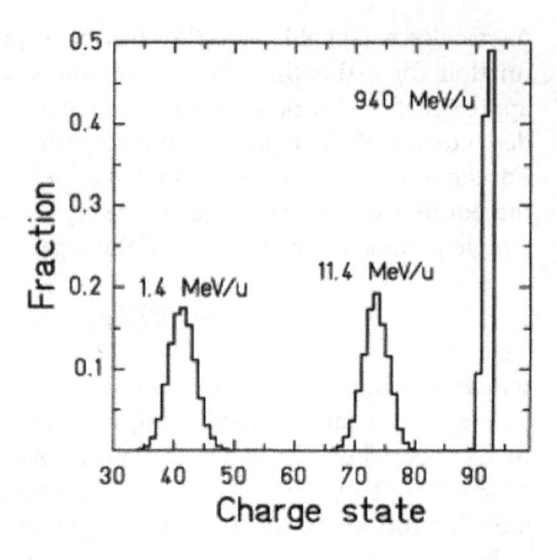

Fig. 11.42 Measured equilibrium charge state distributions at different energies for ^{238}U behind foil strippers.

charge states due to the strongly reduced influence of density effects compared to solid strippers [195]. Since electron capture cross sections of the heavy ions in the low-Z gases are considerably suppressed, in particular hydrogen promises higher equilibrium charge states as compared to nitrogen which is routinely used at the UNILAC gas stripper [201].

In synchrotrons and storage rings, charge-changing reactions between beam ions and the residual gas in the vacuum chamber can lead to the immediate loss of the ion. These processes are not significant in linear and cyclotron accelerators, because of the short acceleration path length there. However, in synchrotrons for partially stripped ions and in storage rings, these processes can limit the intensity and life time of the beam. In order to determine the lifetime, it is necessary to know the charge-changing cross-sections for both the capture and the loss of electrons during the interaction of the ions with the residual gases.

For practical purposes, the one-electron-capture cross-section σ_c, for example, is usually estimated by applying a semi-empirical formula of Schlachter et al., derived from experimental data in gases from H_2 to Xe [202]. It describes most of the experimental data for σ_c in the low- and medium-energy range within a factor of two. For the electron-loss cross-section σ_l, simple semi-formulas cannot be found for a comparable wide energy range.

In recent years methods calculating charge-changing cross-sections have been developed taking into account the electronic structure of projectile ions and target atoms. As an example calculated cross-sections of U^{39+} + Ar as a function of the ion energy are shown in Fig. 11.43 in comparison with experimental data [203–206]. Figure 11.43 shows quite well the general energy dependence of cross-sections of partially ionized highly charged heavy ions: The probability of electron capture saturates at very low velocities and decreases steeply at energies above a few 100 keV/u. Loss cross-sections increase with increasing energy, pass through

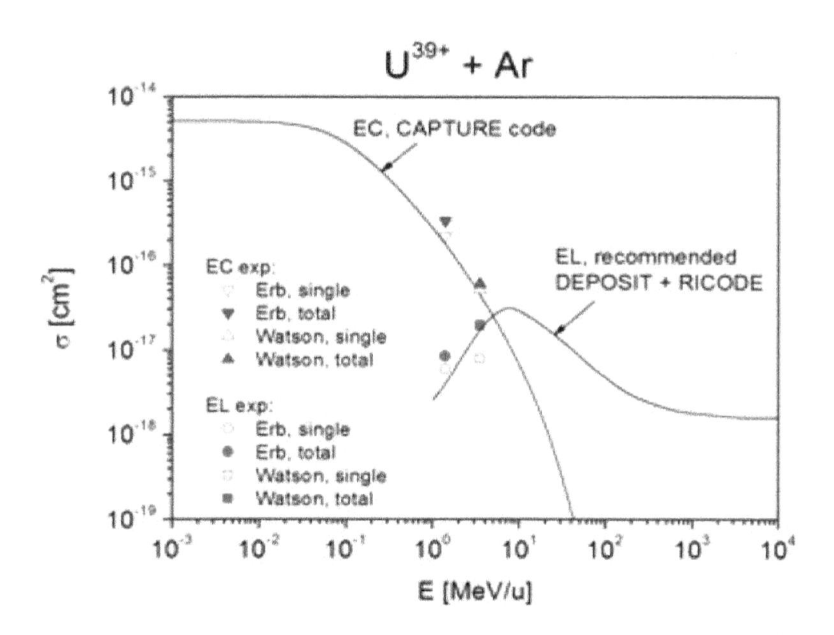

Fig. 11.43 Calculated cross sections (solid line) for electron capture (EC) and loss (EL) for uranium 39+ ions in argon gas as a function of the collision energy [190, 191]. Experimental single-electron capture and loss cross-sections (open symbols) are given to show their contribution to the total cross-sections (solid symbols) [192, 193]. Solid curves: EC calculations using the CAPTURE code, and EL calculations using the RICODE and DEPOSIT codes, respectively. The DEPOSIT code is described in [190].

a maximum at intermediate energies, and reach a lower nearly constant value at relativistic energies.

For heavy gases, the cross-sections can be orders of magnitudes larger than for helium or hydrogen, especially the capture cross-sections. This is important with respect to the desorption of heavy gas molecules from vacuum chamber walls induced by impinging heavy ions (see section "Heavy Ion-Induced Desorption"). Cross-sections for electron capture and loss in the medium-energy range have been measured at RIKEN [207].

Charge changing processes can also occur due to ion-electron recombination during electron cooling of an ion beam. Significant recombination processes under these conditions, that means near zero relative energy $E_{rel} = 0$, are Radiative Recombination (RR) and Dielectric Recombination (DR). Unexpectedly high recombination rates have been observed in merged-beam experiments using a cold dense electron target ($n_e = 4 \times 10^8$ cm^{-3}) in an experiment with 6.3 MeV/u U^{28+} at the GSI Unilac [208], indicating that the lifetime of a stored electron-cooled U^{28+} beam would be only seconds. Later, in a series of experiments at CERN's Low Energy Antiproton Ring (LEAR) Pb^{53+} ions were stored and cooled

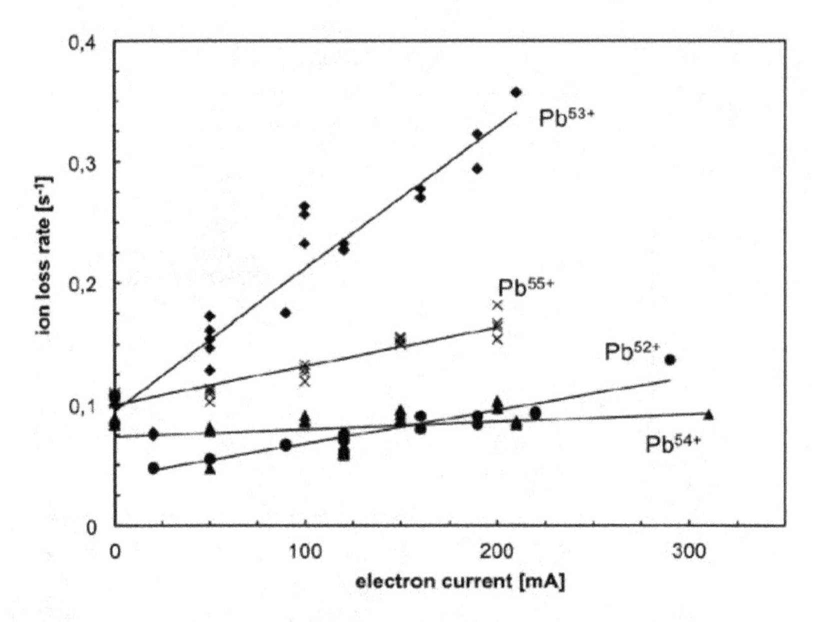

Fig. 11.44 Beam decay rates $1/\tau$ as a function of the electron cooler current for 4.2 MeV/u lead ions with charge states 52+ to 55+ in LEAR [209]. The decay rate for an electron cooler current of zero is determined by the residual gas pressure in the ring, which varies for the different runs. [Data for the graph have been provided from CERN, C. Carli]

in preparation for further acceleration [209]. The recombination rates obtained from the observed lifetimes of a few seconds were by far larger (by a factor of about 50) than the calculated RR. Neighbouring charge states Pb^{52+} and Pb^{54+}/Pb^{55+} behaved quite differently and provided sufficient time for cooling without extensive beam losses (Fig. 11.44). Experiments at the Test Storage Ring (TSR) at the Max-Planck-Institut für Kernphysik (MPIK) in Heidelberg investigated $Au^{49+,50+,51+}$ ions with an energy of 3.6 MeV/u; these are isoelectronic with $Pb^{52+,53+,54+}$. In the TSR-experiments recombination rates were not determined indirectly from lifetime measurements; rather, they were measured as a function of the relative energy in the electron-ion-center-of-mass frame [210]. An extremely sharp recombination peak was found for Au^{50+} at $E_{rel} = 0$. The enhancement factor was about 60 for Au^{50+} in comparison to RR theory. Obviously, for this ion DR resonances dominate the recombination at relative energy zero. For Au^{49+}, the maximum recombination rate at $E_{rel} = 0$ is lower by roughly a factor of 10.

Apparently, the huge recombination rates observed for Au^{50+} and Pb^{53+} are caused by the individual electronic structure of these ions, which happens to support dielectronic recombination resonances at very low relative energies, leading to much shorter lifetimes under cooling conditions. Unfortunately, so far theory is not able to predict DR rates for these complex multi-electron systems.

11.5.2.2.3 Heavy Ion-Induced Desorption

During high-intensity, heavy-ion operation of several particle accelerators worldwide, dynamic pressure build-ups of several orders of magnitude have been observed. The pressure increase is caused by lost beam ions that impact the vacuum chamber walls at a grazing angle. Ion-induced desorption, which has been observed at BNL, CERN, and GSI, can seriously limit beam intensity and lifetime in synchrotrons or storage rings, because mainly heavy gas molecules are desorbed, resulting in large cross-sections for projectile charge changing processes. In the past several years, experiments have been performed at several laboratories to study the observed dynamic vacuum degradations; it is important to understand and overcome this problem for present and future heavy ion accelerators [211]. The ion-induced desorption yield is defined as

$$\eta = \frac{\#\text{desorped molecules}}{\#\text{incident ions}}. \tag{11.91}$$

Experimental studies of the desorption yield for energetic ions on stainless steel at room temperature indicate a scaling law $\eta \sim (dE/dx)^n_{\text{el}}$, where $(dE/dx)_{\text{el}} \sim Z^2/A$ is the electronic energy loss, Z is the charge number and A the mass number. In the 5–100 MeV/u energy range and for perpendicular incidence, $n \approx 3$ was obtained for uranium ions. Unfortunately, the measured desorption yields at 100 MeV/u were $\eta \approx 100$ for uranium and $\eta \approx 4$ for argon (Fig. 11.45) [212]. As a consequence, uncontrolled beam loss must be minimized to avoid pressure bumps caused by lost heavy ions. Ions lost due to collisions with the residual gas should hit only low-desorption materials or dedicated collimators and catchers [213]. Vacuum pressures in the range of 10^{-11} mbar and lower are required for highly charged and partially ionized heavy ion beams in order to control these processes. Therefore, the vacuum requirements and beam-loss collimation issues in high-intensity heavy ion synchrotrons and storage rings differ greatly from those in proton facilities.

11.5.2.3 Ion Accelerator Facilities

In this article is not possible to describe all existing accelerator facilities for stable or radioactive ion beams, or those under construction. A few examples have been selected and described. The focus will be on facilities that are considered by the scientific community especially important for heavy ion nuclear physics in the long term, as described in the IUPAP Report 41 and the NuPECC Long Range Plan 2010 [157, 214]. The facilities described here have been categorized on the basis of the main production methods for radioactive ions: In-flight and ISOL facilities.

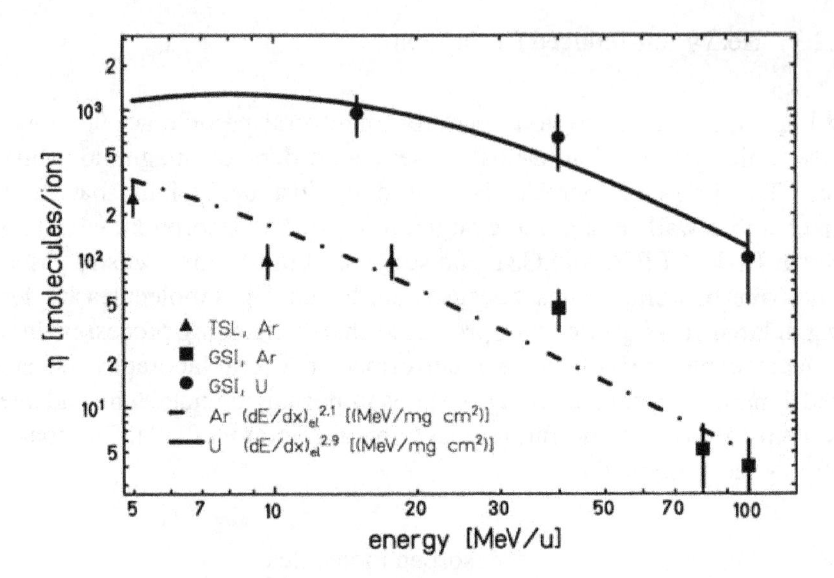

Fig. 11.45 Desorption yields for argon and uranium ions impacting stainless steel at a perpendicular angle, shown here as a function of the projectile energy. The solid lines represent the calculated electronic energy loss $(dE/dx)^n_{\text{el}}$ for argon and uranium to the power of $n = 2.1$ and $n = 2.9$, respectively, adapted to the experimental values (left) [212]. [Reprinted with permission from J. Vac. Sci. Technol. A 27(2) (2009) 245, H. Kolmus, A. Krämer, M. Bender, M.C. Bellachioma, H. Reich-Sprenger, E. Mahner, E. Hedlund, L. Westerberg, O.B. Malyshev, M. Leandersson, E. Edqvist, Copyright (2009), American Vacuum Society]

11.5.2.3.1 In-Flight Facilities

The in-flight method takes advantage of reaction kinematics to efficiently separate short-lived nuclei, from the limits of stability to lifetimes in the μs range and even to the level of a single ion. At medium and high energies (approximately several 10 to several 100 MeV/u), the radioactive isotope beams produced are forward-directed at a velocity approaching beam velocity. Due to interaction with the target, the emittance is larger than that of the projectile beam. Therefore, a large acceptance separator system is needed for the fragment beam separation. The advantage of this rare isotope-production process is its independence of chemical properties; it can be used with isotopes from all elements.

In a new generation of in-flight facilities, powerful and versatile heavy ion accelerators, as projectile sources for the production of exotic nuclei, are combined with large-acceptance electromagnetic separators and different high-resolution systems, such as high-resolution spectrometers, storage rings, and ion traps. Post-acceleration up to more than 10 MeV/u for stopped radioactive isotopes is also included to obtain high-quality exotic beams at low energies. The different in-flight scenarios are described in [215].

In the following sections, examples of large scale cyclotron-, synchrotron-, and linear-accelerator facilities as the primary projectile source for in-flight-produced nuclei are briefly described and discussed. The production targets, electromagnetic separators, and experimental set-ups (ion traps, high-resolution spectrometers, etc.) are not addressed here. However, storage rings and re- or post-accelerators are included where they are part of the overall facility.

RIKEN Radioactive Isotope Beam Facility (RIBF)

The history of cyclotron accelerators at RIKEN Nishina Center, Wako, Japan, began before World War II [216]. The era of heavy ion beams began in 1966 with a 160-cm cyclotron. Work with radioactive ion beams began in 1986 when the RIKEN K540 Ring Cyclotron (RRC) [217] went into operation along with the Projectile Fragment Separator (RIPS). A K70 AVF cyclotron was built for light ion injection [218] and the existing variable-frequency linac (RILAC) [219] was used as injector for heavy ions. The RILAC has been in operation since 1980 and has successfully accelerated light ions to energies of 4 MeV/u and heavy ions to 0.8 MeV/u.

The RIKEN accelerator facility was expanded beginning in 1997 and became the Radioactive Isotope Beam Facility (RIKEN RIBF). The aim of the expansion was to improve experimental conditions for research with radioactive isotopes [220]. In the first phase of the expansion project, a new multi-stage acceleration system was built based on the existing RIKEN accelerators RILAC/AVF and RRC. To these were added three booster cyclotrons, the fixed-frequency Ring Cyclotron fRC/K = 570 [221], the Intermediate Ring Cyclotron IRC/K = 980 [209], and the Superconducting Ring Cyclotron SRC/K = 2600 [222]. Figure 11.46 shows a schematic diagram of the facility, including the key parameters of the accelerator stages. The main accelerator is the SRC, with six sector magnets. Figure 11.47

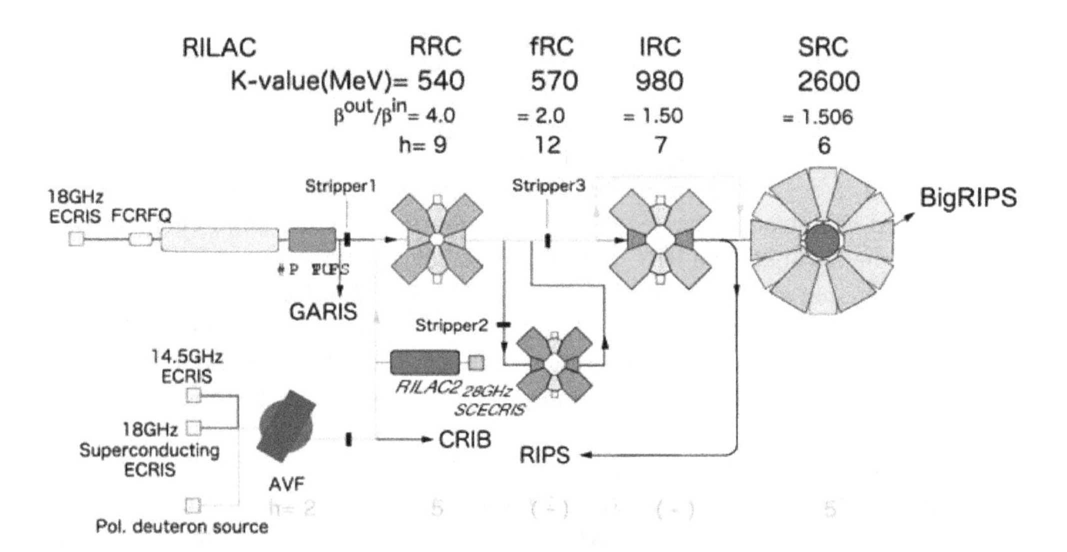

Fig. 11.46 Schematic diagram of the RIKEN RIBF showing the key parameters of the accelerator stages (see article) [220]. The injectors RILAC, RILAC2, and AVF are used for different acceleration modes of the four booster cyclotrons: RRC, fRC, IRC, and SRC.

Fig. 11.47 Picture of the RIKEN Superconducting Ring Cyclotron SRC

shows a picture of this facility. When these accelerators are combined in a cascade, with strippers in between, all ions of all elements can be accelerated up to at least 70% of the speed of light. The accelerated stable ion beams pass a thin low-Z target. Radioactive ion beams are produced there using projectile fragmentation and sometimes in-flight fission as well. The beams are then selected in the following BigRIPS spectrometer [223].

RIBF is operated in three different acceleration modes [224]. The first one uses RILAC, RRC, IRC, and SRC to accelerate medium-mass ions. The beam energy can be varied in a wide range below 400 MeV/u by varying the RF frequency of the linac and cyclotrons. The second acceleration mode is the fixed-energy mode, which uses the fRC between RRC and IRC. The beam energy from the SRC is fixed at 345 MeV/u, due to the fixed frequency operation of the fRC. This mode is used to accelerate heavy ions such as uranium and xenon. The third mode uses the AVF cyclotron as the injector and two boosters, the RRC and SRC. This mode is only used for light ions such as deuterons and nitrogen, also with variable RF frequency in the SRC.

The RIKEN RIBF started operation in 2006 and has been used to accelerate beams from the full mass range from deuteron to uranium. The design energy of 345 MeV/u for uranium was achieved in 2007. Since then, experimental studies on rare isotopes have been carried out. For example, 45 new neutron-rich isotopes were created in 4 days [225], halo structure and large deformation were found in the medium-mass nuclei far from the stability line [226, 227], and decay half-lives of very neutron-rich isotopes were measured to study cosmic r-process [228].

Recently the main focus of accelerator development has been on improving the heavy ion beam intensities by optimizing the accelerator tuning and by reducing matching and transmission losses in the individual stages. Overall transmission for light ions is now approaching 100%. Construction of a second linac for the RRC (RILAC2) began in 2008, fed from a new superconducting 28-GHz ECR ion source, to provide an independent choice of ion beams for rare isotope physics and super-heavy element research [224].

GSI Accelerator Facility (Unilac, SIS18, ESR)

The accelerator facility at the GSI Helmholtzzentrum for heavy ion research, Darmstadt, Germany, (see Fig. 11.52 below), consists of the universal heavy ion linear accelerator (Unilac), the heavy ion synchrotron SIS18 (Schwer-Ionen-Synchrotron, 18 Tm), and the Experimental Storage Ring (ESR, 10 Tm). Construction of the Unilac started in 1970 [229, 230]. The first heavy ions were accelerated in 1975, and uranium ions in 1976 [231]. The Unilac can deliver beams with energies >12.5 MeV/u for all elements up to uranium. Upgrade measures were performed for beam energy in the 1980s and for beam intensity in the 1990s [232]. The present layout of the Unilac is shown in Fig. 11.48. The Unilac pre-stripper linac consists of a 36-MHz IH-type RFQ and an IH-drift tube linac [235], followed by a nitrogen gas-jet stripper and an achromatic charge-analysis system at 1.4 MeV/u [233]. The high-current pre-stripper linac is fed from a Penning ion source for high-duty-cycle (~25%) operation or from high current sources of the MEVVA or CORDIS type for synchrotron injection (low-duty-cycle ~ 1%) [223]. In the low-duty-cycle mode, ions with *A*/*q*-ratio up to 60 (U^{4+}) can be accelerated there [236].

Switching between the different ion sources with up to 50 Hz enables the acceleration of up to three different ion species; this way, experiments at the Unilac, the SIS18, and in ESR can be performed with different ion species in parallel [237].

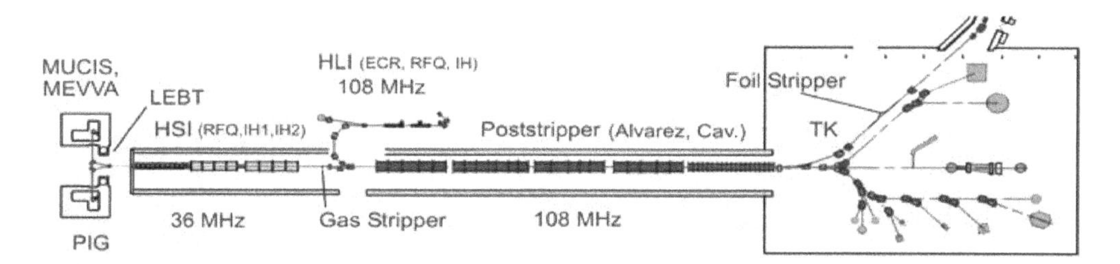

Fig. 11.48 Layout of the Unilac: The pre-stripper linac consists of a 36-MHz high-current IH-type RFQ and an IH-drift tube linac, followed by a nitrogen gas-jet stripper and a charge-analysis system at 1.4 MeV/u [233]. The post-stripper linac consists of four 10-m 108-MHz Alvarez drift-tube sections (up to 11.4 MeV/u) and a chain of single gap cavities either for post-acceleration (up to about 12.5 MeV/u for heavy ions) or for interpolation of the energy steps (from 2.6 to 11.4 MeV/u) of the Alvarez sections. In addition to the high current IH-pre-stripper linac, there is a high-charge-state 1.4 MeV/u injector linac (108 MHz), equipped with an ECR ion source which is used for direct beam delivery (e.g. of U^{28+}-ions) without stripping into the Alvarez post-stripper linac [234].

Construction of the synchrotron storage ring facility [225] started in 1985; it went into operation in 1990 [226, 227]. The synchrotron SIS18 delivers U^{72+} beams (foil stripped after the Unilac at 11.4 MeV/u) at energies of up to 1 GeV/u; U^{92+}-ions (using stripping at 1 GeV/u and re-injection into the SIS18 via the ESR) are delivered at energies of up to 1.4 GeV/u. Light ion beams ($q/A = 1/2$) can be accelerated at energies of up to 2 GeV/u. Beam intensities currently range from 10^{10} ions (for heavy elements) to 10^{11} ions (for light elements) per synchrotron cycle (1 Hz). The Unilac and the SIS18 will be used as the injector system in the Facility for Antiproton and Ion Research (FAIR) (Fig. 11.52 below). Intensities are being increased using a current upgrade program for injection in FAIR (see section "Facility for Antiproton and Ion Research (FAIR) at GSI").

Experiments with in-flight-produced isotopes are performed both at Unilac and SIS18 energies. Super-heavy elements, produced with low-energy Unilac beams using fusion reactions in very thin heavy targets, move forward with the center-of-mass energy. They are analyzed in an electromagnetic separator (SHIP) before being stopped in a detector, or investigated at rest in an ion trap, or separated using chemical methods [148]. Fast radioactive isotopes produced by means of the projectile fragmentation of SIS18 beams in a thin target are analyzed in the FRagment Separator (FRS) [238]. The isotopes thus analyzed can either be transferred to the ESR (Fig. 11.49) or used for investigations of nuclei with lifetimes as short as microseconds in high-resolution spectrometers.

The nuclear fragments emerge from the fragmentation target with increased momentum-spread in all phase spaces, but with comparable average velocities. Therefore, many isotopes with the same magnetic rigidity may be injected into the ESR, where all components of the mixed beam are phase-space-cooled to the

Fig. 11.49 Photo of the experimental storage ring (ESR) at GSI, a unique tool for studying in-fight-produced radioactive ion beams with or without beam cooling, using stochastic and electron beam cooling or the isochronous operation mode (see text). [Photo from GSI Helmholtzzentrum für Schwerionenforschung, A. Zschau]

same velocity, with extremely small velocity spread [239]. Therefore, the frequency spectrum of the beam noise (Schottky spectrum) from the multi-component beam stored in the ESR consists of clearly separated lines. The frequency differences between the lines are determined only by the mass-to-charge ratio A/q. The difference in the revolution frequency (or revolution time) between ions of varying A/q ratios is given by

$$\frac{\Delta f}{f} = -\frac{\Delta T}{T} = -\frac{1}{\gamma_t^2}\frac{\Delta(A/q)}{A/q} + \left(1 - \frac{\gamma^2}{\gamma_t^2}\right)\frac{\Delta v}{v}. \tag{11.92}$$

where γ_t is the transition energy and $\Delta v/v$ is the velocity spread in the ion beam; the velocity spread can be reduced by stochastic or electron cooling. The best results in terms of accuracy and resolution are achieved with electron cooling at very low beam intensities, and when the number of total stored ions within a small A/q-interval is below 1000 (this minimizes the counteracting effect of intra-beam scattering ($\sim q^4/A^2$) on the cooling process). Figure 11.50 shows the equilibrium-momentum spread as a function of the number of fully stripped uranium ions in the ESR storage ring at GSI. Above a threshold of roughly 1000 ions, the equilibrium-momentum spread is defined by the balance of intra-beam scattering and electron cooling. Below this threshold, intra-beam scattering is suppressed because the beam ions form a longitudinal string, in which the ions are no longer able to pass each other. In this ordered state, the equilibrium is determined by the balance of cooling and machine noise [240, 241]. For electron cooling, the precision of mass measurements performed using the Schottky Mass Spectroscopy (SMS) is around 1×10^{-6} or better. The resolution power of this method is illustrated by Fig. 11.51, which shows the separation of mass-resolved ^{52}Mn isomers [155, 242].

The cooling times at high energies restrict the application of electron cooling to nuclei with lifetimes of about 10 s. For shorter lifetimes, stochastic pre-cooling must be applied to the "hot" beams, with typical time constants of roughly 1 s [243].

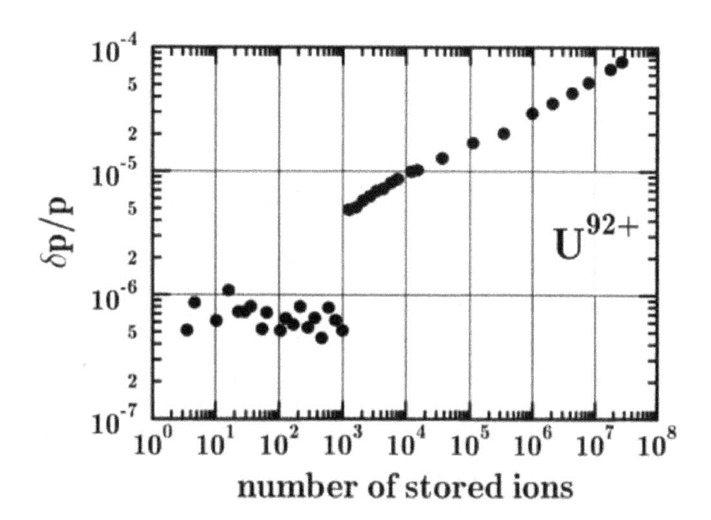

Fig. 11.50 Experimental momentum spread plotted against the number of stored ions in the ESR for fully stripped electron cooled uranium ions at 360 MeV/u.

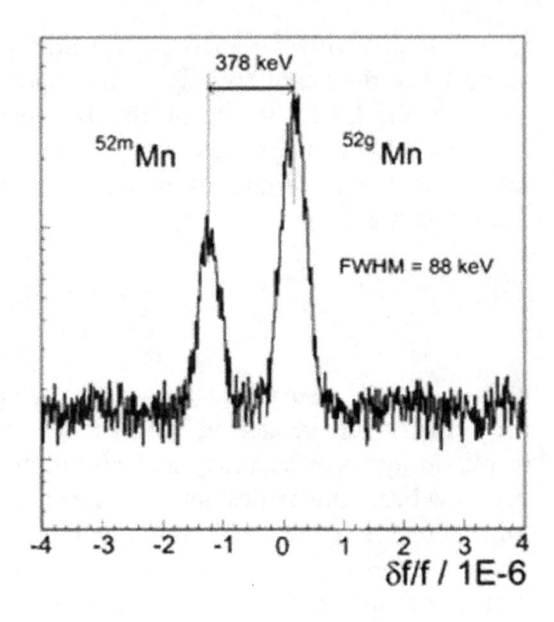

Fig. 11.51 High-resolution mass spectra of electron cooled ^{52}Mn ions at the ground (right) and isomeric (left) states in the ESR.

Operation in the isochronous mode with $\gamma = \gamma_t$ makes possible the investigation of short-lived nuclei with lifetimes of as little as few milliseconds [244].

For atomic and nuclear physics experiments with low energy, electron cooled ion beams in 2012 the CRYRING storage ring was delivered from Stockholm to GSI. Since 2017 the modified CRYRING is operational at GSI. In combination with the ESR and later also with the new FAIR facility CRYRING is the only facility world-wide that provides low-energy highly charged stable beams and beams of rare isotopes with a free choice of the charge state, including bare ions [245].

Facility for Antiproton and Ion Research (FAIR) at GSI

With the new international Facility for Antiproton and Ion Research (FAIR), shown in Fig. 11.52, the research possibilities in nuclear physics with radioactive ions will be expanded considerably at GSI, and the range of accessible isotopes further extended [246, 247]. However, this is only one part of the physics research to be performed at the new facility. Hadron physics, the study of highly compressed nuclear matter, and atomic physics will be additional fields of research [248].

The FAIR facility in the Modularized Start Version (MSV) [247] will consist of six circular accelerators (SIS18, SIS100, CR, HESR, ESR and CRYRING), of two linear accelerators (p-Linac, UNILAC) and of about 1.5 km of beam lines (see Fig. 11.52). The existing Unilac-SIS18 combination will be used as the injector for the new superconducting synchrotron SIS100 (100 Tm), which will have five times the circumference of the SIS18. For radioactive-isotope research much higher primary beam intensities will be reached. To give one example, in the case of uranium U^{28+} will be accelerated both in the SIS18 and SIS100, without stripping after the Unilac in order to avoid the related beam losses. In addition, the ramping and cycling rate of SIS18 will be increased (2.7 Hz). The achievable energy for U^{28+} will be 2.7 GeV/u

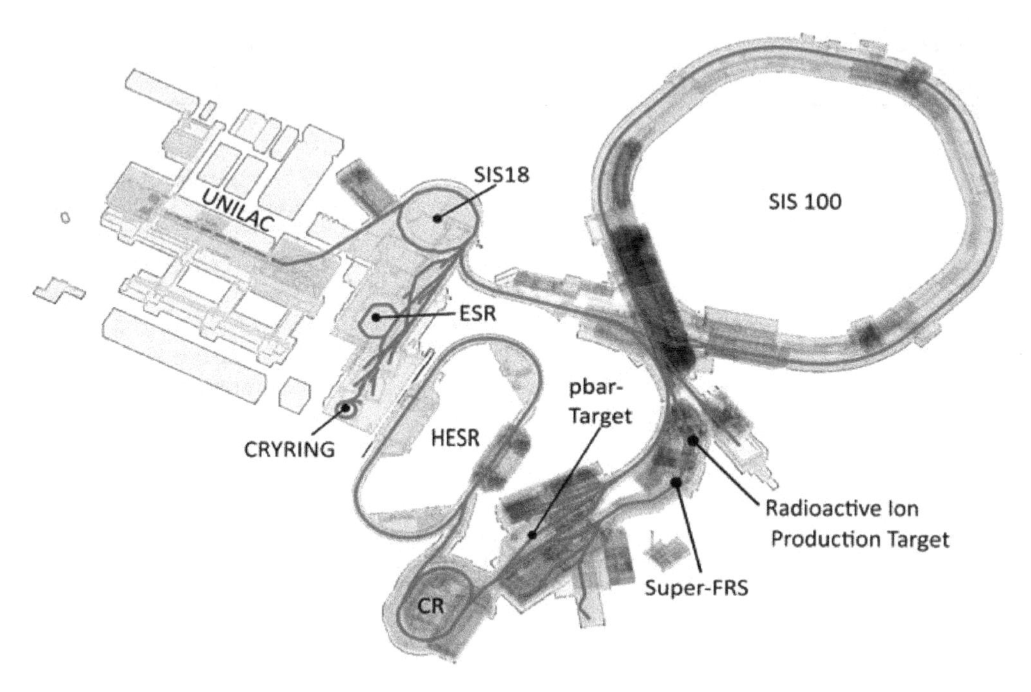

Fig. 11.52 Left: Layout of the existing GSI facilities with the accelerators Unilac, SIS18, the Experimental Storage Ring ESR and the CRYRING storage ring. Right: Layout of the planned new FAIR facilities, with the existing accelerators (Unilac and SIS18) acting as the injector system: the diagram shows the new Superconducting Synchrotrons SIS100 (100 Tm), the Collector Ring (CR), the Superconducting FRagment Separator (Super-FRS), the antiproton production target, the additional proton-injector linac, and the High Energy Storage Ring (HESR) (see text). [Courtesy of GSI Helmholtzzentrum für Schwerionenforschung, C. Pomplun]

and up to 11 GeV/u for fully stripped U^{92+} beams in the SIS100 [249]. With a new 70 MeV proton linac injector for the SIS18, intense proton beams with energies of up to 29 GeV will be provided from the SIS100 for antiproton production [250].

The ion beam from SIS100 can either be transferred to different experimental set-ups or be sent through a production target for fast radioactive-isotope beams. Together with a new large acceptance fragment separator (SuperFRS) behind the SIS100, it is expected that radioactive-isotope intensities will be reached that are 10,000 times the current level [251]. The SIS100 RF-system is also designed to compress the accelerated heavy ion or proton beams into short bunches (to ~60 ns in the case of heavy ions and to ~25 ns in the case of protons). That is required for the production and subsequent storage and efficient cooling of "hot" rare isotope and antiproton beams in the following CR cooler-storage ring [252]. The main task of the collector ring (CR) is stochastic cooling of radioactive ions or antiproton beams from the production targets. In addition, this ring offers the possibility for mass measurements of short-lived ions, by operating in isochronous mode. For research with high-energy antiprotons up to 14 GeV and with heavy-ions, a high-energy storage ring (HESR) will be available, equipped with a high-energy electron cooler (up to 8 MeV), a stochastic cooling system, internal target, and an associated

Table 11.12 Key parameters and features of the synchrotrons and cooler/storage rings [247]

Ring	Circumference [m]	Energy [GeV/u]	Specific features
SIS-100 Synchrotron	1083	2.7 for U^{28+}, 29 for p	Fast ramped (up to 4 T/s) superferric magnets up to 2 T, extraction of short (60 ns), single pulses of up to 5×10^{11} U^{28+} and 4×10^{13} p (25 ns) or slow extraction.
CR Collector Ring	215	0.740 for A/Z = 2.7, 3 for antiprotons	Large aperture. Fast stochastic cooling of radioactive ion beams and antiprotons. Isochronous mode for mass measurements of short-lived nuclei.
HESR High Energy Storage Ring	575	14 for anti-protons, heavy ions (50 Tm)	Stochastic cooling of antiprotons up to 14 GeV. Electron cooling up to 14 GeV. Internal pellet or cluster target.

detector set-up [253]. The key parameters and features of the synchrotrons and cooler/storage rings are given in Table 11.12.

The FAIR facility is presently under construction. The CRYRING storage ring has already started its operation. The upgraded SIS18 will be available 2018. Start of commissioning of the SIS100, the production targets and the storage rings CR and HESR is expected for 2024/2025.

National Superconducting Cyclotron Laboratory (NSCL) at Michigan State University (MSU)

The first superconducting cyclotron, the K500 (K = 500) was built and went into operation at MSU in 1982 [254, 255]. A second cyclotron, the K1200, with greater bending power and hence higher energy beams (K = 1200) was commissioned in 1988 [256]. It was capable accelerating fully stripped light ions with $N = Z$ to 200 MeV/u and heavy ions to approximately 50 MeV/u, depending on the charge state. At these energies it was possible to convert the primary beam into radioactive secondary (sometimes called rare isotope) beams using projectile fragmentation or projectile fission. The A1200 separator was constructed for cyclotron beam analysis and for production and separation of radioactive fragment beams [257]. This device allowed radioactive beams to be delivered to all experimental set-ups and made possible a successful research program with radioactive-isotopes. The Coupled Cyclotron Facility (CCF) was built to increase the intensities (by several orders of magnitude) and energies for heavy ion beams. The project included an upgrade to the K500 cyclotron and its use as an injector for the K1200 cyclotron, along with a

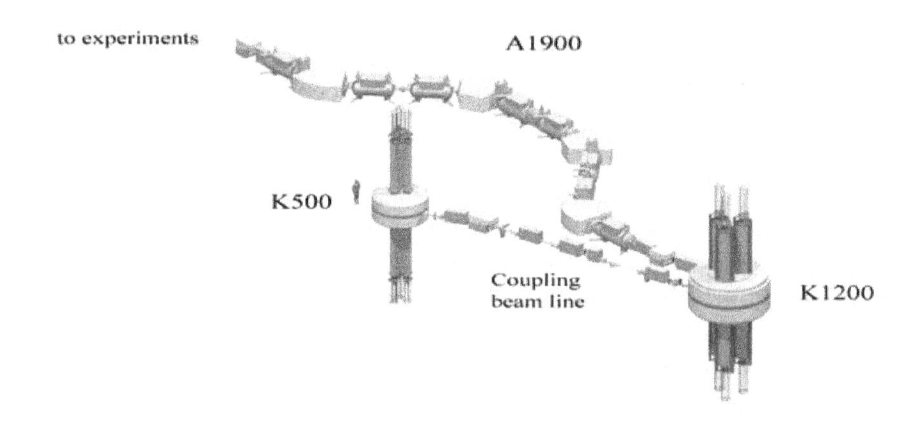

Fig. 11.53 The NSCL coupled cyclotron K500+K1200 facility with the superconducting large-acceptance A1900 fragment separator [246]. To show scale, the size of a person is shown near the K500 cyclotron (Courtesy of NSCL, Michigan State University, B. Sherrill)

new A1900 fragment separator. The CCF went into operation in 2001 [258, 259]. The program is centered on experimentation with radioactive ion beams produced by the A1900. Nearly 1000 different radioactive beams have been produced and used for experiments at the CCF since its inception.

Figure 11.53 shows the MSU-NSCL accelerator facility with the superconducting A1900 fragment separator with a collection efficiency near to 100% as compared to a few % for the A1200 system [260]. The maximum energy of this coupled cyclotron facility is limited to 200 MeV/u for $q/A = ½$ by the focussing in the K1200. Energies up to 80 MeV/u can be delivered for the heaviest ions. High transmission efficiencies allow 0.7–1.0 kW beams to be routinely delivered for experiments at the NSCL. Net beam transmission measured from just before the K500 to extracted beam from the K1200 can be about 30% depending on the ion used (factoring out the unavoidable loss due to the charge stripping foil in the K1200) [261].

The facility is currently being expanded for the investigation of radioactive isotopes at low energies; this will be done by stopping the in-flight-produced isotopes in a cryogenic gas stopping system and re-accelerating them in a compact linac (ReA3) [261, 262]. The ReA3 linear accelerator will provide low-energy radioactive isotope beams of high beam quality. The stopped ions will be re-ionized in an Electron Beam Ion Trap (EBIT), then re-accelerated in a room-temperature RFQ and a superconducting linac built of $\lambda/4$-resonators (QWR). ReA3 beam energies range from 0.3 to 6 MeV/u for $A < 50$ and up to 3 MeV/u for uranium. A later upgrade to 12 MeV/u (ReA12) will be done by expanding the ReA3 linac with QWRs. That will be carried out prior to completion of the FRIB project described in the following section.

MSU Facility for Rare Isotope Beams (FRIB)

The major new initiative in the US in radioactive beams is the Facility for Rare Isotope Beams (FRIB). FRIB is a research facility for the creation and utilization of radioactive isotope beams based on the concept of a high-intensity (400 kW) and high-energy (200 MeV/u) heavy-ion linac. It will include the capability to provide radioactive ion beams at all energies from thermal to nearly 200 MeV/u. Originally, the Rare Isotope Accelerator (RIA) with 100 kW and 400 MeV/u beams was proposed in 2003. In the years that followed an alternative design, FRIB, based on a lower energy (200 MeV/u) but higher intensity (400 kW) heavy-ion driver linac, was developed. Michigan State University was selected near the end of 2008 to build FRIB and the project attained DOE Critical Decision 1, CD-1, in 2010 [263].

Figure 11.54 shows the overall layout and the topology of the FRIB facility [264]. The heavy ion driver linac is fed by ECR ion sources. The low-energy beam transport (LEBT), the RFQ structure, and the medium-energy beam transport (MEBT) to the first section of the main linac are capable of transporting and accelerating heavy ion beams containing more than one charge state, such as 33+ and 34+ for uranium. The linac segment 1 consists of two types of superconducting quarter-wave resonators (QWR) operating at 80.5 MHz with $\beta_{\text{opt}} = 0.041$ and 0.085 to increase uranium-beam energy to 17.5 MeV/u and higher for lighter ions [265].

A stripping section will be installed after linac section 1 to increase the charge states of the ions and hence the acceleration efficiency. An isochronous charge

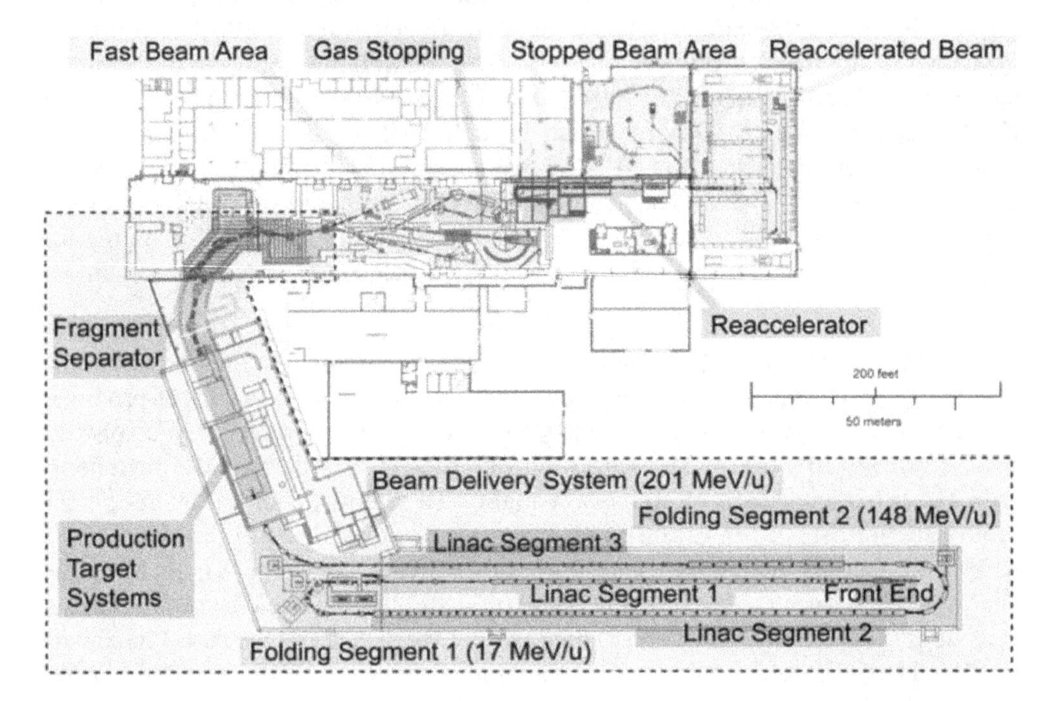

Fig. 11.54 Layout of the MSU FRIB project (Courtesy of FRIB, Michigan State University, B. Sherrill)

analysis system will make possible the selection of several charge states for further acceleration. Additional RF-cavities will help to optimize the longitudinal matching to linac section 2. This system will provide efficient acceleration of all beams.

The post-stripper linac will be built with two types of Half-Wave Resonators (HWR) with $\beta_{opt} = 0.285$ and 0.53 operating at a frequency of 322 MHz. The four cavity types will be able to efficiently cover the full velocity range from $\beta \sim 0.025$ to $\beta \sim 0.57$ ($E/A \geq 200$ MeV/u). Transverse focusing in the linac structures will be performed using superconducting 9T solenoids. With this condition, linac sections 2 and 3 will be able to accept several charge states (e.g. five charge states from 77+ to 81+ for uranium), and hence most of the intensity of the stripped beam charge state distribution for further acceleration. For purposes of hands-on machine maintenance, the specification for uncontrolled beam loss has been set to 1 W/m.

The high-power beam will be transferred to RIB production target systems, which are followed by a large-acceptance fragment separator with three stages of separation. The experimental area will reuse the existing NSCL experimental facilities. The fragments will be either transported to experimental set-ups for fast radioactive beams or slowed down using a gas-stopping system. Stopped isotopes can be investigated in traps or transferred to a charge breeder and then re-accelerated to low or medium energies up to 12 MeV/u (ReA12) [262].

The cryogenic facility, ECR, and RF support facilities are located above the linac tunnel, which according to plans will be about 13 m underground.

The project includes several upgrade options. Space will be left in the linac tunnel to make possible the addition of cryomodules; this will allow the beam energy to be upgraded to 400 MeV/u. In addition, the facility will include a path for a beam line to an optional ISOL production area that might be added in the future. There will also be space to add a light-ion injector and fast beam-switching system to make simultaneous multiple use of the FRIB possible. Construction of FRIB started in 2014. The project should be completed by 2022, with a possible early completion in 2021.

Cyclotron Facility at the Flerov Laboratory for Nuclear Reactions (FLNR) Dubna

The Laboratory for Nuclear Reactions, now named Flerov Laboratory for Nuclear Reactions (FLNR) in Dubna, Russia, was founded in the Joint Institute for Nuclear Research (JINR) in Dubna in 1957 [266]. For more than 40 years, classical and isochronous cyclotrons have been used there for the acceleration of heavy ions for nuclear physics research and heavy ion beam applications. Two isochronous cyclotrons (U400 and U400M) are used for heavy ion nuclear physics. The U400 (K = 625) has been in operation since 1978. In 1996 it was upgraded; the Penning ion source was replaced by an ECR ion source [267]. The synthesis of super heavy elements, predominantly using ^{48}Ca-beams, was and is still the main research approach there [268]. Steady improvement of ^{48}Ca-beams allowed to successfully synthesise new elements from Z = 114 to 118. A review of the discovery of super-heavy nuclei at FLNR, JINR is presented in [149].

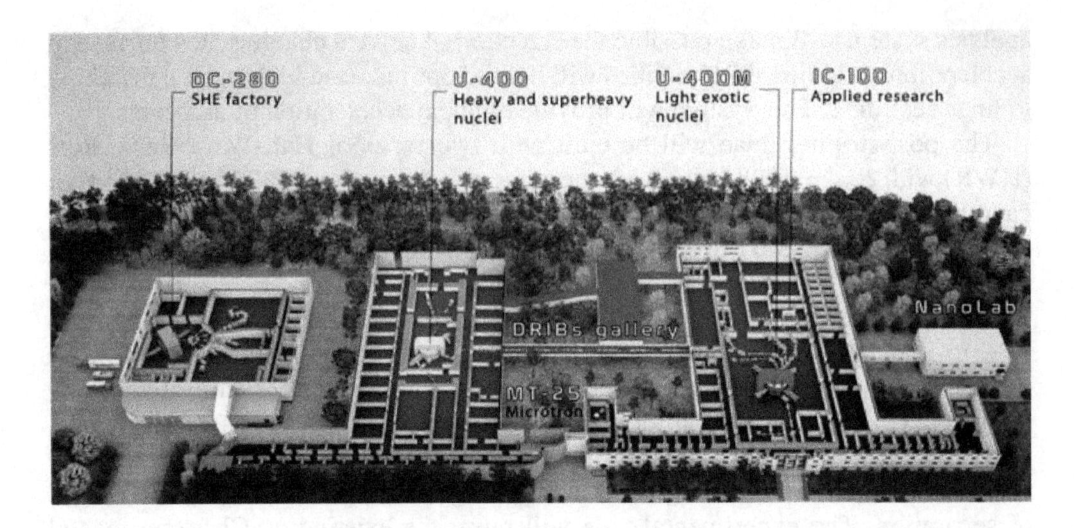

Fig. 11.55 Layout of the Dubna Radioactive Ion Beam (DRIBsIII) facility at the Flerov Laboratory for Nuclear Research (FLNR). It shows from left to right the new DC280 cyclotron, the U400 cyclotron, the U400M, and the beam line (DRIBs gallery) for transport of RIBs from the U400M to the U400 cyclotron, used for the acceleration of RIBS (see text).

The U400M (K = 550) has delivered light ion beams in the energy range of up to 50 MeV/u since the beginning of the 1990s. Originally, it has mainly been used for the production of light radioactive isotopes such as ^{6}He or ^{8}He using the ISOL method. These ions are transferred to an ECRIS for charge breeding, and then transported with keV/u energies via a 120-m beam transport system to the U400, where they are post-accelerated to MeV/u energies [269]. Upgrades of the U400M in the past years allowed accelerating ions of very heavy elements e. g. bismuth up to energies of 15 MeV/u [270]. A new cyclotron DC280 (K = 280) has been built, which will be the main accelerator of a Super Heavy Element Factory in the future, expected delivering first beams for experiments in 2018. Fed by normal and superconducting (18 GHz) ECR-ion sources, it should provide ten times higher beam intensities than the cyclotrons used so far for masses up to A = 50. The layout of the FLNR accelerator complex is shown in Figs. 11.55 and 11.56.

Heavy Ion Research Facility in Lanzhou (HIRFL)

One new and rather versatile accelerator combination of cyclotrons and synchrotron-storage rings is the Heavy Ion Research Facility in Lanzhou (HIRFL), China [271]. Its first stage consists of a superconducting ECR ion source, a compact sector-focusing injector cyclotron (K = 69), and a main sector-cyclotron accelerator SSC (K = 450). The SSC went into operation in 1988. The construction of a new heavy ion synchrotron (CSRm) (with the SSC as injector), combined with a fragment separator, and an experimental storage ring (CSRe) for maximum uranium beam energies of 500 MeV/u, started in 1999; the facility went into operation in 2007.

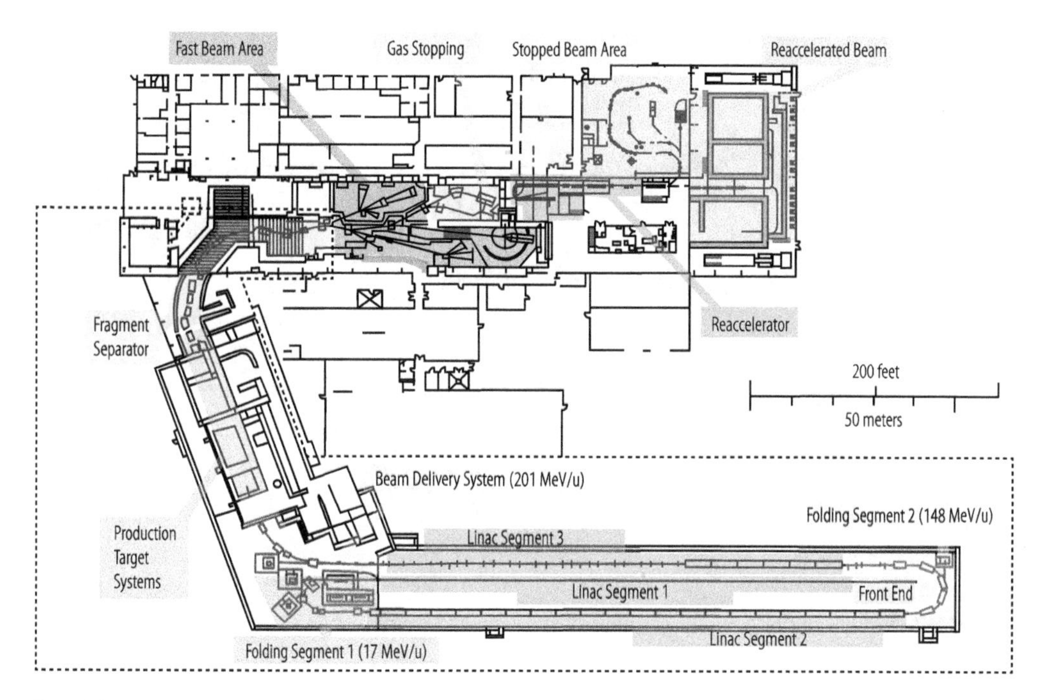

Fig. 11.56 Picture of the U400 cyclotron at the Flerov Laboratory for Nuclear Research (FLNR) in Dubna, used for super-heavy element synthesis.

The synchrotron and storage rings are both equipped with electron-cooler devices. The HIRFL facility is used for in-flight production of exotic isotopes at medium and high energies, and other basic and applied science, as well (Fig. 11.57).

Other In-Flight Facilities

A number of other facilities exist which can or could produce radioactive isotopes using the in-flight method. As can be concluded from Fig. 11.41, cyclotrons with K-values $\geq$150 (this includes all superconducting cyclotrons) can, using modern ECR ion sources, deliver sufficient beam energy for heavy ions. Again we refer to the IUPAP Report 41 and the NuPECC Long Range Plan 2010 [157, 214].

11.5.2.3.2 ISOL Facilities

Two examples of ISOL facilities have been selected and will be described here; both involve the post-acceleration of radioactive isotope beams. The first is the ISAC facility at TRIUMF, Vancouver, which like ISOLDE at CERN uses a high-energy proton beam to produce radioactive isotopes in a thick target. It has recently been expanded, the existing linac post-accelerator being upgraded to achieve higher energies. The second one, SPIRAL2 at GANIL, is an extension of the existing in-flight/ISOL facility GANIL-SPIRAL. It involves an additional new powerful

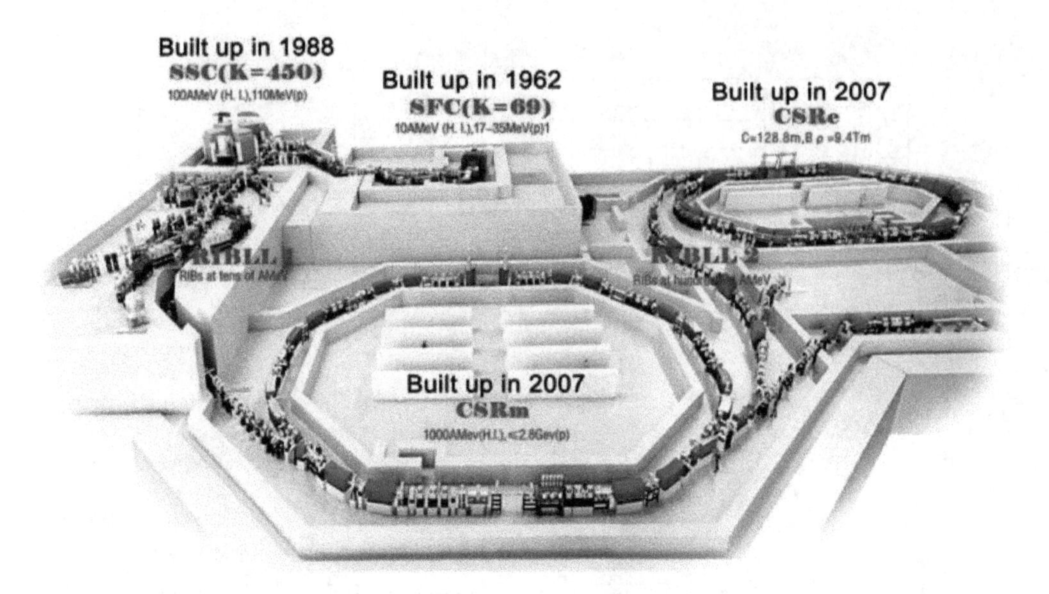

Fig. 11.57 Layout of the Accelerator facility at the Institute of Modern Physics (IMP) in Lanzhou.

medium-energy light ion linac driver to produce much higher intensities of exotic isotopes than is currently possible, and which will make accessible new regions of exotic nuclei.

ISAC at TRIUMF

TRIUMF has operated a 500 MeV H^- cyclotron since 1974. The TRIUMF facility (Fig. 11.58) was expanded in 1995 with the addition of a radioactive beam facility, ISAC [272]. The radioactive species at ISAC are produced using the ISOL method with a 500 MeV proton beam of up to 100 μA bombarding a thick target. After production the species are ionized, mass-separated and sent to either a low-energy area or pass through a string of linear accelerators to feed experiments at higher energies. The first beams from ISAC, now named ISAC-I, were available in 1998, while the first accelerated beams were delivered in 2001 to a medium-energy area and used chiefly for nuclear astrophysics [273]. The TRIUMF ISAC-II superconducting linac, proposed in 1999, was designed to raise the energy of radioactive ion beams above the Coulomb barrier to support nuclear physics at TRIUMF. The first stage of this project, Phase I, commissioned in 2006, involved the addition of 20 MV of superconducting linac [274]. Phase II of the project consisting of an additional 20 MV of superconducting linac was installed and commissioned in 2010 [275].

Typically, 1+ beams are produced in the on-line source, but an ECR ion source was installed in 2009 to act as a charge-state booster (CSB) to raise the q/A ratio of low-energy high-mass beams so that they could be accelerated through the ISAC

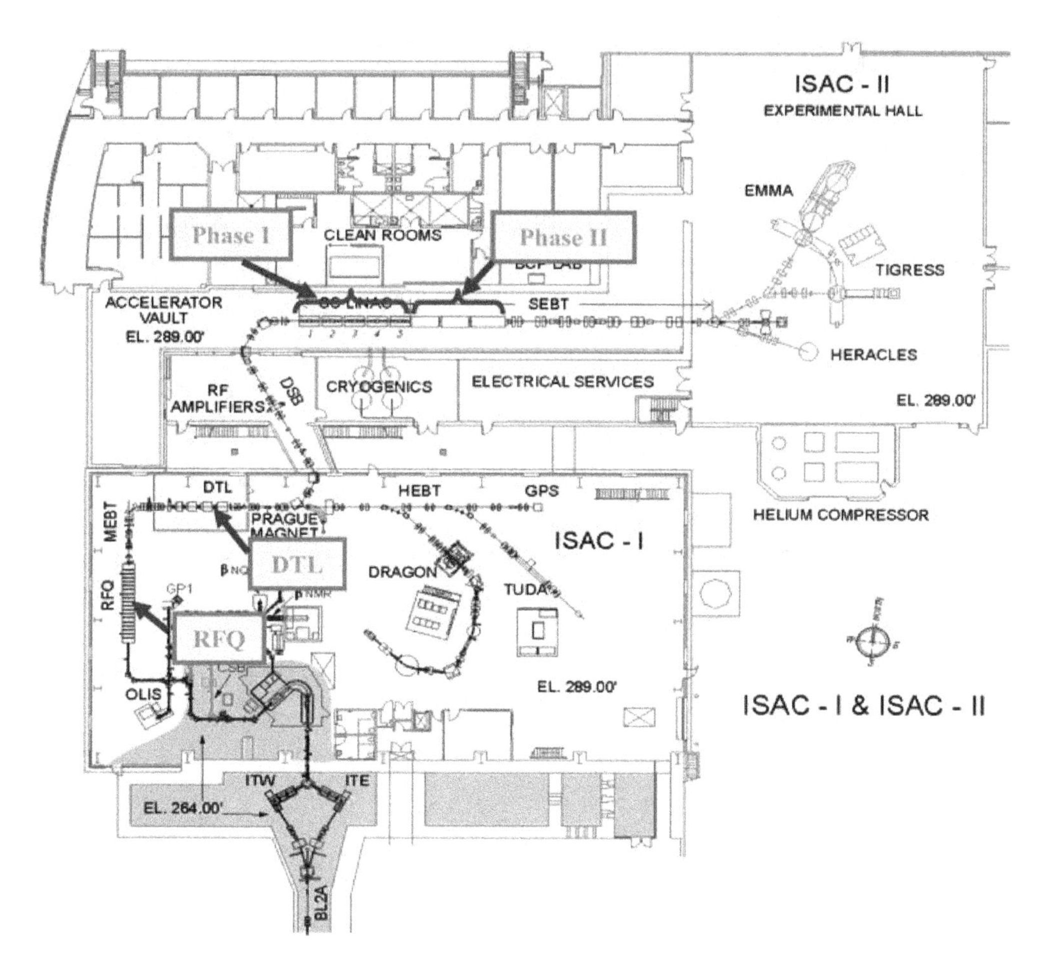

Fig. 11.58 The ISAC facility at TRIUMF showing the RFQ, DTL in ISAC-I and the two installation phases of the superconducting heavy ion linac in ISAC-II. [Courtesy of ISAC/TRIUMF, Vancouver, R. E. Laxdal]

accelerators. An off-line ECR ion source provides stable beams for accelerator commissioning and tuning as well as for the science program.

The ISAC-I accelerator chain includes a four-vane split-ring structure RFQ (35.4 MHz), which accelerates beams of $A/q \leq 30$ from 2 keV/u to 153 keV/u [276]. The post-stripper, a 106 MHz variable-energy drift tube linac (DTL), accelerates ions of $2 \leq A/q \leq 6$ to a final energy between 0.153 MeV/u and 1.53 MeV/u. The variable-energy DTL is based on a unique separated-function approach with five independent interdigital H-mode (IH) structures [277]. Both the RFQ and DTL have been used since 2001 to reliably provide a variety of radioactive and stable ions.

The ISAC-II superconducting linac is composed of bulk niobium, quarter wave resonators (QWR) for acceleration, and superconducting solenoids for periodic transverse focusing, housed in several cryomodules. The Phase-I linac consists of 20 QWR housed in five cryomodules. The first eight cavities have a geometric

Fig. 11.59 The ISAC-II Superconducting linac. [Courtesy of ISAC/TRIUMF, Vancouver, R. E. Laxdal]

$\beta = 0.057$ and the remainder a geometric $\beta = 0.071$. The cavities operate at 106 MHz. The Phase-II upgrade also consists of 20 QWR; they are housed in three cryomodules. These bulk niobium cavities have a geometric $\beta = 0.11$ and resonate at 141.44 MHz. One 9T superconducting solenoid is installed in the middle of each of the eight cryomodules in close proximity to the cavities. The ISAC-II SC-linac is shown in Fig. 11.59 [278].

The performance of the SC-linac is quoted in terms of the peak surface field achieved at a cavity RF power of 7 W. The Phase-I section has operated at Ep = 33 ± 1 MV/m since 2006 with little sign of degradation. The Phase-II section averaged Ep = 26 MV/m in the first 6 months of operation [279].

GANIL-SPIRAL2 Accelerator Facility

Construction of the cyclotron facility at the French national heavy-ion laboratory (Grand Accelerateur National d'Ion Lourd GANIL) in Caen, France, started at the end of the 1970s. It consisted of two compact cyclotrons C01 and C02, (K = 30 each) and two separated-sector cyclotrons CSS1and CSS2, (K = 380 each). The facility went into operation in 1982. The first beams for experiments were delivered in 1983 [280]. Using one of the compact cyclotrons as the injector for the CSS1 and a stripper before CSS2, it can achieve ion energies of up to 95 MeV/u for light ions and up to 24 MeV/u for uranium beams. Stable beams of intermediate energies in the range of $\leq$1 MeV/u to 13 MeV/u are used after C01 or C02, and CSS1 for atomic physics, biology, and solid state physics. Upgrades to improve beam energies and intensities have been performed since mid-1980s, with the special aim of improving the situation for exotic-beam experiments. This has been done by introducing ECR ion sources, by modifying the injection systems into the C01/C02 injector cyclotrons, and by improving performance of components for the operation with beam powers of up to several kW in and after the CSS2 [281, 283]. In-flight technology was used to produce exotic beams [283].

Since 2001, when a fifth cyclotron, CIME, went into operation, the Isotope Separation On-Line (ISOL) technique was also used to produce rare isotopes [284]. The exotic isotopes, which were produced in thick targets using high power ion beams of up to 3 kW from the CSS2 cyclotron, were transferred into a charge breeding ECR ion source, and post-accelerated in the compact cyclotron CIME (K = 265). Depending on the isotope, energies in the range from 1.2 to 25 MeV/u are delivered from CIME. These beams are transferred to the existing experimental facilities. An overview of the existing GANIL-SPIRAL1 accelerator and experiment facility displays Fig. 11.60 (right) [285]. A review on stable and radioactive beams produced there is given in [286].

To improve the possibilities for heavy-ion nuclear physics research both for stable and for radioactive ions an expansion of the GANIL facilities has been proposed named SPIRAL2. The SPIRAL2 project at GANIL was approved in 2005 [287, 288]. This new facility will provide intense beams of neutron rich exotic nuclei (10^{+6} to 10^{+11}) by the ISOL method in the mass range from $A = 60$ to $A = 140$. The facility will be composed of a flexible high power superconducting

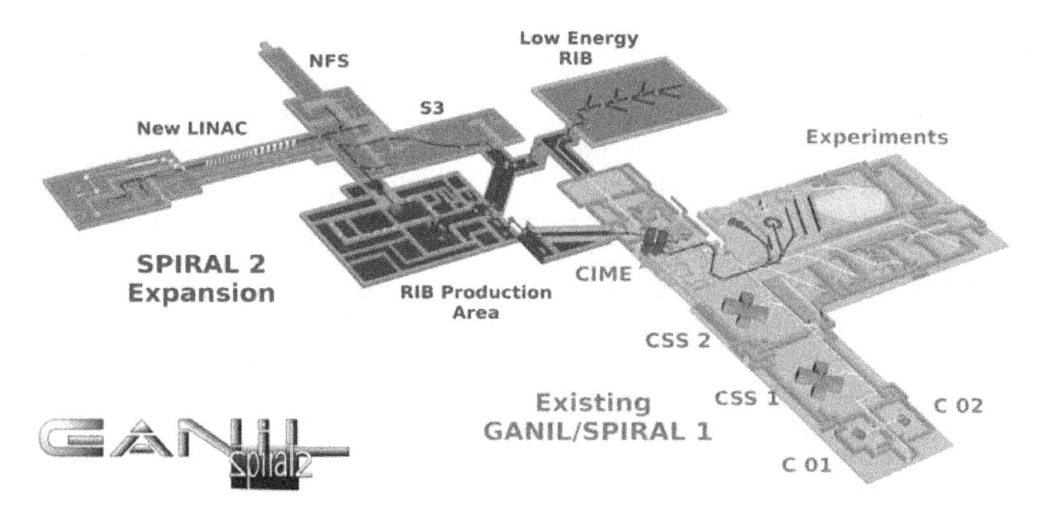

Fig. 11.60 Layout of the existing GANIL-SPIRAL1 accelerators and the experiment facilities (right), and of the new accelerator and experiment facilities under construction for SPIRAL2 (left): the existing GANIL-SPIRAL1 facility builds on the two injector cyclotrons C01/C02 the two sector cyclotrons CSS1 and CSS2 with stripper in between and the cyclotron CIME for acceleration of exotic beams. SPIRAL2 is based on a new superconducting linear ion accelerator as particle source for different production methods of radioactive isotopes (RIB) and stable ion beams. [Courtesy of GANIL/SPIRAL2, Caen, T. Junquera]

linear driver accelerator (5 mA/40 MeV deuterons, 5 mA/33 MeV protons, and 1 mA/14.5 MeV/u heavy ions with $q/A = 1/3$), a dedicated building for the production of radioactive isotopes (RI), the existing cyclotron CIME for the re-acceleration, and new experimental facilities. The high power deuteron beam is directed on a carbon converter target, producing fast neutrons. The main process for the production of the RI is based on the fast neutron induced fission in uranium carbide targets. Figure 11.60 (left) [285] shows an overview of the GANIL-SPIRAL2 facility [287, 289].

The new linac is composed of two families of 88 MHz superconducting QWR ($\beta = 0.07$, $\beta = 0.12$), which permits the acceleration of the different ions and energies mentioned before. The basic principle of this design is to install the resonators in separate cryomodules (one QWR per cryomodule in the $\beta = 0.07$ section and two QWR per cryomodule in the $\beta = 0.12$ section). Between each cryomodule beam focusing is performed by means of 2 room temperature quadrupoles with short vacuum/diagnostics boxes in between to allow for flexible beam optics and tuning.

Phase 1 of SPIRAL2 includes construction of buildings, which started in 2011, installation of the driver linac, the high energy beam lines and first experimental equipment in 2012. Proton beams were accelerated in the RFQ in 2015. The phase 2 contains the radioactive isotope production cave, low energy experiments for RIB and the connection to GANIL/SPIRAL1 facilities. Construction is expected to start in 2014. As a long term option a heavy ion source with $q/A = 1/6$ will be added to the linac. It will accelerate heavy ions up to 8.5 MeV/u.

Other ISOL-Facilities

There is a number of additional accelerator facilities used for the production of radioactive isotopes with the ISOL method building on different accelerator combinations and using different production methods. Spontaneous fission products from californium are used as radioactive isotope source in the frame of the CARIBU project at Argonne, in order to reach new areas of the nuclide chart, and accelerated in the superconducting linac ATLAS [290]. Upgrades are going on at existing ISOL facilities as for example at the ISOLDE facility at CERN [291]. In the frame of this article it is not possible to address more. Therefore, it is again referred to the compendium of nuclear physics facilities included in the IUPAP Report 41 [157] and in the NuPECC Long Range Plan 2010 [214].

11.5.2.3.3 Conclusions

Since about 20 years there is a growing worldwide demand for in-flight and in ISOL facilities for the production of radioactive isotopes. Developments in the fields of ion sources and superconducting accelerator technologies have facilitated access of many nuclear physics laboratories to exotic beams. The new facilities will further increase the production rates and the measurement precision for exotic ions.

In-flight facilities have advantages in the field of nuclear physics, whereas ISOL facilities have their strengths in the field of nuclear astrophysics [292]. A new generation of in-flight facilities builds on very powerful driver accelerators covering a broad range of light to heavy primary ion beams. For new and expanding ISOL facilities with superconducting post-accelerators, target handling, charge breeding, acceleration, and beam handling of radioactive ions are important issues. ISOL-post-accelerators are operated typically in the same energy range as machines for super-heavy element synthesis.

References

1. J.D. Jackson: *Classical Electrodynamics*, Wiley, 1962.
2. R. Feynman: *Feynman Lectures on Physics*, Vol. 2, eq. (21.1) ff.
3. G. Geloni, et al.: *Undulator radiation in a waveguide*, Nucl. Instrum. Meth. A 584 (2008) 219.
4. A. Hofmann: *The Physics of Synchrotron Radiation*, Cambridge Univ. Press, 2004.
5. M. Sands: *The Physics of Electron Storage Rings*, SLAC Report 121, SLAC, Stanford, 1970.
6. J.A. Clarke: *The Science and Technology of Undulators and Wigglers*, Oxford Publ., 2004.
7. Schmüser, P., Dohlus, M., Rossbach, J., Behrens, C.: *Free-Electron Lasers in the Ultraviolet and X-Ray Regime*, Springer, 2014.
8. G. Stupakov, D. Zhou: *Analytical theory of coherent synchrotron radiation wakefield of short bunches shielded by conducting parallel plates*, Phys. Rev. Accel Beams 19 (2016) 044402.
9. M. Abo-Bakr, et al.: *Steady-State Far-Infrared Coherent Synchrotron Radiation detected at BESSY II*, Phys. Rev. Lett. 88(25) (2002) 254801.
10. S. Wesch, et al.: *A Multi-Channel THz and Infrared Spectrometer for Femtosecond Electron Bunch Diagnostics by Single-Shot Spectroscopy of Coherent Radiation*, Nucl. Instrum. Meth. A 665 (2011) 40.
11. S. Reiche: *GENESIS 1.3: a fully 3D time-dependent FEL simulation code*, Nucl. Instrum. Meth. A 429 (1999) 243.
12. I. Zagorodnov, *Numerical modeling of collective effects in free electron laser, in Proceedings of 11th International Computational Accelerator Physics Conference, Rostock-Warnemünde, Germany* (JACoW, Geneva, Switzerland, 2012) p. 81.
13. J.B. Murphy, C. Pellegrini: *Introduction to the physics of free electron lasers*, Laser Handbook, Vol. 6, (1990), p. 115.
14. E.L. Saldin, E.A. Schneidmiller, M.V. Yurkov: *The Physics of Free Electron Lasers*, Springer (2000).
15. Z. Huang, K.-J. Kim: *A Review of X-Ray Free-Electron Laser*, Phys. Rev. ST Accel. Beams 30 (2007) 034801.
16. Z. Huang, P. Schmüser: *Handbook of Accelerator Physics and Engineering*, A.W. Chao, M. Tigner (eds.), World Scientific, 1999.
17. J.M.J. Madey: *Relationship between mean radiated energy, mean squared radiated energy and spontaneous power spectrum in a power series expansion of the equations of motion in a free-electron laser*, Nuovo Cimento B 50 (1979) 64.
18. FEL at JLAB: see www.jlab.org/FEL
19. P. Emma, et al.: *First Lasing of the LCLS X-Ray FEL at 1.5 Å*, PAC2009 Proc., 2009.
20. A.M. Kondratenko, E.L. Saldin: *Generation of Coherent Radiation by a Relativistic Electron Beam in an Undulator*, Part. Accel. 19 (1980) 207.
21. R. Bonifacio, C. Pellegrini, M. Narducci: *Collective instabilities and high-gain regime in a free electron laser*, Opt. Comm. 50 (1984) 373.

22. K.-J. Kim: *Three-Dimensional Analysis of Coherent Amplification and Self-Amplified Spontaneous Emission in Free Electron Lasers*, Phys. Rev. Lett. 57 (1986) 1871.
23. M. Xie: *Exact and variational solutions of 3D eigenmodes in high gain FELs*, Nucl. Instrum. Meth. A 445 (2000) 59.
24. E.L. Saldin, et al.: *FAST a three-dimensional time-dependent FEL simulation code*, Nucl. Instrum. Meth. A 429 (1999) 233.
25. W. Fawlay: *A User Manual for GINGER and Its Post-Processor XPLOTGIN*, Report LBNL-49625, Lawrence Berkeley Laboratory (2002).
26. G. Moore: *The high-gain regime of the free electron laser*, Nucl. Instrum. Meth. A 239 (1985) 19.
27. R. Ischebeck, et al.: *Study of the transverse coherence at the TTF free electron laser*, Nucl. Instrum. Meth. A 507 (2000) 175.
28. P.B. Corkum: *Plasma perspective on strong field multiphoton ionization*, Phys. Rev. Lett. 71 (1993) 1994.
29. K.J. Schafer, et.al.: *Above threshold ionization beyond the high harmonic cutoff*, Phys. Rev. Lett. 70 (1993) 1599.
30. J. Feldhaus, et al.: *Possible application of X-ray optical elements for reducing the spectral bandwidth of an X-ray SASE FEL*, Opt. Comm. 140 (1997). 341.
31. L.H. Yu: *Generation of intense uv radiation by subharmonically seeded single-pass free-electron lasers*, Phys. Rev. A 44 (1991) 5178.
32. G. Stupakov: *Using the Beam-Echo Effect for Generation of Short-Wavelength Radiation*, Phys. Rev. Lett 102 (2009) 074801.
33. E.L. Saldin, et al.: *Self-amplified spontaneous emission FEL with energy-chirped electron beam and its application for generation of attosecond x-ray pulses*, Phys. Rev. ST Accel. Beams 9 (2006) 050702.
34. W. Henning, C. Shanks: *Accelerators for Americas Future*, US. Dep. Energy, 2010.
35. K. Bethge, in: Advances of accelerator physics and technologies, H. Schopper (ed.), World Scientific, 1993.
36. M.R. Cleland: CAS-CERN Accelerator School and KVI specialized CAS course on small accelerators, 2005, p. 383.
37. A. Zaitsev, I.A. Dobrinets, G. Vins; HPHT-Treared Daimonds, Springer-Verlag, ISBN: 9783662506462, 2016.
38. J. Lutz et al.: Semiconductor Power Devices: Springer Verlag, ISBN 978-3-642-11125-9, 2011.
39. J.R. Tesmer, et al.: Handbook of modern ion beam materials analysis, Materials Research Society, 1995.
40. T. Schulze-König, et al.: Nucl. Instrum. Meth. B 268 (2010) 891.
41. U. Zoppi, et al.: Radiocarbon 49 (2007) 171.
42. W. Kutschera: private communication.
43. J.J. Nelson, D.J. Muehlner, in: Magnetic bubbles, H. Jouve (ed.), Acad. Press, London, 1986.
44. F. Joliot, I. Curie, *Artificial Production of a New Kind of Radio-Element,* Nature, Volume 133, Issue 3354, (1934) 201-202
45. E. Fermi, *Radioactivity Induced by Neutron Bombardment*, Nature 133, (1934) 757-757
46. O. Chievitz, G. Hevesy, *Radioactive indicators in the study of phosphorous in the metabolism in rats*, Nature 136 (1935) 754
47. J. H. Lawrence, K. G. Scott, and L. W. Tuttle, *Studies on leukemia with the aid of radioactive phosphorus*. Internat. Clin., 3, (1939) 33
48. J. G. Hamilton, R. S. Stone, *The Intravenous and Intraduodenal Administration of Radio-Sodium,* Radiobiology, 28, (1937) 178
49. J. H. Lawrence and R. Tennant, *The comparative effects of neutrons and x-rays on the whole body*, J Exp Med 66, (1937) 667-688
50. T. Ido, C.N. Wan, J.S. Fowler, A.P. Wolf, *Fluorination with F,: a convenient synthesis of 2-deoxy-2-fluoro-D-glucose*. J Org Chem 42 (1977) 2341-2342

51. J.S. Fowler, T. Ido, *Initial and subsequent approach for the synthesis of 18FDG*, Semin Nucl Med. 32(1) (2002) 6-12
52. D. B. Mackay, C. J. Steel, K. Poole, S. McKnight, F. Schmitz, M. Ghyoot, R. Verbruggen, F. Vamecq and Y. Jongen, *Quality assurance for PET gas production using the Cyclone 3D oxygen-15 generator*, Applied Radiation and Isotopes, Volume 51, Issue 4 (1999) 403-409
53. IAEA-DCRP/2006, *Directory of Cyclotrons used for Radionuclide Production in Member States - 2006 Update*, October 2006
54. N. Ramamoorthy, *Production of radioisotopes for medical applications*, Oral presentation, Workshop Physics For Health in Europe, 2-4 February 2010, CERN, Geneva
55. G.O. Hendry et al., "Design and Performance of a H-Cyclotron," Proc. of the 9th Int. Conference on Cyclotrons and their Applications (1981) 125.
56. P. W. Schmor, Review of cyclotrons used in the production of radioisotopes for biomedical applications, Proceedings of CYCLOTRONS 2010, Lanzhou, China
57. C. Oliver, Compact and efficient accelerators for radioisotope production, IPAC2017 - Proceedings, June 2017, ISBN 978-3-95450-182-3
58. R. E. Shefer, R. E. Klinkowstein, M. J. Welch and J. W. Brodack, *The Production of Short Lived PET Isotopes at Low Bombarding Energy with a High Current Electrostatic Accelerator*. Proc. Third Workshop on Targetry and Target Chemistry, T. J. Ruth, ed., Vancouver, B.C., (1989)
59. A.R. Jalilian, J.A. Osso, Production, applications and status of zirconium-89 immunoPET agents, J Radioanal Nucl Chem (2017) 314: 7. https://doi.org/10.1007
60. IAEA, *Cyclotron Produced Radionuclides: Principles and Practice*, Technical Reports Series No. 465, International Atomic Energy Agency, Vienna, 2008
61. *Therapeutic Nuclear Medicine*, Richard P Baum, (Ed.), Springer-Verlag Berlin Heidelberg, 2014, ISBN 978-3-540-36719-2. DOI https://doi.org/10.1007/978-3-540-36719-2
62. Making Medical Isotopes, Report of the Task Force on Alternatives for Medical-Isotope Production, TRIUMF University of British Columbia Advanced Applied Physics Solutions, Inc. (2008)
63. Molybdenum-99 for Medical Imaging, The National Academies Press, Washington, DC, ISBN 978-0-309-44531-3, DOI 10.17226/23563
64. IAEA, Non-HEU production technologies for molybdenum-99 and technetium-99m, Nuclear Energy Series No. NF-T-5.4, Vienna: IAEA, 2013, ISBN 978–92–0–137710–4
65. W. K. H. Panofsky, *Big Physics and Small Physics at Stanford*, Stanford Historical Society, Sandstone and Tile, Volume 14, no. 3 (1990) 1-10
66. D.W. Fry, R.B.R.S. Harvie, L.B Mullett, W. Walkinshaw, *Travelling-wave linear accelerator for electrons*, Nature 160 (1947) 351-353
67. IAEA, DIrectory of RAdiotherapy Centres https://dirac.iaea.org/Query/Countries, as of 30 November 2017 [25] http://iopscience.iop.org/article/10.1088/1361-6560/aa9517
68. B W Raaymakers et al. *First patients treated with a 1.5 T MRI-Linac: clinical proof of concept of a high-precision, high-field MRI guided radiotherapy treatment*, Phys. Med. Biol. (2017) 62 L41
69. G. L. Locher, Biological effects and therapeutic possibilities of neutrons, Am. J. Roentgenol. 36, (1936) 1-13.
70. R.S. Stone, *Neutron therapy and specific ionization*, Am. J. Roentgenol. 59 (1948) 771-785
71. M. Awschalom, et al. *The Fermilab Neutron Radiotherapy Facility*, 1977 PAC, Chicago, IEEE Trans. Nuc. Sci., Vol. NS-24, No. 3, (1977) 1055.
72. R.R. Wilson, *Radiological use of fast protons*, Radiobiology 47, (1946) 487-491
73. R.P. Levy, J.I. Fabrikant, K.A. Frankel, M.H. Phillips, J.T. Lyman, J.H. Lawrence, C.A. Tobias, *Heavy-charged-particle radiosurgery of the pituitary gland: clinical results of 840 patients*, Stereotact. Funct. Neurosurg. 57 (1-2), (1991) 22-35
74. C. A. Tobias, J. H. Lawrence, J. L. Born, R. K. Mccombs, J. E. Roberts, H. O. Anger, B. V. A. Low-Beer, and C. B. Huggins, *Pituitary Irradiation with High-Energy Proton Beams - A Preliminary Report*, Cancer Research 18, No 2, (1958) 121

75. Börje Larsson, Lars Leksell, Bror Rexed, Patrick Sourander, William Mair, and Bengt Andersson, *The High-Energy Proton Beam as a Neurosurgical Tool*, Nature 182, (1958) 1222-1223
76. H.D. Suit, M. Goitein, J. Munzenrider, L. Verhey, K.R. Davis, A. Koehler, R. Linggood, R.G. Ojemann, *Definitive radiation therapy for chordoma and chondrosarcoma of base of skull and cervical spine*, J Neurosurg, 56(3) (1982) 377-385
77. E. Pedroni, R. Bacher, H. Blattmann, T. Böhringer, A. Coray, A. Lomax, S. Lin, G. Munkel, S. Scheib, U. Schneider and A. Tourosvsky, The 200-MeV proton therapy project at the Paul Scherrer Institute: conceptual design and practical realization, Med. Phys. 22 (1995) 37-53.
78. K. Umegaki, K. Hiramoto, N. Kosugi, K. Moriyama, H. Akiyama, S. Kakiuchi, *Development of Advanced Proton Beam Therapy System for Cancer Treatment,* Hitachi Review Vol. 52 No. 4, (2003) 196-201
79. J. R. Castro, *Heavy ion therapy: the BEVALAC epoch. In: "Hadron therapy in oncology"*, U. Amaldi and B. Larsson eds., B., Elsevier, Amsterdam-Lausanne-New York-Oxford-Shannon-Tokyo, (1994) 208-216
80. Y. Hirao et al., Heavy ion synchrotron for medical use, Nucl. Phys. A 538 (1992) 541c.
81. H. Tsujii et al., *Clinical Results of Carbon Ion Radiotherapy at NIRS*, Journal of Radiation Research, Vol. 48, Suppl. A, A1-A13.
82. H. Tsujii et al., *Clinical advantages of carbon-ion radiotherapy*, New J. Phys. 10 (2008)
83. http://www.klinikum.uni-heidelberg.de/
84. S.E. Combs, M. Ellerbrock, T. Haberer, D. Habermehl, A. Hoess, O. Jäkel, A. Jensen, S. Klemm, M. Münter, J. Naumann, A. Nikoghosyan, S. Oertel, K. Parodi, S. Rieken, J. Debus, *Heidelberg Ion Therapy Center (HIT): Initial clinical experience in the first 80 patients*, Acta Oncol. 49(7), (2010) 1132-40
85. T. Haberer, W. Becher, D. Schardt and G. Kraft, Magnetic scanning system for heavy ion therapy, Nuclear Instruments and Methods A 330 (1993) 296.
86. W. Enghardt et al., Charged hadron tumour therapy monitoring by means of PET, Nucl. Instr. Methods A525 (2004) 284;
87. P. Crespo, Optimization of In-Beam Positron Emission Tomography for Monitoring Heavy Ion Tumour Therapy, Ph. D. Thesis, Technische Universitaet Darmstadt (2005).
88. K. Parodi et al., The feasibility of in-beam PET for accurate monitoring of proton therapy: results of a comprehensive experimental study, IEEE Trans. Nucl. Sci. 52 (2005) 778.
89. S. Webb, Intensity-Modulated Radiation Therapy, Institute of Physics Publishing, Bristol and Philadelphia, 2001.
90. Particle Therapy CoOperative Group (PTCOG), www.ptcog.com and ptcog.web.psi.ch.
91. H. Souda, *Facility Set-up and operation*, International Training Course on Carbon-ion Radiotherapy 2016, Chiba/Maebashi, Japan
92. H. Souda, *Carbon Ion Therapy Facilities in Japan*, 2017, Asian Forum for Accelerators and Detectors (AFAD) 2017, Lanzhou, China
93. U. Amaldi, *Particle Accelerators: From Big Bang Physics to Hadron Therapy*, Springer, 2012, ISBN 978-3-319-08870-9, DOI 10.1007/978-3-319-08870-9_4
94. U. Amaldi, G. Magrin (Eds), *The Path to the Italian National Centre for Ion Therapy*, Ed. Mercurio, Vercelli, (2005)
95. E. Feldmeier, T. Haberer, M. Galonska, R. Cee, S. Scheloske, A. Peters, *The First Magnetic Field Control (B-train) to Optimize the Duty Cycle of a Synchrotron in Clinical Operation*, Proceedings of IPAC 2012, New Orleans (Louisiana, USA), THPPD002, pp. 3503-3505
96. Y. Iwata et al. *Multiple-energy operation with extended flattops at HIMAC*, NIM A, 624 (1), 2010, pp. 33-38.
97. K. Crandall, M. Weiss, *Preliminary design of compact linac for TERA, TERA 94/34 ACC 20, September 1994*
98. U. Amaldi et al., *LIBO—a linac-booster for protontherapy: construction and tests of a prototype*, Nuclear Instruments and Methods in Physics Research A, 512 (2004) 521
99. L. Picardi et al., Progetto del TOP Linac, ENEA-CR, Frascati 1997, RT/INN/97-17.

100. A. Garonna, U. Amaldi, R. Bonomi, D. Campo, A. Degiovanni, M. Garlasché, I. Mondino, V. Rizzoglio and S. Verdú Andrés, *Cyclinac medical accelerators using pulsed* C6+/H2+ *ion sources*, J. Inst. 5 C09004 (2010)
101. J. Fourrier et al, Variable energy proton therapy FFAG accelerator, proceedings of EPACS08, 1791-1793.
102. Y. Yonemura et al, Development of RF acceleration system for 150 MeV FFAG accelerator, NIM A 576 (2007) 294-300.
103. D. Trbojevic et al., Design of a non-scaling FFAG accelerator for proton therapy, Proc. Cycl. 2004 (2005) 246–248;
104. S. Antoine et al, Principle design of a protontherapy, rapid-cycling, variable energy spiral FFAG, Nuclear Instruments and Methods A 602 (2009) 293-305.
105. E. Keil, A. M. Sessler and D. Trbojevic, Hadron cancer therapy complex using non-scaling fixed field alternating gradient accelerator and gantry design, Phys. Rev. ST Accel. Beams 10 (2007) 054701.
106. S. Vérdu Andrés, U. Amaldi and A. Faus-Golfe, *Literature review on Linacs and FFAGs for hadron therapy,* Int J Mod Phys. A26, (2011) 1659–1689
107. Carbon ion therapy, Proceedings of the HPCBM and ENLIGHT meetings held in Baden and in Lyon', Radiotherapy and Oncology 73/2 (2004) 1-217.
108. T.R. Bortfeld & J.S. Loeffler *Three ways to make proton therapy affordable*, 28 September 2017, Nature, V 549 (2017) 451-453
109. U. Amaldi, S. Braccini, G. Magrin, P. Pearce and R. Zennaro, patent WO 2008/081480 A1.
110. J. Fuchs et al, Laser-driven proton scaling laws and new paths towards energy increase, Nature Physics 2 (2005) 48-54.
111. S. V. Bulanov, G. A. Mourou and T. Tajima, Optics in the relativistic regime, Rev. Mod. Phys. 78 (2006) 309-372.
112. U. Linz and J. Alonso, What will it take for laser driven proton accelerators to be applied to tumour therapy?, Phys. Rev. ST Acc. Beams 10 (2007) 094801.
113. B.T.M. Willis, C.J. Carlile: Experimental neutron scattering, Oxford University Press, 2009, ISBN 978-0-19-851970-6.
114. F. Mezei: 2011, private communication.
115. Scientific Prospects for Neutron Scattering with support from EC/TMR and ILL, in Autrans, France, 11-13 January 1996.
116. K. van der Meer, et al.: Nucl. Instrum. Meth. B 217 (2004) 202–220.
117. F. Mezei: *Comparison of neutron efficiency of reactor and pulsed source instruments*, Proc. ICANS-XII (Abingdon 1993) (RAL Report No. 94-025), I-137; and F. Mezei: *The raison d'être of long pulse spallation sources*, J. Neutron Research 6 (1997) 3-32.
118. S. Gammino, L. Celona, R. Miracoli, D. Mascali, G. Castro, G. Ciavola, F. Maimone, R. Gobin, O. Delferrière, G. Adroit, F. Senèe: Proc. 19th Workshop ECR Ion Sources (MOPOT012), Grenoble, August 2010, to be published on Jacow.
119. R. Gobin, et al.: *High intensity ECR ion source (H^+, D^+, H^-) developments at CEA/Saclay*, Rev. Sci. Instrum. 73 (2002) 922.
120. M. Eshraqi, G. Franchetti, A.M. Lombardi: *Emittance control in rf cavities and solenoids*, Phys. Rev. ST Accel. Beams 12 (2009) 024201.
121. J. Stovall: *Low and medium energy beam acceleration in high intensity linacs*, Eur. Particle Accelerator Conf., 2004, Lucerne, Switzerland.
122. S. Molloy, et al.: *High precision superconducting cavity diagnostics with higher order mode measurements*, Phys. Rev. ST Accel. Beams 9 (2006) 112802.
123. G.E. McMichael; et al.: *Accelerator research on the RCS at IPNS*, Proc. EPAC'06, MOPCH126 (2006).
124. G. H. Rees: *Status of the SNS (now ISIS)*, Proc. PAC83, IEEE Trans. Nucl. Sci. NS-30(4) (1983) 3044-3048.
125. P. Bryant, M. Regler and M. Schuster, (eds), *The AUSTRON Feasibility Study*, Vienna (1994).
126. JAERI-KEK Project Team, *Accelerator Technical Design Report for J-PARC*, J-PARC 03-01 (2003).

127. Shinian Fu et al, *Accelerator design for China Spallation Neutron Source*, ICFA Beam Dynamics Newsletter, pp. 120-123, April, (2011).
128. J. Wei et al, Injection choice for Spallation Neutron Source ring, Proc. of PAC01, (2001).
129. R. Damburg et al, Chapter 3, in: Rydberg states of atoms and molecules, Cambridge University Press, pp 31-71, (2003).
130. G.H. Rees, *Linac, beam line and ring studies for an upgrading of ISIS*, Internal RAL note GHR1/ASTeC/ December (2009).
131. The ESS (European Spallation Source) Project, Volume III, Technical Report, pp.2-4 to 2-56, (2002).
132. M.A. Plum et al, *SNS ring commissioning results*, Proceedings of EPAC'06, MOPCH131 (2006).
133. C.R. Prior and G.H. Rees, *Multi-turn injection and lattice design for HIDIF,* Nuclear Instruments and Methods, Physics Research A, 415 pages 357-362, (1998).
134. G.H. Rees, *Direct proton injection for high power, short pulse, spallation source rings,* Internal RA note GHR1/ASTeC/ November (2016).
135. G. H. Rees, Non-isochronous & isochronous, non-scaling FFAG designs, Proc. of 18th International Conf. on Cyclotrons & their Applications, Sicily, MOP1197, p.189-192 (2007).
136. S. Machida, Parameters of a DF spiral ring for ISIS Hall, RAL note smb://ISIS/Shares/Accelerator R&D/IBIS Meetings/ ISIS II FFAG Parameters.
137. L.M. Onishchenko: *Cyclotrons*, Phys. Part. Nucl. 39 (2008) 950.
138. E.O. Lawrence, N.E. Edlefsen: *On the production of high speed protons*, Science 72 (1930) 376.
139. H. Willax: *Proposal for a 500 MeV Isochronous Cyclotron with Ring Magnet*, Proc. Intern. Conf. Sector-Focused Cyclotrons (1963) 386.
140. M. Seidel, et al.: *Production of a 1.3 Megawatt Proton Beam at PSI*, Proc. IPAC10, Kyoto, Japan (2010) 1309-1313.
141. L.H. Thomas: *The Path of Ions in the Cyclotron*, Phys. Rev. 54 (1938) 580-598.
142. W. Joho: *High Intensity Problems in Cyclotrons*, Proc. 5th Intern. Conf. Cyclotrons and their Applications, Caen (1981).
143. G. Dutto, et al.: *TRIUMF High Intensity Cyclotron Development for ISAC*, Proc. 17th Intern. Conf. Cyclotrons and Their Applications, Tokyo (2004) 82–88.
144. M. Kase, et al.: *Present Status of the RIKEN Ring Cyclotron*, Proc. 17th Intern. Conf. Cyclotrons and their Applications, Tokyo (2004) 160–162.
145. H. Okuno, et al.: *Magnets for the RIKEN Superconducting RING Cyclotron*, Proc. 17th Intern. Conf. Cyclotrons and their Applications, Tokyo (2004) 373–377.
146. V. Yakovlev et al.: *The Energy Efficiency of High Intensity Proton Driver Concepts*, Proc. IPAC'17, Copenhagen (2017) 4842-4847
147. *List of Cyclotrons*, Proc. 18th Intern. Conf. Cyclotrons and their Applications, Giardini Naxos (2007).
148. S. Hofmann, G. Münzenberg: Rev. Mod. Phys. 72(3) (2000) 733.
149. Yu.Ts. Oganessian, V.K. Utyonkov, Rep. Prog. Phys. 78, (2015) 036301
150. S. Hofmann, *J. Phys. G: Nucl. Part. Phys. 42 (2015) 114001*
151. J.R. Alonso: Proc. EPAC'90, Nice (1990) 95.
152. H.W. Schreuder: Proc. EPAC'90, Nice (1990) 82.
153. I. Tanihata, et al.: Phys. Rev. Lett. 55(24) (1985) 2676.
154. P.G. Hansen, B. Jonson: Europhys. Lett. 4(4) (1987) 409.
155. H. Geissel, G. Muenzenberg, H. Weick: Nucl. Phys. A 701 (2002) 259.
156. U. Koester: Eur. Phys. J. A 15 (2002) 255.
157. http://www.iupap.org/wg/
158. C.E. Anderson, K.W. Ehlers: Rev. Sci. Instrum. 27 (1956) 809.
159. A.S. Pasuyk, Y.P. Tretiakov, S.K. Gorbacher: Dubna-Report 3370 (1967).
160. P. Spaedtke, et al.: Proc. LINAC'96, Geneva (1996) 163.
161. V.A. Monchinsky, L.V. Kalagin, A.I. Govorov: Laser Part. Beams 14 (1996) 439.
162. E.D. Donets: Rev. Sci. Instrum. 69(2) (1998) 614.

162. A. V. Butenko et al., Proc. IPAC'14, Dresden (2014) 2103
163. U. Zoppi, et al.: Radiocarbon 49 (2007) 171.
164. J. Alessi, et al.: Proc. HIAT'09, Venice (2009) 138.
165. C. J. Gardner et al., Proc. IPAC'15, Richmond (2015) 3805
166. F. Wenander: EBIST2010, J. Instrum. 5 (2010) C10004.
167. P. Briand, R. Geller, B. Jacquot, C. Jacquot: Nucl. Instrum. Meth. 131 (1975) 407.
168. R. Geller, B. Jacquot, M. Pontonnier: Phys. Rev. 56(8) (1985) 1505.
169. Y. Jongen and G. Ryckewaert, IEEE Trans. Nucl. Sci. 30(4):2685 (1983)
170. D. Leitner, C. Lyneis, in: Physics and Technology of Ion Sources, I.G. Brown (ed.), Wiley-VCH (2004) 203.
171. L. Sun et al., Proc. LINAC'16, East Lansing (2016) 1028
172. G. Machicaone et al., Proc. ECRIS'14, Nizhny Novgorod (2014) 1
173. Y. Higurashi et al., Proc. ECRIS'16, Busan (2016) 10
174. S. Jeong, Proc. IPAC'16 Busan (2016) 4261
175. J. Benitez et al., Proc. ECRIS'12, Sydney (2012) 153
176. H. W. Zhao et al., Phys. Rev. Accel. Beams **20** 094801 (2017)
177. K. Tinschert et al., Proc. ECRIS'08, Chicago (2008) 97
178. H. Koivisto et al., Proc. HIAT'09, Venice, (2009) 128
179. T. Loew et al., Proc. PAC'07, Albuquerque (2007) 1742
180. S. L. Bogomolov et al., Proc. EPAC'98, Stockholm (1998) 1391
181. S. Gammino et al., Proc. Cyclotrons'01, East Lansing (2001) 223
182. P. Sortais et al., Rev. Sci. Instrum. **75** 1610 (2004)
183. D. Leitner et al., Proc. ECRIS'08, Chicago (2008) 2.
184. T. Lamy, J. Angot, C. Fourel, Proc. HIAT'09, Venice (2009) 114
185. L. Maunoury et al., Proc. ECRIS'16, Busan (2016) 35
186. L. Sun et al., Proc. LINAC'16, East Lansing (2016) 1027
187. S. Gammino, ISIBHI collaboration, Proc. Cyclotrons'07, Giardini-Naxos, Sicily (2007) 256
188. D. Leitner, C. Lyneis, Physics and Technology of Ion sources, 223, Ed. I. G. Brown, Wiley-VCH
189. P. Spädtke et al., Rev. Sci. Instrum. 83 02B720 (2012)
190. M. Chanel, Nucl. Instrum. Meth. A 532 (2004) 137
191. R. Hollinger et al., Rev. Sci. Instrum. 79 (2008) 02C703
192. R. Hollinger et al., Proc. RUPAC'12, Saint-Petersburg (2012) 436
193. N. Bohr: Kgl. Danske. Videnskab. Selskab. Mat.- Fys. Medd. 18(8) (1948).
194. H.H. Heckmann, E.L. Hubbard, W.G. Simon: Phys. Rev. 129(3) (1963) 1240.
195. H.-D. Betz: Rev. Mod. Phys. 44 (1972) 465.
196. P. Strehl, in: Handbook of Accel. Phys. Eng., A.W. Chao, M. Tigner (eds.), World Sci. (2006) 603.
197. C.D. Moak, et al.: Phys. Rev. Lett. 18(2) (1967) 41.
198. J.P. Rozet, C. Stéphan, D. Vernhet: Nucl. Instrum. Meth. B 107 (1996) 67.
199. A. Leon, et al.: Atomic Data & Nuclear Data Tables 69 (1998) 217.
200. C. Scheidenberger, et al.: Nucl. Instrum. Meth. B 142 (1998) 441.
201. W. Bath et al., Phys. Rev. Accel. Beams 18 (2015) 040101
202. A.S. Schlachter, et al.: Phys. Rev. A 27(11) (1983) 3372.
203. V.P. Shevelko, et al.: J. Phys. B 37 (2004) 201.
204. V.P. Shevelko, et al.: Nucl. Instrum. Meth. B 269 (2011) 1455.
205. W. Erb: GSI-Report GSI-P-7-78, (1978).
206. A.N. Perumal, et al.: Nucl. Instrum. Meth. B 227 (2005) 251.
207. H. Okuno, et al.: Phys. Rev. ST Accel. Beams 14 (2011) 033503.
208. A. Mueller, et al.: Phys. Ser. T 37 (1991) 62.
209. J. Bosser, et al.: Part. Accel. 63 (1999) 171; S. Baird, et al.: Phys. Lett. B 361 (1995) 184.
210. O. Uwira, et al.: Hyp. Interact. 108, (1997) 149.
211. E. Mahner: Phys. Rev. ST Accel. Beams 11 (2008) 104801.
212. H. Kolmus, et al.: J. Vac. Sci. Technol. A 27(2) (2009) 245.
213. C. Omet, H. Kollmus, H. Reich-Sprenger, P. Spiller: Proc. EPAC'08 Genoa (2008) 295.

214. http://www.nupecc.org
215. H. Geissel, G. Münzenberg, H. Weick: Nucl. Phys. A 701 (2002) 259.
216. http://www.rarf.riken.go.jp/rarf/acc/history.html
217. H. Kamitsubo: Proc. Cyclotrons'84, East Lansing (1984) 257.
218. A. Goto, et al.: Proc. Cyclotrons'89, Berlin (1989) 51.
219. M. Odera, et al.: Nucl. Instrum. Meth. 227 (1984) 187.
220. Y. Yano: Proc. Cyclotrons'04, Tokyo (2004) 18A1; Y. Yano: Nucl. Instrum. Meth. B 261 (2007) 1009.
221. N. Inabe, et al.: Proc. Cyclotrons'04, Tokyo (2004) 200; T. Mitsumoto, et al.: Proc. Cyclotrons'04, Tokyo (2004) 384.
222. H. Okuno, et al.: IEEE Trans. Appl. Supercond. 17 (2007) 1063.
223. T. Kubo: Nucl. Instrum. Meth. B 204 (2003) 97.
224. O. Kamigaito, et al.: Proc. Cyclotrons'10, Lanzhou (2010) TUM2CIO01.
225. T. Ohnishi, et al.: J. Phys. Soc. Jpn. 79 (2010) 073201.
226. T. Nakamura, et al.: Phys. Rev. Lett. 103 (2009) 262501.
227. P. Doornenbal, et al.: Phys. Rev. Lett. 103 (2009) 032501.
228. S. Nishimura, et al.: Phys. Rev. Lett. 106 (2011) 052502.
229. Ch. Schmelzer, D. Böhne: Proc. Prot. Lin. Accel. Conf. NAL (1970) 981.
230. D. Böhne: Proc. Prot. Lin. Accel. Conf. Los Alamos (1972) 25.
231. D. Böhne: Proc. PAC'77, IEEE Trans. Nucl. Sci. 14(3) (1977) 1070.
232. N. Angert: Proc. PAC'83, IEEE Trans. Nucl. Sci. 30(4) (1983) 2980.
233. L. Dahl: Proc. HIAT'09, Venice (2009) 193.
234. N. Angert, et al.: Proc. EPAC'92, Berlin (1992) 167.
235. U. Ratzinger: Proc. LINAC'96, Geneva (1996) 288.
236. J. Ohnishi, et al.: Proc. Cyclotrons'04, Tokyo (2004) 197.
237. J. Glatz: Proc. LINAC'86, SLAC-Rep. 303, Stanford (1986) 302.
238. H. Geissel, et al.: Nucl. Instrum. Meth. B 70 (1992) 286.
239. B. Franzke, et al.: Proc. EPAC'98, Stockholm (1998) 256.
240. R. Hasse: Phys. Rev. Lett. 83, 3430 (1999).
241. M. Steck, et al.: Phys. Rev. Lett. 77 (1996) 3803; M. Steck, et al.: Proc. PAC'01, Chicago (2001) 137.
242. H. Irnich, et al.: Phys. Rev. Lett. 75(23) (1995) 4182.
243. F. Nolden, et al.: Proc. EPAC'00 Vienna (2000) 1262.
244. M. Hausmann, et al.: Nucl. Instrum. Meth. A 446 (2000) 569.
245. M. Lestinsky et al.: Physics book: CRYRING@ESR, Eur. Phys. J. Spec. Top. 225, 797 (2016).
246. FAIR Technical Design Reports, GSI, Darmstadt (2008).
247. O. Kester, et al.: Proc. IPAC'15, Richmond (2015) 1343.
248. W. Henning: Proc. EPAC'04 Lucerne (2004).
249. P. Spiller, et al.: Proc. EPAC'08, Genoa (2008) 298.
250. U. Ratzinger, et al.: Proc. LINAC'06, Knoxville (2006) 526.
251. H. Geissel, et al.: Nucl. Instrum. Meth. B 204 (2003) 71.
252. M. Steck, et al.: Proc. PAC'09, Vancouver (2009) 4246.
253. R. Toelle, et al.: Proc. PAC'07, Albuquerque (2007) 1482.
254. H.G. Blosser: Proc. Cyclotrons'78, Bloomington (1978), Nucl. Sci. NS-26(2) (1979) 2040.
255. H. Blosser, et al.: Proc. Cyclotrons'86, Tokyo (1986) 157.
256. J.A. Nolen, et al.: Proc. Cyclotrons'89, Berlin (1989) 5.
257. B.M. Sherrill, et al.: Nucl. Instrum. Meth. 56-57(2) (1991) 1106.
258. F. Marti, et al.: Proc. Cyclotrons'01, East Lansing (2001) 64.
259. P. Miller, et al.: Proc. Cyclotrons'04, Tokyo (2004) 62.
260. D.J. Morrissey, et al.: Nucl. Instrum. Meth. B 126 (1997) 316.
261. J. Stetson, et al.: Proc. Cyclotrons'10, Lanzhou (2010) MOA1CIO01.
262. O. Kester, et al.: Proc. SRF'09, Berlin (2009) 57.
263. R.C. York, Proc. PAC'09, Vancouver (2009) 70.

264. http://frib.msu.edu/
265. X. Wu, et al.: Proc. PAC'09, Vancouver (2009) 4947.
266. http://flerovlab.jinr.ru/flnr
267. G. Gulbekyan, et al.: Proc. Cyclotrons'95, Cape Town (1995) 95.
268. Yu. Oganessian: Eur. Phys. J. A 42 (2009) 361.
269. V.V. Bashevoy, et al.: Proc. Cyclotrons'01, East Lansing (2001) 387.
270. G. Gulbekyan et al. Proc. Cyclotrons'16, Zurich (2016), 278
271. W. Zhan, et al.: Proc. Cyclotrons'07, Catania (2007) 110.
272. P.W. Schmor, et al.: Proc. LINAC'04, Lübeck (2004) 251.
273. R.E. Laxdal, et al.: Proc. PAC'01, Chicago (2001) 3942.
274. R.E. Laxdal: Proc. LINAC'06, Knoxville (2006) 521.
275. V. Zvyagintsev, et al.: Proc. RuPAC'10, Protvino (2010) 292.
276. R. Poirier, et al.: Proc. LINAC'00, Monterey (2000) 1023.
277. R.E. Laxdal, et al.: Proc. LINAC'00, Monterey (2000) 97.
278. R.E. Laxdal, et al.: Proc. PAC'05, Knoxville (2005) 3191.
279. R.E. Laxdal: priv. commun.
280. A. Joubert, et al.: Proc. Cyclotrons'84, East Lansing (1984) 3.
281. J. Ferme: Proc. Cyclotrons'86, Tokyo (1986) 24.
282. E. Baron, et al.: Proc. Cyclotrons'95, Cape Town (1995) 39.
283. E. Baron, et al.: Nucl. Instrum. Meth. A 362 (1995) 90.
284. M. Lieuvin, et al.: Proc. Cyclotrons'01, East Lansing (2001) 59.
285. By courtesy of GANIL/SPIRAL2, T. Junquera.
286. F. Chautard, et al.: Proc. EPAC'04, Lucerne (2004) 1270.
287. F. Chautard: Proc.Cyclotrons'10, Lanzhou (2010) MOM2CIO02.
288. M. Lewitowicz: *Acta Physica Polonica B 40* (*2009*) 811.
289. T. Junquera (SPIRAL2 Team): Proc. Linac'08, Victoria (2008) 348.
290. R.C. Pardo, et al.: Proc. PAC'09, Vancouver (2009) 65.
291. M. Pasini (HIE-ISOLDE design team): Proc. SRF'09, Berlin (2009) 924.
292. I. Tanihata: Nucl. Instrum. Meth. B 266 (2008) 4067.

12

Future Directions

C. Joshi, A. Caldwell, P. Muggli, S. D. Holmes, and V. D. Shiltsev

12.1 Plasma Accelerators

C. Joshi · A. Caldwell · P. Muggli

12.1.1 Introduction

The charge separation between electrons and ions that exists within an electron plasma density wave can create large electric fields. In 1979 Tajima and Dawson first recognized that the longitudinal component of the field of a so-called "relativistic" wave (one propagating with a phase velocity close to c), could be used to accelerate charged particles to high energies in a short distance [1]. The accelerating gradient of such a plasma wave, E_0, can be approximated—assuming a total separation of

Coordinated by C. Joshi, A. Caldwell

C. Joshi
Los Angeles, CA, USA
e-mail: joshi@ee.ucla.edu

A. Caldwell (✉) · P. Muggli
Max-Planck-Institut, Munich, Germany
e-mail: caldwell@mpp.mpg.de; muggli@mpp.mpg.de

S. D. Holmes · V. D. Shiltsev
Fermi National Accelerator Laboratory, Batavia, IL, USA
e-mail: shiltsev@fnal.gov

electrons and ions in such a wave with wavelength $\lambda_p = 2\pi c/\omega_p$—as

$$E_o \approx \frac{m_e c \omega_p}{e},$$
$$E_o\,(\text{eV/m}) \approx 96.2\sqrt{n_o\,(\text{cm}^{-3})}. \tag{12.1}$$

Here λ_p is the plasma wavelength, $\omega_p = \sqrt{4\pi e^2 n_o/m_e}$ is the plasma frequency and n_o is the background plasma density, c is the speed of light, m_e and e are the mass and charge of an electron, respectively. The majority of plasma accelerators have been operated at plasma densities $10^{14}\ \text{cm}^{-3} < n_o < 10^{20}\ \text{cm}^{-3}$ giving $10^9 < E_o(\text{eV/m}) < 10^{12}$. Such ultrahigh accelerating gradients are the principal attraction of plasma accelerators.

In a laser-plasma accelerator (LPA), an accelerating structure is created as an intense laser pulse propagates through the plasma driving an electron plasma wave also known as a *wake*. Specifically it is the ponderomotive force, F_p, of the intense laser that displaces the electrons from the heavier ions creating the charge separation and accelerating *wakefield*. The ponderomotive force is related to E_l or A_l, the electric field and vector potential of the laser respectively via $F_p \propto \nabla E_l^2 \propto \nabla A_l^2$. Often the normalized vector potential of the laser driver, a_o, is used to characterize the strength or magnitude of the laser driver and subsequent wake that is driven. The normalized vector potential of the laser is given by

$$a_o = \frac{eA_l}{m_e c^2} = \frac{eE_l}{\omega_o m_e c} \simeq 8.6 \times 10^{-10} \lambda_o\,(\mu\text{m}) \sqrt{I_o\,(\text{W/cm}^2)}, \tag{12.2}$$

where I_o is the focused intensity of the laser pulse, and λ_o and ω_o are the vacuum laser wavelength and frequency respectively. In laser wakefield accelerators, laser intensities of $10^{17} < I_o(\text{W/cm}^2) < 10^{20}$ are commonly used. For $a_o \approx 1$ and laser pulse duration on the order of half-a-plasma wavelength (typically 50–100 fs), wakes with accelerating fields of approximately 100 GeV/m are produced in a $10^{18}/\text{cm}^3$ density plasma. This gradient is more than three orders of magnitude larger than the accelerating gradient in a conventional RF driven accelerator.

It was recognized in the 1980s that wakes in plasmas could also be driven by a relativistic beam of electrons [2]. Given an intense enough beam of electrons, the plasma is both created [3] and excited by the passage of the bunch. Recently, driving the wake with a proton bunch has also been suggested [4].

In a beam-driven plasma wakefield accelerator (PWA), it is the space charge force of a highly relativistic beam of particles which excites the wakefield. For such a bunch, the electric field seen by the plasma electrons is in the transverse direction, and the plasma electrons initially move away from the beam axis, while the more massive ions remain effectively frozen. These transversely expelled electrons are attracted back towards the beam axis by the space charge force of the ions, creating a cavity with very strong electric fields. An appropriately timed witness bunch can be placed in a region of very strong longitudinal component of the electric field of this cavity and accelerated. See Fig. 12.1. The cavity also provides a radial force that

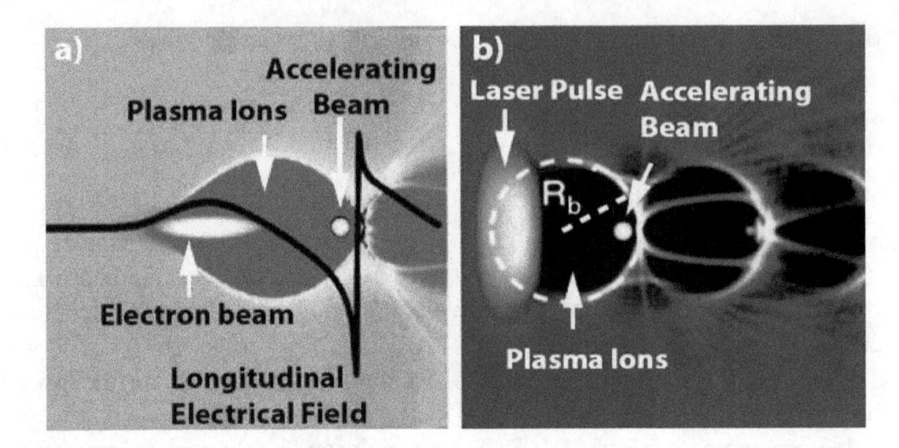

Fig. 12.1 Schematic of the accelerating structures produced in a plasma in the case of (**a**) electron beam-driven plasma wakefield accelerator and (**b**) laser wakefield accelerator (LWFA), both operating in the blow-out regime. In this regime all the plasma electrons are expelled by the electron beam or the laser pulse thereby creating a "cavity" containing only plasma ions. The extremely nonlinear longitudinal electric field generated in both cases is shown as a black curve in (**a**). In the LWFA case the laser spot size is matched to the plasma to produce a near spherical ion cavity with radius R_b equal to the laser spot size. The blow-out regime is attractive because the transverse focusing field increases linearly and the longitudinal accelerating field is constant with radius from the axis

Table 12.1 This table summarizes the scaling laws for the three distinct LWFA regimes characterized by the laser a_o, normalized spot-size $k_p w_o$ and longitudinal electric field amplitude ε_{LW} $(m_e c\omega_p/e)$

Regime	a_o	$k_p w_o$	$\epsilon_{LW}(E_o)$	$k_p L_d$	$k_p L_{pd}$	λ_W	γ_ϕ	$\Delta W(mc^2)$
Linear	<1	2π	a_o^2	$\frac{\omega_o^2}{\omega_p^2}$	$\frac{\omega_o^2}{\omega_p^2}\frac{\omega_p\tau'}{a_o^2}$	$\frac{2\pi}{k_p}$	$\frac{\omega_o}{\omega_p}$	$a_o^2\frac{\omega_o^2}{\omega_p^2}$
1D Nonlinear	>1	2π	a_o	$4a_o^2\frac{\omega_o^2}{\omega_p^2}$	$\frac{1}{3}\frac{\omega_o^2}{\omega_p^2}\omega_p\tau'$	$\frac{4a_o}{k_p}$	$\sqrt{a_o}\frac{\omega_o}{\omega_p}$	$4a_o^2\frac{\omega_o}{\omega_p^2}$
3D Nonlinear	>2	$2\sqrt{a_o}$	$\frac{1}{2}\sqrt{a_o}$	$\frac{4}{3}\frac{\omega_o^2}{\omega_p^2}\sqrt{a_o}$	$\frac{\omega_o^2}{\omega_p^2}\omega_p\tau'$	$\sqrt{a_o}\frac{2\pi}{k_p}$	$\frac{1}{\sqrt{3}}\frac{\omega_o}{\omega_p}$	$\frac{2}{3}\frac{\omega_o^2}{\omega_p^2}a_o$

Here $k_p L_d$ and $k_p L_{pd}$ are the normalized dephasing and pump depletion lengths respectively, λ_W is the plasma wavelength, γ_ϕ is the Lorentz factor associated with the phase velocity of the wake, and $\Delta W(mc^2)$ is the dephasing limited net energy gain in a single stage of a LWFA. τ' is the characteristic laser pulse duration given approximately by $\tau' = \frac{4}{3}\frac{1}{\omega_p}\sqrt{a_o}$. This table has been reproduced from [70]

keeps the witness bunch from expanding radially. A conceptual e^-e^+ linear collider based on plasma acceleration envisions multiple plasma wakefield stages powered by either a laser or a particle beam driver. Each stage adds typically 10 GeV to the accelerating beam. Using a multi-TeV class proton beam as a driver, it may be possible to accelerate an electron beam to energies of interest (TeV) in a single stage.

12.1.2 Physical Concepts

We now discuss some of the important physical concepts in plasma accelerators. These concepts will determine the necessary laser/beam and plasma conditions for the efficient acceleration of electrons. Unless specified otherwise the expressions given in this section are in the limit of small amplitude or sinusoidally varying (linear) wakes.

12.1.2.1 Phase Velocity v_ϕ

The phase velocity, v_ϕ, of the accelerating wake structure is equal to the group velocity of the driver. In the case of a LPA, v_ϕ of the wakefield is equal to $v_g = c\left(1 - \omega_p^2/\omega_o^2\right)^{1/2}$ of the laser pulse as it propagates through the plasma. Often it is also useful to define a relativistic Lorentz factor of the wake $\gamma_\phi = \left(1 - v_\phi^2/c^2\right)^{-1/2} \simeq \omega_o/\omega_p$. For a PWA driven by a highly relativistic electron bunch $v_\phi \approx c$.

12.1.2.2 Dephasing Length L_d

In a LPA, electrons that are accelerated by the wakefield can gain enough energy as to move faster than the phase velocity of the accelerating wake structure. The difference in velocity causes a relative slip in position between the electron beam and the accelerating phase of the wakefield. Eventually the electron beam will slip out of the accelerating phase and into the decelerating phase of the wakefield and begin to lose energy. The length over which the electron beam can gain energy from the accelerating phase of the wakefield is known as the dephasing length, L_d. Within the wake, the accelerating phase extends over approximately half a plasma wavelength and the dephasing length is approximated by assuming that the velocity of the relativistic electrons equals c. The dephasing length is then given by $L_d = c\lambda_p/\{2(c - v_\phi)\}$. This length limits the maximum energy gain that an electron experiences in a LPA.

In a PWA driven by an electron beam, where $v_\phi \approx c$ energy gain is not limited by dephasing. However, in a proton driven PWA the dephasing between the drive and witness bunch must be considered. The phase slippage between the drive and witness bunch has been estimated as [5]

$$\delta \approx \frac{\pi L}{\lambda_p}\left[\frac{1}{\gamma_f \gamma_i}\right], \tag{12.3}$$

where L is the distance traveled, and γ_i and γ_f are the initial and final Lorentz factors of the particles in the driving bunch. With an initial proton energy of 1 TeV and final energy of 0.5 TeV, it is nevertheless possible to have dephasing lengths of many hundreds of meters.

In either LPA or PWA cases, it is conceivable to control the plasma wavelength by adjusting the density of the plasma, thus in part or fully compensating for the phase slippage.

12.1.2.3 Pump Depletion Length L_{pd}

In a LPA, the laser pulse is depleted of its energy as it excites the wake. Additionally, laser light that is not coupled into driving the wakefield is lost due to diffraction/refraction within the plasma. The depletion of energy from the laser pulse limits the distance over which the laser remains intense enough to drive a wakefield. The characteristic distance over which the laser pulse is depleted of it energy is defined as the pump depletion length L_{pd}. In order to gain further energy, two or more laser-plasma accelerator stages must be combined in series. This is known as staging. In a PWA if $\gamma_f \ll \gamma_i$, then the pump depletion length is approximately $\gamma_i mc^2/E_-$, where E_- is the decelerating field of the wake.

12.1.2.4 Injection and Trapping

Once a plasma accelerator has been created, electrons can be *injected* into the accelerator to gain energy. In a LPA, the background plasma typically provides the source for injected electrons. In order to gain a significant amount of energy these injected plasma electrons must be *trapped* by the wakefield. An electron is considered to be trapped once it has gained enough energy to have a longitudinal velocity equal to the phase velocity of the wakefield. After an electron is trapped it can remain in the accelerating phase of the wakefield and continue to gain energy until it travels a dephasing length. Next different methods of electron injection are introduced.

In a PWA, a relativistic witness beam is injected at the appropriate phase with respect to the drive beam in order to experience the accelerating phase of the wakefield. Since both the witness beam and the wake propagate at c, there is no relative phase slippage.

Self-Injection

In a thermal plasma, (such as those typically created using lasers with joules of energy and pulse durations on the ps-ns time scale), some electrons in the tail of the electron distribution function may have sufficiently high momentum so as to reside on a "trapped-orbit". This occurs when the electron velocity is equal to v_ϕ in the plasma wake potential.

In a cold plasma, plasma electrons from outside the wake can cross into or be self-injected into the accelerating phase of the wake. If the wake amplitude is large enough, these injected electrons can quickly gain enough momentum from the accelerating field to move with the wake and become trapped [6, 7]. Self-injection process can be facilitated in a density downramp [8] or at a sudden transition between high and low density plasma regimes [9].

Downramp Injection
If the drive laser pulse or the particle bunch traverses across a plasma downramp [9] some of the plasma electrons forming the sheath around the plasma ions of a 3D nonlinear wake (discussed later) can be injected into the wake. This happens because the wavelength of the wake increase as the drive bunch goes from higher to lower density causing some electrons to cross the sheath and gain sufficient longitudinal velocity to move synchronously with the wake as they reach the back of the wake. These electrons can also have a very small transverse emittance because their transverse velocity is reduced to near zero by the strong defocusing field they feel as they converge on the axis of the wake.

Colliding-Pulse Injection
In a LPA, if a second laser pulse is collided with the laser pulse that induces the wake, then the ponderomotive force associated with the beating of the two pulses can impart enough longitudinal momentum to the initially untrapped electrons so that they are injected into the separatrix of the wakefield to be trapped [10, 11].

Ionization Injection
Electrons can be injected into a fully formed wake via ionization-trapping [12, 13]. A binary mixture of atoms is ionized by the field of the laser or space charge of the driving bunch. The minority atoms in this mixture typically have a step in the ionization potential such that the inner electrons are ionized close to the wake axis, near the peak field of the laser pulse or beam driver. These electrons experience a much greater wake potential so as to be trapped by the wake.

External Injection
In analogy to a conventional traveling wave accelerating structure, a witness beam of electrons or other charged particles may be pre-accelerated and injected into a plasma accelerating structure to increase their energy [14]. Within the wake the magnitude of accelerating field varies with space. Therefore, to minimize the difference in the accelerating gradient that is experienced, the injected electron beam should ideally have a longitudinal length that is much less than the plasma wavelength of the accelerator. For a plasma density of 10^{18} cm^{-3}, $\lambda_p \sim 30$ μm ($\sim$100 fs). Additionally, the external electron beam must be synchronized to the accelerating structure such that it is injected into the correct accelerating and focusing phases of the wakefield.

12.1.2.5 Net Energy Gain ΔW

For a LPA, The net energy gain of an electron ΔW, can be estimated as

$$\Delta W = eE_{LW}L_{acc} = \left(\frac{E_{LW}}{E_o}\right)\left(\frac{\omega_p}{c}L_{acc}\right)mc^2. \tag{12.4}$$

Here E_{LW} is the local longitudinal accelerating field seen by the electron of the wake, and L_{acc} is the distance over which the electron interacts with the accelerating field.

12.1.2.6 Beam Loading

The accelerated particles can modify the wakefield in a process known as beam loading [15, 16]. It can be understood as destructive interference between the fields of the wake and fields of the accelerated particles. Beam loading places limitations on the number of particles, the peak current, the energy spread and duration of the accelerated particles. The plasma-accelerator can be optimized for one or more of these parameters. The maximum number N_o of electrons that can be loaded into the accelerating phase of the wake is [15]

$$N_o = 5 \times 10^5 \,(E_1/E_o)\, n_o^{1/2} \left[\text{cm}^{-3}\right] S \left[\text{cm}^2\right], \tag{12.5}$$

where $S \gg \pi/k_p^2$ is the cross sectional area of the wake. As $N \rightarrow N_o$ the energy spread $\Delta\gamma/\gamma \rightarrow 1$. From energy conservation arguments the beam loading efficiency scales as $(N/N_o)(2 - N/N_o)$.

12.1.2.7 Drive Pulse Evolution

In a LPA, plasma can act as a lens to focus the spot size of the laser in a process known as relativistic self-focusing [17]. Relativistic self-focusing of the laser pulse occurs when the power of the laser pulse exceeds P_c, the critical power for self-focusing given by

$$P_c(GW) \simeq 17.4\frac{\omega_o^2}{\omega_p^2}. \tag{12.6}$$

The laser pulse will continue to focus until all the electrons are expelled from within the spot size of the laser and there is no longer a gradient in the index of refraction in the radial direction. The ratio of the laser power, P, to P_c gives a good indication of how strongly the laser pulse will be focused by the plasma.

The laser pulse can also be compressed longitudinally in time by the density gradient of a wake via photon acceleration and deceleration [18, 19]. If the laser pulse width is on the order of λ_p the entire pulse will be compressed and this can lead to an increase the laser intensity and a_o [20]. However, if the laser pulse width is much greater than a plasma wavelength, the laser pulse will be compressed within each wake driving a laser-plasma accelerator in the self-modulated regime.

In a PWA, the particles in a symmetric drive pulse lose their energy to the wake at a different rate. For ultra-relativistic electrons, the pulse shape of the drive beam does not evolve longitudinally since there is insignificant relative motion between the particles. But for a proton drive beam the spatial spread of particles d due to momentum spread will induce a lengthening of the bunch. The effect can be evaluated for vacuum propagation as

$$d \approx \frac{L}{2\Delta\gamma^2} \approx \left(\frac{\sigma_p}{p}\right) \frac{M_p^2 c^4}{p^2 c^2} L, \tag{12.7}$$

where M_p is the proton mass, and p is the proton momentum. Given a 1 TeV proton beam, a 10% momentum spread leads to a growth of ~0.1 μm/m. Large relative momentum spreads will still allow for long plasma acceleration stages provided the drive beam is relativistic.

12.1.2.8 Guiding

The length of the accelerating structure can limit the energy gain of the particles. For a LPA the length of the accelerator (assuming Gaussian optics) is approximately equal to π times the Rayleigh length of the focused laser pulse, $z_R = \pi w_o^2/\lambda_o$, where w_o is the spot size of the laser. Often, in order to reach the intensities necessary to drive large amplitude wakefields, the laser must be focused to a spot sizes on the order of ~10 μm. This limits the accelerating structure length to ~1 mm and subsequently severely limits the energy gain of electrons if the L_d is longer than this.

To overcome this limitation, a channel can be created that has an appropriate radial plasma density profile such that the diffraction of the laser pulse is minimized. This allows the spot size of the laser pulse to remain small, or 'guided', over the length of the channel that is much greater than a Rayleigh length. For a channel with a radial parabolic electron density profile, a change in electron density of $\Delta n = 1/\left(\pi r_e w_o^2\right)$ is required to optical guide the laser pulse, where here $r_e = e^2/m_e c^2$ is the classical electron radius. Two main methods have been used to guide short laser pulses.

In the first method, an electrical discharge is struck across a tube, or capillary containing hydrogen gas creating a plasma. The electron temperature close to the capillary wall is lower than that near the center of the discharge. Therefore over a short amount of time (~100 ns), hotter electrons located on axis of the discharge will diffuse radially, creating an appropriate density channel that can guide a laser pulse [21].

A second method, known as self-guiding, relies on matching the ponderomotive force of the laser, which expels the plasma electrons and drives the wake, to the space-charge attraction force that the plasma ions exert on the expelled electrons. When this is achieved, the radial plasma density profile of the wake that is created has the appropriate shape and depth to minimize the diffraction of the laser pulse [22, 23].

Both of these techniques have been used to extend the length of a laser-plasma accelerator from hundreds of microns to centimeter scale lengths [24–26].

In a PWA, if the density-length product of the plasma is large enough, the electron drive pulse can be guided by the transverse focusing force provided by the plasma ions. If the beam density $n_b > n_o$, the beam electrons can blow-out all the plasma electrons creating an ion channel. In this blow-out regime, the beam envelope is described by the differential equation [27]:

$$\sigma_r'' + \left[K^2 - \epsilon_{Nr}^2/\gamma^2\sigma_r^4(z)\right]\sigma_r(z) = 1, \tag{12.8}$$

where $K = \omega_p/(2\gamma)^{1/2}c$ is the restoring force provided by the plasma ions or equivalently the betatron wavenumber $k_\beta = \omega_\beta/c$ where ω_β is the betatron frequency. The beam is said to be matched to the plasma if $\beta_{beam} = 1/K = \beta_{plasma}$. In this case the beam propagates through the plasma with a constant radius [28]. The matched beam radius r_b is found by letting $\sigma_r''(z) = 0$ in this equation giving $r_b = (\epsilon_N/\gamma k_\beta)^{1/2}$. The beam now propagates through the plasma until it is either pump depleted or its front is slowly eroded away by finite emittance of the particles.

12.1.2.9 Head Erosion

In a LPA, the front portion of the laser pulse will diffract at a rate of some fraction of c/ω_p per z_R unless the pulse is guided in a pre-formed plasma channel [23, 29]. This head erosion will eventually limit the distance over which a wake can be excited. Similarly, in a PWA, the head of the drive bunch propagates in a region of no wakefield, so that the beam emittance will eventually erode the drive bunch [30]. A lower emittance drive beam can reduce head erosion as an obstacle limiting the energy gain, thus extracting more energy from the bunch. Proton beams have larger emittances than electron beams, therefore longer beams, lower plasma densities and thus long plasma cells are envisaged for efficient energy extraction. Therefore strong magnetic focusing of the proton drive bunch along the length of the plasma channel is likely necessary.

12.1.2.10 Instabilities, Scattering, and Radiation Loss

Generally short laser/beam drivers are relatively immune to laser-plasma instabilities because both the driver and the wake are continuously entering undisturbed plasma at $\sim c$. Nevertheless, there are two instabilities that can grow the emittance of the beam being accelerated. These are laser and electron beam hosing instability [31] and plasma ion motion [32].

In the hosing instability any transverse offset of the electron beam from the axis of propagation of the wake or any head-to-tail tilt can grow. The growth rate depends on the initial offset of a particular slice, how far the slice is from the front of the bunch and how far the bunch has propagated into the plasma. In the nonlinear 3D regime the instability involves a coupling of the centroid of the wake cavity and the bunch. In plasma accelerators the ions are usually assumed to be immobile because they are far more massive than the electrons. However if the accelerating bunch density exceeds the plasma density by a factor (mi/me), the focusing force exerted by the electrons on the plasma ions becomes so large that the plasma ions implode inward toward the axis. This can locally alter the linear focusing force provided by the plasma ions and affect the emittance of the accelerating electrons.

Coulomb scattering in the plasma can increase the emittance of the witness bunch. In plasma accelerators, typical values of n_o will be in the range of $10^{14} - 10^{17}$ cm^{-3}, and both the radiation length and the mean-free-path are orders of magnitude larger than the expected plasma length on the order of 100 m. A 1 TeV proton beam in Li vapor of density 1×10^{15} atoms/cm^3 gives a transverse growth rate of the proton beam of less than 0.01 μm/m due to multiple scattering, which is small compared to the size of the drive bunch. Multiple scattering for high energy electrons will also be small.

In addition to these instabilities, as particles are accelerated they will oscillate transversely emitting betratron radiation. The energy lost to radiation per unit is given by [33, 34]:

$$\frac{dW_{loss}}{dz} = \frac{1}{3} r_e \gamma_b^2 \omega_\beta^2 K_w^2. \tag{12.9}$$

Here $K_w^2 = \gamma_b \omega_\beta r_o / c$ is the effective wiggler strength of the ion column. At extremely high energies, the betatron radiation loss rate will equal the rate particles gain energy. This ultimately limits the maximum energy gain in a plasma accelerator.

12.1.3 Beam Driven Plasma Wakefield Accelerators

12.1.3.1 Electron Beam Driven PWA

The basic concept of the plasma wakefield accelerator involves the passage of an ultra-relativistic charged particle bunch through a stationary plasma. The plasma may be formed by ionizing a gas with a laser or through field-ionization by the Coulomb field of the relativistic bunch itself. This second method allows the production of meter-long, dense ($10^{16} - 10^{17}$cm^{-3}) plasmas suitable for the PWA and greatly simplifies the experimental set-up. In single bunch experiments [35] carried out with ultra-short electron bunches, the head of the bunch creates the plasma and drives the wake. The wake produces a high-gradient longitudinal field that in turn accelerates particles in the back of the bunch. The system effectively operates as a transformer, where the energy from the particles in the bulk of the bunch is transferred to those in the back, via the plasma wake. The physics is unchanged if there are two bunches rather than a single bunch; energy from the leading drive bunch is transferred to a trailing witness bunch. The maximum energy which can be given to a particle in the witness bunch in a PWA is limited by the transformer ratio, defined as

$$R = \frac{\Delta W_{max}^{witness}}{\Delta W_{max}^{drive}} \leq 2 - \frac{N_{witness}}{N_{drive}} , \tag{12.10}$$

which is at most two for longitudinally symmetric drive bunches and an unloaded wake ($N_{witness} \ll N_{drive}$) [36]. Here ΔW_{max} is the change in energy of the particles in the drive/witness beam and N is the number of particles in the witness/drive beam. This upper limit can in principle be overcome by nonsymmetric bunches [37]. Another option currently under study is to use an appropriately phased train of bunches so that the wakefield is increased with each bunch [38]. According to linear plasma theory [39], maximum accelerating field of the wake is given by [40]:

$$eE_{linear} \simeq 100 \frac{N}{2 \times 10^{10}} \frac{20}{\sigma_z\ (\mu\text{m})} \ln \sqrt{\frac{2.5 \times 10^{17}}{n_o\ (\text{cm}^{-3})} \frac{10}{\sigma_r\ (\mu\text{m})}} \text{GeV/m}, \tag{12.11}$$

where N is the number of particles in the electron bunch and σ_z is the bunch length, and σ_r is the spot size. Linear theory is valid when the normalized charge per unit length of the beam, $\Lambda = \frac{n_b}{n_o}(k_p \sigma_z) \simeq 2.5 \left(\frac{N}{2\times 10^{10}}\right)\left(\frac{20}{\sigma_z(\mu\text{m})}\right)$ is less than 1. Equation (12.11) indicates that generating large gradient wakefields requires short, high density electron bunches. For high Λ the nonlinear equivalent of Eq. (12.11) should be used [41].

In a PWA operating in the so-called blowout regime, a short but high-current electron bunch, with density n_b larger than the plasma density n_p, expels all the plasma electrons from a region surrounding the beam as shown in Fig. 12.1. The expelled plasma electrons rush back in because of the restoring force of the relatively immobile plasma ions and thus generate a large plasma wakefield. This wakefield has a phase velocity equal to the beam velocity ($\approx c$ for ultra-relativistic beams). Since the electrons in the bunch are ultra-relativistic, there is no relative phase slippage between the electrons and the wake over meter-scale plasmas.

When the wakes are in the blowout regime, shaped bunches can be used to transfer a substantial amount of energy from the drive bunch to the wake and optimize the energy extraction efficiency from the wake to the trailing bunch while not further increasing its energy spread. It can be shown that particles in a trapezoid shape drive bunch—with beam current increasing from the front to the back—loose energy at a near constant rate. A trailing bunch with a reverse shape can load the wake in such a way that nearly all the particles gain energy at the same rate. Significant beam loading will occur [42] when the loaded charge

$$Q(nC) > [(0.047mcp)/eEs]\,(1016/np\,(cm - 3))\,1/2(kpRb)4 \qquad (12.12)$$

Here, we take the normalized electric field eEs/mcp = kpRb/2 as the electric field seen by the accelerating electrons, kp−1 is the plasma skin depth, and Rb is the blowout radius of the wake.

12.1.3.2 Short Proton and Positron Beam Driven PWA

In contrast to plasma wakes driven by bunches of electrons, only limited investigations of the plasma wave excitation by a positively charged driver exist [42–46]. For a positively charged driver (such as a proton bunch or a positron bunch), plasma electrons are attracted towards the axis of the driver, overshoot and setup the accelerating wake structure [47, 48]. For positively charged drivers with $n_b/n_o \ll 1$ the electric field distribution of the wake is the same as that for the negative driver but shifted in phase. However, for a positively charged driver driver with $n_b/n_o \gtrsim 1$ the wake behavior is significantly different than for a negatively charged driver with the same beam density. Due to the radial symmetry, for the wakefield driven by a strong positively charged driver ($n_b/n_o \geq 1$), there is an electron density enhancement on-axis and effective increase of the local plasma frequency. As a result, the proton or positron driver beams must be even shorter in order to excite the plasma wake resonantly. However, given that protons can be accelerated to the TeV regime in conventional accelerators it is conceivable to accelerate electron bunches in the wake of the proton bunch up to several TeV (e.g., in the wake of a Large Hadron Collider (LHC) proton beam) in a single plasma stage.

12.1.3.3 Long Proton Beam Driven PWA

Proton bunches available today are in principle very interesting to drive wakefields because, besides being relativistic, they also have a large population and thus carry large amounts of energy (10s to 100s of kJ). A proton bunch can therefore drive wakefields over a very long single plasma and lead to very large energy gain of a witness bunch. It was shown through simulations [4] that a 10 GeV electron bunch injected into wakefields driven by a 100 μm-long, 10 kA, 1 TeV proton bunch can reach an energy of ~0.5 TeV in a single plasma, only ∼300 m-long. This corresponds to an average accelerating gradient larger than 1 GeV/m in a plasma with density of 6×10^{14} cm^{-3}. We note here that in these simulations the proton bunch length σ_z is shorter and the width σ_r is narrower than the wakefields period λ_p, a condition necessary to effectively drive wakefields. These two conditions are usually expressed as $k_p\sigma_z \leq 1$ and $k_p\sigma_r \leq 1$.

Proton bunches produced for example by the CERN Super Proton Synchrotron (SPS) or Large Hadron Collider (LHC) are long: σ_z~6–12 cm.

Compression of proton bunches produced by these synchrotrons is in principle possible [49]. However, it would first require means to impart a %-level correlated energy chirp along the already high-energy bunch. Second, the magnetic compressor (chicane) would have to be rather long (km), again because of the high energy of the particles. Therefore, producing short proton bunches such as the one used in the simulations of [4] remains a challenge.

Matching the plasma density to the long SPS proton bunch ($k_p\sigma_z \cong 1$) would mean using a low density plasma ($\sim 10^{11}$cm^{-3}) and driving low amplitude wakefields (~10s of MV/m range from E_0 in Eq. 12.11).

However, proton bunches can be focused to small transverse sizes (e.g., $\sigma_r = 200$ μm). Choosing the plasma density to reach $k_p\sigma_r \cong 1$, leads to a much larger plasma density (7×10^{14}cm^{-3}) and also much larger possible accelerating field (E_0 ~ 2.5 GV/m, Eq. 12.11). This choice of $k_p\sigma_r \cong 1$ means $k_p\sigma_z \gg 1$ since $\sigma_z \gg \sigma_r$. Such bunches are subject to a symmetric transverse self-modulation process [50]. The self-modulation process transforms the long continuous bunch into a train of short bunches separated by the wakefields period. The short bunches are formed by the action of the periodically focusing/defocusing transverse wakefields. The train can then resonantly drive wakefields to amplitudes on the order of the 2.5 GV/m predicted by Eq. 12.11. Furthermore, the process can be seeded, so that it becomes a well-controlled, seeded self-modulation (SSM) process [51]. In this case the beam-plasma system is turned into an amplifier of seed wakefields, with well determined final fields amplitude *and phase*. The process is also weakly sensitive to variations of the input bunch parameters (±5%) [52]. The SSM process makes it possible to deterministically inject an electron bunch where wakefields reach their peak amplitude and into the range of accelerating and focusing phase of the wakefields (corresponding to region $\sim\lambda_p/4$ long), a large number of periods ($\sim\sigma_z/\lambda_p$) behind the seed point [53]. Simulation results show that with this scheme, electrons can be accelerated to multi-TeV energies in a plasma a few km-long [54].

12.1.4 *Laser-Driven Plasma Accelerators*

12.1.4.1 Plasma Beat Wave Accelerator (PBWA)

Two co-propagating long ($c\tau \gg \lambda_p$) but moderate intensity ($a_o < 1$) laser beams with frequencies ω_1 and ω_2 can resonantly excite a relativistic plasma wave when $\omega_1 - \omega_2 \cong \omega_p$ [55]. Here τ is the pulse width of the laser pulse. For a square pulse, the amplitude of the plasma wave grows linearly in time and eventually saturates due to change in plasma frequency caused by relativistic mass increase of the plasma electrons. For laser pulses with $a_o \ll 1$, the maximum wave amplitude is given by $E_1/E_o = (4a_{o1}a_{o2}/3)^{1/3}$ and occurs when $\omega_1 - \omega_2 = \omega_p(1 - (a_{o1}a_{o2})^{2/3}/8)$. The relativistic plasma wave is prone to modulational instability [56], and mode coupling

[57]. For linear wakes, electrons need to be externally injected for acceleration [58, 59].

12.1.4.2 Self-Modulated Laser-Wakefield accelerator (SM-LWFA)

An accelerating wakefield can be driven by a single long laser pulse ($c\tau > \lambda_p$) via the Raman forward-scattering (RFS) instability [60] if $\omega_o > 2\omega_p$ and the laser power $P \cong P_c$ the critical power for relativistic self-focusing. In this process the laser pulse amplitude is modulated at ω_p by the creation of frequency sidebands at $\omega_o \pm \omega_p$. The spatiotemporal gain for RFS instability [61] has been estimated as $G = e^g(2\pi g)^{-1/2}$, where g is given by

$$g = \frac{a_o}{\sqrt{2}}\left(1 + \frac{a_o^2}{2}\right)\left(\frac{\omega_p^2}{\omega_o^2}\right)\frac{\omega_o}{c}\sqrt{x\phi}, \tag{12.13}$$

where x is the length of the plasma, and ϕ/c is the time over which the plasma experiences a constant laser intensity. The normalized plasma wave amplitude is $\delta n/n_o = \alpha_n G$, where α_n is the initial wave amplitude associated with the thermal noise of the plasma. When $\alpha_n G \approx 1$ the wave amplitude becomes large enough to self-trap and accelerate electrons [62, 63].

12.1.4.3 Laser Wakefield Accelerator (LWFA)

A laser wakefield accelerator (LWFA) is a laser-plasma accelerator which is created directly by the ponderomotive force of a single short ($c\tau \sim \lambda_p$), intense laser pulse. With the advent of Ti:sapphire laser systems, such ultra-short, ultra-intense laser pulses are readily available. An advantage of using such short laser pulses to drive an accelerating wake is that they are relatively immune to most laser-plasma instabilities. Additionally, these short, intense laser pulses can drive wakefields with large amplitudes, that are capable of self-trapping and accelerating electrons to high energies [64–67]. As a result of this, laser wakefield accelerators are the focus of the majority of current laser-plasma accelerator research. Next, details on a few specific regimes in LWFA research are presented.

12.1.5 LWFA Regimes

12.1.5.1 Linear Regime

The wake induced by a short laser pulse ($c\tau \sim \lambda_p$) is said to be in the linear regime if $\delta n/n_o < 1$, where δn is the change in electron density associated with the wake. Laser drivers with modest intensities $a_o < 1$ and $P/P_c \leq 1$, excite wakefields in the linear regime. In the linear regime the electron density response and the longitudinal electric field of the wake vary sinusoidally. The transverse and longitudinal fields are $\pi/2$ out of phase with one another. Thus there is a quarter wavelength region where the fields are both accelerating and focusing for the particles.

12.1.5.2 Nonlinear Regime

A nonlinear wake is created when electron density perturbation of the wake is on the order of or greater than the background plasma density ($\delta n/n_o \gtrsim 1$) and the longitudinal accelerating field of the wake becomes larger than E_o. Such nonlinear wakes are typically created by using a laser driver with an $a_o \gtrsim 2$ or with a beam driver with $n_b/n_o \gg 1$ and a $k_p\sigma_r \ll 1$. A characteristic of a nonlinear wake is a "saw toothed" shape longitudinal electric field profile. Additionally, the plasma wavelength of a nonlinear wake increases as the a_o of the laser (wake amplitude) is increased. A nonlinear wake is said to be in the 1D regime if the normalized laser pulse spot is broad, i.e., $k_p w_o \gg 1$. However, when $k_p w_o \sim 1$, and $a_o > 1$ one approaches the 3D-nonlinear regime.

The electron density response and corresponding fields of nonlinear wakes driven by laser pulses in 3-D regime are difficult to study analytically. To gain a better understanding in this regime, particle in cell (PIC) codes have been used to model the LWFA in this regime. Early work in this regime focused on simulating and developing scalings to describe the dynamics of a singular nonlinear wake driven in the so called bubble regime by a laser pulse with an $a_o \geq 2\omega_o/\omega_p$ [68, 69]. More recently a nonlinear regime in which a laser pulse with an $2 \leq a_o \leq 2\omega_o/\omega_p$ is used to drive a periodic nonlinear wake, in the so called blowout regime, has been studied with simulations, a phenomenological theory and experimental investigation [23, 25, 26, 41, 70–73]. In both the bubble and blowout regimes, the ponderomotive force of the laser pushes out all the electrons from within the first period of the wake creating an ion bubble into which background plasma electrons are self-injected, trapped and accelerated.

An advantage of operating in the blowout regime is that there exits a matched self-guiding condition, in which the intensity, spot size and pulse width of the laser pulse are matched to the plasma density, such that the driven wake serves to minimize the diffraction of the laser pulse [70]. This allows a strong and stable wake to be sustained over tens of Rayleigh lengths and can increase the interaction length between trapped electrons and the accelerating field of the wake. The matched self-guiding condition is valid when the laser $a_o \gtrsim 2$ and $c\tau \sim \lambda_p/2$. When these conditions are met, the matching condition is given by

$$k_p w_o \simeq k_p R_b = 2\sqrt{a_o}, \tag{12.14}$$

where R_b is the blowout radius of the spherically shaped wake. Remarkably, even when the spot size and or pulse width are not precisely matched to the plasma density, the laser pulse evolves within the plasma towards the matched spot size and pulse width, via relativistic self-focusing and longitudinal pulse compression [67, 72, 73].

Simulations indicate that in the blowout regime self-trapping of background electrons will occur for large amplitude wakefields when the normalized blowout radius $k_pR_b \approx 4 - 5$.

12.1.5.3 Scaling Laws

The scaling laws for a LWFA in the regimes which were discussed above are now presented.

12.1.6 Status

12.1.6.1 LWFA and PWA

The development of laser-plasma accelerators has closely followed the progress of high-power, short pulse lasers [74]. Prior to 1990, laser-plasma accelerator experiments focused primarily on the PBWA [14] or SM-LWFA concepts [54, 56]. These demonstrated the existence of greater than many GeV/m gradients and acceleration of self-injected as well as externally injected electrons [14, 69]. The accelerated particles typically had an exponentially falling energy spectrum [57]. In the mid 1990s, the chirped pulse amplification (CPA) scheme was implemented using titanium-sapphire as a gain medium and enabled the development of terawatt-class, sub-100 fs laser systems. Such short pulses (cp) were ideal for exciting wakes in underdense plasmas. Through better control of plasma density and length, and laser pulse parameters, electron bunches with a significant charge and relatively narrow (¡20%) energy spread were observed [58–60]. The electron beam energy spread and charge were controlled further by employing the colliding pulse injection technique [10, 11]. The CPA technique has extended the peak power reach of the Ti-sapphire lasers to multi-petawatt (PW) levels. For instance at Lawrence Berkeley National Laboratory (LBNL) a petawatt laser facility became operational as part of the Berkeley Lab Laser Accelerator (BELLA) project [74]. Using a plasma-channel to extend the acceleration distance, narrow energy electron bunches with energy up to 4.3 GeV have been observed [24] at LBNL. Similar peak power laser facilities have either come on line (for instance at GIST (Korea)) or expected to soon be operational in several institutes worldwide. Most of the LPA experiments are being conducted using the highly nonlinear 3D (sometimes called the blow-out or bubble) regime using either self-injected or ionization injected charge into the plasma wake. For example electrons were accelerated to beyond 1.4 GeV using ionization injection into a LPA operating in the nearly matched, self-guided blowout regime [70]. At present, several groups are exploring this regime to obtain electron beams for the generation of betatron and undulator radiation [71–73]. In past few years much effort has been devoted to controlling the charge, minimizing the transverse emittance and energy spread of electrons created from laser wakefield accelerators [74]. New techniques for beam injection such as downramp injection [74] and transverse collide pulse injection [74] have been proposed for generating electron beams with sub 100 nm transverse emittances. Possible applications for such beams might be fifth generation light sources (ultra compact XUV and X-FELs) and high-energy electron therapy of cancerous tumors. As in the past, continued progress in this field will be tied to progress in making lasers more reliable, efficient and cheaper. Additionally, to extend the interaction between the electron beam and the LPA progress must be made in making reproducible, meter-scale plasma sources.

Experiments on beam-driven PWA have been going on at Argonne National Laboratory (ANL) [75], Fermilab [76], KEK [77], Brookhaven National Laboratory (BNL) [78] and at SLAC [79]. New facilities such as FLASHForward at GSI, Germany are being commissioned. With the exception of the SLAC experiments, all other work has hitherto used modest energy electron beam (¡100 MeV) to study both acceleration and focusing effect of plasmas on beams. Experiments at SLACs FFTB facility were initially carried out using 28 GeV electron and positron beams. These used 4 ps long beams containing more than 10 kA current to systematically study beam propagation, focusing, betatron radiation emission, and acceleration in meter-scale, laser-ionized plasmas [80, 81]. These were followed by experiments that used a sub-50 fs beam to both produce the plasma via tunnel ionization and excite the wakefield in much denser, meter-scale vapor columns. The front portion of the bunch was used to excite the field while the particles in the back of the same bunch were used to sample to accelerating wake. An example of data on changes in electrons energy in the above mentioned PWA experiment at SLAC is shown in Fig. 12.2. Most of the initially 28.5 GeV electrons in the pulse are seen to lose energy but electrons in the back of the same pulse are accelerated by the wake as its electric field changes sign. The energy gain is seen to increase with distance. When the electron beam energy was increased to 42 GeV some electrons were seen to gain more than 42 GeV in just 85 cm, implying that a gradient of 50 GeV/m was sustained throughout the plasma [82].

In 2006 the FFTB facility was replaced by a new experimental facility called FACET for Advanced Acceleration research [83]. Between 2019 and 2016 FACET provided electron and positron capability with a driver beam-witness beam structure so that high gradient acceleration of a significant number of charged particles with a narrow energy spread could be demonstrated. The drive and witness beams were only about 50 fs long and were separated by about 300 fs. In order to preserve its narrow energy spread the witness beam had to beam-load the wake so that particles in the witness beam flattened the electric field of the wake thereby experiencing a nearly uniform accelerating field. With this loading, the energy contained in the wake is efficiently transferred to the accelerating particles. In a landmark experiment carried out at FACET, it was shown that the efficiency of transferring drive bunch energy to the core of the accelerated bunch was up to 30% [84]. In a follow up campaign an energy gain of 9 GeV for a bunch containing 30 pC of charge with a 5% energy spread in a 1.2 m long plasma was observed [85]. Using the energy and

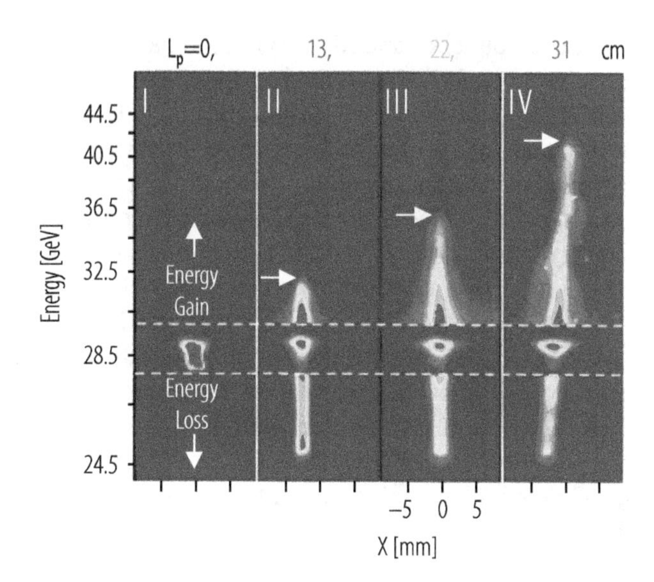

Fig. 12.2 Changes in electron energy in the PWA experiment at SLAC

the spot size changes experienced by different slices of the drive bunch itself, the PWA cavity in the nonlinear blowout regime was shown to have a longitudinal and transverse field structure that in principle will accelerate electrons without emittance growth, as long as the electrons (to be accelerated) are matched in and out of the plasma with minimal energy spread [86].

The plasma wake produced by an electron bunch cannot be used to accelerate a positron beam when the wake is in the nonlinear blow out regime because the plasma ions strongly defocus the positrons. Until recently it was not very clear how efficient positron acceleration at a high gradient could be carried out using highly nonlinear plasma wakes. Experiments at FACET showed that a certain positron beam current profile can lead to a loaded wake where the longitudinal electric field reverses sign (from decelerating to accelerating) in the middle of the single drive bunch [87]. This happens because the presence of the positrons pulls in the plasma electrons towards the axis. These plasma electrons can cross the axis in the middle of the drive bunch. Most of the electrons overshoot and set up a bubble like wake cavity but a significant fraction of the electrons are confined by the back of the positron beam close to the axis. This flattens the wake shape by beam loading [87]. A significant amount of positron charge can now be accelerated at the same electric field gradient producing a well-defined narrow energy spectrum. The energy extraction efficiency is similar to the electron bunch acceleration case described above.

In 2016, FACET ceased operation to make way for the LCLS II facility that will occupy the first 1 km space of the original SLAC linac tunnel. A new facility for advanced accelerator research, known as FACET II, is being constructed between the LCLS II linac and the LCLS linac. Together with the BELLA laser the FACET II facility [88] will arguably be the backbone of research facilities for short drive pulse driven plasma accelerator research for the next decade. The foremost physics challenge is the generation of collider quality transverse and longitudinal emittance bunches and identifying the factors that cause emittance growth in plasma accelerators [88]. Transverse emittance growth may occur as a result of the hosing instability and also ion motion [31, 32]. Even if the electrons have a very small emittance inside the wake, such a beam must be extracted and matched either to another plasma acceleration stage or to conventional magnetic optics [89–91] to avoid emittance growth. Longitudinal emittance will depend on how small the energy spread of the beam can be, which in turn depends on optimizing the beam loading to give a constant accelerating field. Continued development of diagnostic techniques to visualize the fields of the highly nonlinear wake and the injected electrons is also needed. Issues of generating asymmetric emittance flat beams and spin polarized beams are still open questions. Creative solutions for the generation and acceleration of collider quality positron bunches using plasmas are needed. In the near future it is highly likely that a single stage of LWFA and a PWA will produce 10 GeV bunches with a percent level energy spread and a high efficiency.

12.1.6.2 Proton-Driven Plasma Wakeeld Acceleration

The AWAKE experiment at CERN studies the driving of plasma wakefields with proton bunches [53, 90, 91]. AWAKE uses the (6–12)cm-long bunch delivered by the SPS with $(1\text{–}3)\times10^{11}$, 400 GeV protons focused to a transverse rms size of $\sigma_r = 200$ μm. The plasma density is chosen so that the accelerating field can reach ~1 GV/m, i.e.: $n_0 > 10^{14}$ cm^{-3} corresponding to $\lambda_p < 3$ mm. Since the proton bunch is long when compared to the wakefield period, the experiment relies on the seeded self-modulation (SSM) process [51] to reach these wakefield amplitudes.

A 10 m-long rubidium source was developed to produce a column of vapor with a very uniform density ($\delta n_{Rb}/n_{Rb} < 0.2\%$) and with sharp density ramps (<10 cm) at the entrance and exit [92]. A 450 mJ, 100 fs laser pulse propagating within with the proton bunch creates a sharp (< 100fs), relativistic ionization front that seeds the SSM process. With full ionization of the only electron of the rubidium atom outer shell, the plasma density longitudinal profile is identical to that of the rubidium vapor. The plasma radius is on the order of 1 mm. The plasma density is adjustable from 1×10^{14} to 10×10^{14} cm^{-3}.

The occurrence of the SSM on the proton bunch is observed with three diagnostics [51]: a two-screen method [93], the time resolved emission of the optical transition radiation (OTR) emitted by the protons [94], and spectral analysis of the coherent transition radiation (CTR) emitted by the bunch train [95].

Preliminary experimental results obtained recently show clear evidence of the occurrence of SSM [51]. These results also show that the SSM leads to stable excitation of wakefields and that the process corresponds to the amplification of the seed wakefields with a final phase that is very weakly dependent on the variations of the bunch initial parameters. This was also shown in numerical simulations [52]. Excitation of seed wakefields was demonstrated with a low energy electron bunch [96].

First acceleration experiments of low energy electrons (~15 MeV) externally injected into the wakefields are currently underway. In these experiments the electron bunch is purposely made longer than the plasma period in order to ease the temporal synchronization requirements between the witness electron bunch and the wakefields. Numerical simulation results show that a fraction of these electrons could emerge from the plasma with an energy in the GeV range and with a finite final energy spread ($\delta E/E \sim 10\%$) [90].

Future experiments will use a short witness electron bunch ($\sigma_z < \lambda_p$) that will load the wakefields in order to minimize the final the energy spread and at the same time preserve the emittance of the a large fraction of the bunch population [97] while gaining a few GeVs of energy.

The application of the proton-driven plasma wakefield accelerator is to fixed target experiments and to a possible very high energy electron/proton collider [98]. Electron/proton collision applications ease the requirements on the accelerated electron parameters since proton beams are not focused as tightly as beams of an electron/positron collider and do not require production of a high quality positron beam. Also, electron/proton collisions are used to study QCD physics in which interaction cross-sections tend to increase and not decrease with collision energy.

Self-modulation experiments with low energy electron bunches are also performed at DESY-Zeuthen [99].

12.2 Muon Collider

S. D. Holmes · V. D. Shiltsev

Both e^+e^- and $\mu^+\mu^-$ colliders have been proposed as possible candidates for a lepton collider to complement and extend the reach of the Large Hadron Collider (LHC) at CERN. The physics program that could be pursued by a new lepton collider (e^+e^- or $\mu^+\mu^-$) with sufficient luminosity would include understanding the mechanism behind mass generation and electroweak symmetry breaking; searching for, and possibly discovering, super symmetric particles; and hunting for signs of extra space-time dimensions and quantum gravity. However, the appropriate energy reach for such a collider is currently unknown, and will only be determined following initial physics results at the LHC. It is entirely possible that such results will indicate that a lepton collider with a collision energy well in excess of 1 TeV will be required to illuminate the physics uncovered at LHC. Such a requirement would require consideration of muons as the lepton of choice for such a collider.

The lifetime of the muon, 2 μs in the muon rest frame, is just long enough to allow acceleration to high energy before the muon decays into an electron, a muon-type neutrino and an electron-type antineutrino ($\mu^- \rightarrow e^- \nu_\mu \overline{\nu}_e$). However, constructing and operating a muon based collider with useable luminosity requires surmounting significant technical challenges associated with the production, capture, cooling, acceleration, and storage of muons in the required quantities and with appropriate phase space densities. Over the last decade there has been significant progress in developing the concepts and technologies needed to produce, capture, cool, and accelerate muon beams with high intensities of the order of $O(10^{21})$ muons/year. These developments have established a multi-TeV Muon Collider (MC) in which μ^+ and μ^- are brought to collision at high luminosity in a storage ring as a viable option for the next generation lepton-lepton collider for the full exploration of high energy physics in the era following the LHC discoveries.

Muon colliders were proposed by Budker [100] in 1969 and later conceptually developed by a number of authors and collaborations (see comprehensive list of references in [101]). Figure 12.3 presents a possible layout on the Fermilab

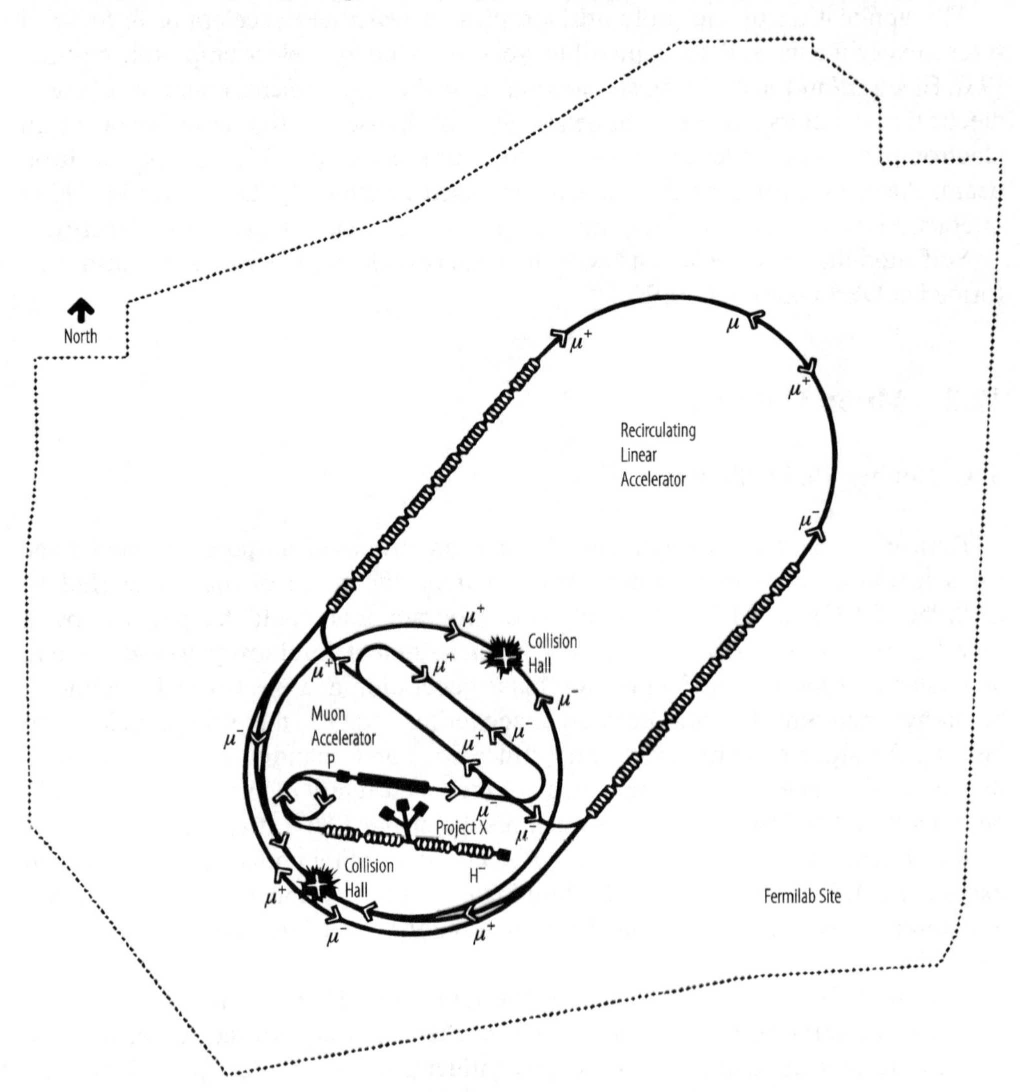

Fig. 12.3 Schematic of a 4 TeV Muon Collider on the 6 × 7 km FNAL site

Table 12.2 The parameters of the low- and high-energy Muon Collider options

Parameter	Higgs factory	Low *E*	High *E*
Center-of-mass energy [TeV]	0.126	1.5	6
Luminosity [cm^{-2} s^{-1}]	$0.005{\cdot}10^{34}$	$4.5{\cdot}10^{34}$	$7{\cdot}10^{34}$
Number of bunches	1	1	1
Muons/bunch [10^{12}]	2	2	2
Circumference [km]	0.3	2.8	6.3
Focusing at IP β_*/σ_z [mm]	25/5	10/10	10/5
Beam energy spread dp/p (rms) [%]	0.003	0.1	0.10
Ring depth [m]	~10	13	~150
Proton driver pulse rate [Hz]	30	12	15
Proton driver power [MW]	≈4	≈4	≈2
Transverse emittance ε_T [π μmrad]	300	25	25
Longitudinal emittance ε_L [π mmrad]	1	72	72

site of a MC that would fully explore the physics responsible for electroweak symmetry breaking. Such a MC requires a center-of-mass energy ($\sqrt{s}$) of a few TeV and a luminosity in the 10^{34} cm^{-2} s^{-1} range (see Table 12.2 for the list of parameters). The MC consists of a high power proton driver based, e.g., on the "Project X" SRF-based 8 GeV 2–4 MW H^- linac [102]; pre-target accumulation and compressor rings where very high intensity 1–3 ns long proton bunches are formed; a liquid mercury target for converting the proton beam into a tertiary muon beam with energy of about 200 MeV; a multi-stage ionization cooling section that reduces the transverse and longitudinal emittances and creates a low emittance beam; a multistage acceleration (initial and main) system—the latter employing Recirculating Linear Accelerators (RLA) to accelerate muons in a modest number of turns up to 2 TeV using superconducting RF technology; and, finally, a roughly 2-km diameter Collider Ring located some 100 m underground where counter-propagating muon beams are stored and collide over the roughly 1000–2000 turns corresponding to the muon lifetime.

12.2.1 Technical Motivations

Synchrotron radiation (proportional to the fourth power of the Lorentz factor γ^4) poses severe limitations on multi-TeV e^+e^- colliders, namely they must have a linear, not circular, geometry. Practical acceleration schemes then require a facility tens of kilometers long. Furthermore, beam-beam effects at the collision point induce the electrons and positrons to radiate, which broadens the colliding beam energy distributions. Since $(m_\mu/m_e)^4 = 2 \times 10^9$, all of these radiation-related effects can be mitigated by using muons instead of electrons. A multi-TeV $\mu^+\mu^-$ collider can be circular and therefore have a compact geometry that will fit on existing accelerator sites, and may be significantly less expensive than alternative machines.

The center-of-mass energy spread for a 3-TeV $\mu^+\mu^-$ collider, $dE/E < 0.1\%$, is an order of magnitude smaller than for an e^+e^- collider of the same energy. Additionally, the MC needs lower wall plug power and has a smaller number of elements requiring high reliability and individual control for effective operation [103].

An additional attraction of a MC is its possible synergy with the Neutrino Factory concept [104]. The front-end of a MC, up to and including the initial cooling channel, is similar (perhaps identical) to the corresponding Neutrino Factory (NF) front-end [105]. However, in a NF the cooling channel must reduce the transverse emittances (ε_x, ε_y) by only factors of a few, whereas to produce the desired luminosity, a MC cooling channel must reduce the transverse emittances (vertical and horizontal) by factors of a few hundred and reduce the longitudinal emittance ε_L by a factor $O(10)$. Thus, a Neutrino Factory could offer the opportunity of a staged approach to a Muon Collider, and also the opportunity of shared R&D.

12.2.2 Design Concepts

Since muons decay quickly, large numbers of them must be produced to operate a muon collider at high luminosity. Collection of muons from the decay of pions produced in proton-nucleus interactions results in a large initial phase volume for the muons, which must be reduced (cooled) by a factor of 10^6 for a practical collider. Without such a cooling, the luminosity reach will not exceed $O(10^{31}\ \text{cm}^{-2}\ \text{s}^{-1})$, a substantial limitation on the discovery reach of the MC. The technique of ionization cooling [106] is proposed for the $\mu^+\mu^-$ collider [107, 108]. This technique is uniquely applicable to muons because of their minimal interaction with matter.

Ionization cooling involves passing the beam through some material absorber in which the muons lose momentum essentially along the direction of motion via ionization energy loss, commonly referred to as dE/dx. Both transverse and longitudinal momentum are reduced via this mechanism, but only the longitudinal momentum is then restored by reacceleration, leaving a net loss of transverse momentum (transverse cooling). The process is repeated many times to achieve a large cooling factor. The energy spread can be reduced by introducing a transverse variation in the absorber density or thickness (e.g., a wedge) at a location where there is dispersion (a correlation between transverse position and energy). This method results in a corresponding increase of transverse phase space and represents in an exchange of longitudinal and transverse emittances. With transverse cooling, this allows cooling in all dimensions. The cooling effect on the emittance is balanced against stochastic multiple scattering and Landau straggling, leading to an equilibrium emittance.

Theoretical studies have shown that, assuming realistic parameters for the cooling hardware, ionization cooling can be expected to reduce the phase space volume occupied by the initial muon beam by a factor of 10^5–10^6. A complete

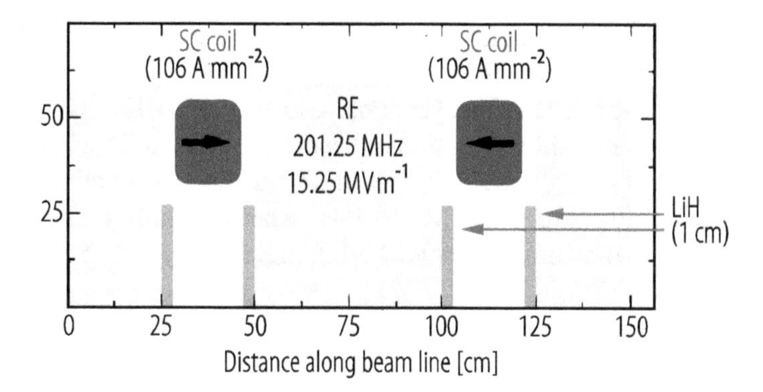

Fig. 12.4 Cooling-channel section. Muons lose energy in lithium hydride (LiH) absorbers (*blue*) that is replaced when the muons are reaccelerated in the longitudinal direction in radio frequency (RF) cavities (*green*). The few-Tesla superconducting (SC) solenoids (*red*) confine the beam within the channel and radially focus the beam at the absorbers. Some representative component parameters are also shown (from [101])

cooling channel would consist of 20–30 cooling stages, each stage yielding about a factor of 2 in 6D phase space reduction—see Fig. 12.4.

Such a cooling method seems relatively straightforward in principle, but has proven quite challenging to implement in practice. One of the main issues is breakdown suppression and attainment of high accelerating gradients in normal-conducting RF cavities immersed in strong magnetic fields. The International Muon Ionization Cooling Experiment (MICE [109]) at RAL (UK) was set to test an ionization cooling channel cell consisting of a sequence of LiH absorbers and 201 MHz RF cavities within a lattice of solenoids that provide the required focusing in a 200 MeV muon beam [110]. The initial results indicate anticipated significant emittance reduction $O(10\%)$ in the "no re-acceleration" configuration [111] and, therefore, can be considered as the first experimental proof of the ionization cooling concept.

12.2.3 Technology Development

Multi-MW target R&D has greatly advanced in recent years, and has culminated in the Mercury Intense Target experiment (MERIT [112]) which has successfully demonstrated a Hg-jet injected into a 15 T solenoid and hit by an intense proton beam from the CERN PS. A high-Z target is chosen to maximize $\pi^{\pm}$ production. The solenoid radially confines essentially all the $\pi^{\pm}$ coming from the target. The Hg-jet choice avoids the shock and radiation damage related target-lifetime issues that arise in a solid target. The jet was viewed by high speed cameras which enabled measurement of the jet dynamics. MERIT results suggest this technology could support beam powers in excess of 4 MW. More advanced solutions for multi-MW targets are under considerations, too, such as granular waterfall targets [113].

Significant efforts are presently focused on high gradient normal conducting RF cavities operating in multi-Tesla magnetic fields as required in the bunching, phase rotation, and cooling channel designs. Closed 805 MHz RF cells with thin Be windows have initially shown significant reduction of maximum RF gradient in a 3 T field—12 MV/m vs. 17 MV/m specified. Further R&D as part of the U.S. based Muon Accelerator Program (MAP) has experimentally demonstrated some 50 MV/m gradients in the RF cavities with high pressure hydrogen gas [114] and in the Be-coated vacuum cavities [115].

Several self-consistent concepts based on different technologies have recently emerged for the MC six-dimensional cooling channel which plays a central role in reaching high luminosity. To achieve the desired mixing of transverse and longitudinal degrees of freedom, the muons must either pass through a series of wedge absorbers in a ring [116] or be put onto a helical trajectory, e.g., as in a "Helical Cooling Channel" [117] or a "FOFO-snake" [118]. The design simulations of the channels are not yet complete and the main challenges are attainment of sufficiently large dynamic apertures, taking into account realistic magnetic fields, RF cavities and absorbers, optimization of the B-fields in RF cavities and technological complexity. The design of the final cooling stages is particularly challenging as it requires very high solenoid fields (up to ~30 T have been considered [119]). The final MC luminosity is proportional to this field. High-field superconducting magnets for the collider ring and for the cooling have been actively developed [120], including feasibility studies of a high temperature superconductor (HTS) option for the 25–50 T final cooling solenoids [121].

A Recirculating Linac with SC RF cavities (e.g. 1.3 GHz ILC-like cavities) is a very attractive option for acceleration of muons from the low energies emerging from the cooling sections to the energy of the experiments. The recirculating linac offers small lengths and low wall plug power consumption but requires small beam emittances.

Recently, realistic collider ring beam optics has been designed which boasts a very good dynamic aperture for about $dP/P = \pm 0.5\%$ and small momentum compaction [122, 123]. The distortions due to the beam-beam interaction will need to be studied as well as practical issues of the machine-detector interface.

Representative performance parameters for a multi-TeV Muon Collider are given in Table 12.2. These parameters are based on the design concepts described above and represent reasonable extrapolations of technologies currently under development. The luminosities displayed are appropriate for the physics research programs that would be undertaken at such a facility [124–127].

12.2.4 Advanced Muon Collider Concepts

In the last few years several advanced muon collider concepts were proposed. An alternative low emittance muon source based on near-threshold production of muons in the reaction $e^+e^- \rightarrow \mu^+\mu^-$ was considered in [128]. The scheme relies on availability of high intensity beam of 45 GeV positrons hitting solid, liquid or crystal target of Be, C or diamond. The resulting emittance of the muon beam is very small and allows direct acceleration with extensive ionization cooling. Synchrotron radiation of high-energy muons channelling in between crystal planes results in very small emittances, too, and opens opportunities for crystal-based muon colliders. Given natural advantages of muons, such as absence of nuclear interaction characteristic of protons and greatly reduced synchrotron radiation compared to electrons, the muons are particle of choice for ultra-high gradient acceleration in crystals, originally proposed in [129]. Such colliders with gradients $O(0.1–1\ \mathrm{TeV/m})$ can potentially reach c.o.m energies hundreds of times higher than in the LHC collisions, though, by necessity, with lower luminosities due to practical limits on the facility total electrical power consumption $O(100\ \mathrm{MW})$ [130]. Of course, significant R&D is needed to demonstrate feasibility of the channelling acceleration in crystals or, as a first step, in carbon nanotubes [131].

References

1. Tajima, T., Dawson, J.M.: Phys. Rev. Lett. 43 (1979) 267.
2. Chen, P., et al.: Phys. Rev. Lett. 54 (1985) 693.
3. O'Connell, C.L., et al.: Phys. Rev. ST Accel. Beams 9 (2006) 101301.
4. Caldwell, A., et al.: Nature Phys. 5 (2009) 363.
5. Ruth, R., et al.: Particle Accelerators 17 (1985) 171.
6. Tsung, F., Narang, R., Mori, W.B., Joshi, C., Fonseca, R.A., Silva, L.O.: Phys. Rev. Lett. 93 (2004) 185002.
7. Mangles, S.P.D., et al.: IEEE Trans. Plasma Sci. 36 (2008) 1715.
8. Geddes, C.G.R., Nakamura, K., Plateau, G.R., Toth, Cs., Cormier-Michel, E., Esarey, E., Schroeder, C.B., Cary, J.R., Leemans, W.P.: Phys. Rev. Lett. 100 (2008) 215004.
9. Suk, H., Barov, N., Rosenzweig, J.B., Esarey, E.: Phys. Rev. Lett. 86 (2001) 1011.
10. Esarey, E., Hubbard, R.F., Leemans, W.P., Ting, A., Sprangle, P.: Phys. Rev. Lett. 76 (1997) 2682.
11. Faure, J., et al.: Nature 444 (2006) 737.
12. Pak, A., Marsh, K.A., Martins, S.F., Lu, W., Mori, W.B., Joshi, C.: Phys. Rev. Lett. 104 (2010) 025003.
13. Oz, E., et al.: Phys. Rev. Lett. 98 (2007) 084801.
14. Clayton, C.E., et al.: Phys. Rev. Lett. 70 (1993) 37.
15. Katsouleas, T., Wilks, S., Chen, P., Dawson, J.M., Su, J.J., et al.: Part. Accel. 22 (1987).
16. Tzoufraz, M., et al.: Phys. Plasmas 16 (2009) 056705.
17. Guo-Zheng Sun, Ott, E., Lee, Y.C., Guzdar, P.: Phys. Fluids 30 (1987) 526.
18. Wilks, S.C., Dawson, J.M., Mori, W.B.: Phys. Rev. Lett. 61 (1988) 337.
19. Esarey, E., Ting, A., Sprangle, P.: Phys. Rev. A 42 (1990) 3526.
20. Faure, J., et al.: Phys. Rev. Lett. 95 (2005) 205003.
21. Butler, A., Spence, D.J., Hooker, S.M.: Phys. Rev. Lett. 89 (2002) 185003.
22. Lu, W., Huang, C., Zhou, M., Mori, W.B., Katsouleas, T.: Phys. Rev. Lett. 96 (2006) 165002.
23. Ralph, J.E., Marsh, K.A., Pak, A.E., Lu, W., Clayton, C.E., Fang, F., Mori, W.B., Joshi, C.: Phys. Rev. Lett. 102 (2009) 175003.
24. Leemans, W.P., et al.: Nature Phys. 2 (2006) 696.
25. Ralph, J.E., et al.: Phys. Plasmas 17 (2010) 056709.
26. Kneip, S., et al.: Phys. Rev. Lett. 103 (2009) 035002.

27. Clayton, C.E., et al.: Phys. Rev. Lett. 88 (2002) 154801.
28. Muggli, P., et al.: Phys. Rev. Lett. 93 (2004) 014802.
29. Esarey, E., et al.: IEEE Trans. Plasma Sci. 24 (1996) 252.
30. Zhou, M.: UCLA Ph.D. Thesis (2008).
31. Huang, C., et al.: Phys. Rev. Lett. 99 (2007) 255001.
32. Rosenweig, J.B., et al.: Phys. Rev. Lett. 95 (2005) 195002.
33. Wang, S., et al.: Phys. Rev. Lett. 88 (2002) 135004.
34. Johnson, D., et al.: Phys. Rev. Lett. 97 (2006) 175003.
35. Hogan, M., et al.: Phys. Rev. Lett. 95 (2005) 054802.
36. Chen, P., et al.: Phys. Rev. Lett. 56 (1986) 1252.
37. Bane, K., et al.: IEEE Trans. Nucl. Sci. NS-32 (1985) 3524.
38. Muggli, P., et al.: Phys. Rev. ST Accel. Beams 13 (2010) 052803.
39. Lu, W., et al.: Phys. Plasmas 12 (2005) 63101.
40. Lu, W., et al.: Phys. Rev. Lett. 96 (2006) 165002.
41. Lu, W., et al.: Phys. Plasmas 13 (2006) 56709.
42. M. Tzoufras et al.: Phys. Rev. Lett. 101(2008) 145002.
43. Blue, B., et al.: Phys. Rev. Lett. 90 (2003) 214801.
44. Blue, B., et al.: Laser Part. Beams 21 (2003) 497.
45. Lee, S., et al.: Phys. Rev. E 64 (2001) 04550.
46. Lotov, K.V., et al.: Phys. Plasmas 14 (2007) 023101.
47. Muggli, P., et al.: Phys. Rev. Lett. 101 (2008) 055001.
48. Wang, X., et al.: Phys. Rev. Lett. 101 (2008) 124801.
49. G. Xia et al.: Proceedings IPAC2010, Kyoto, Japan, June 2010, p. 4395
50. Kumar, N., Pukhov, A., Lotov, K.: Phys. Rev. Lett. 104 (2010) 255003
51. P. Muggli et al., Plasma Physics and Controlled Fusion, 60(1) 014046 (2017).
52. N. Savard et al.: in Proc. North American Particle Accelerator Conf. (NAPAC'16), Chicago, IL, USA, Oct. 2016, paper WEPOA01, pp. 684, 2017, M. Moreira, Phys. Rev. Accel. Beams 22, 031301 (2019)
53. AWAKE Collaboration: *Plasma Phys. Control. Fusion* **56** (2014) 084013
54. A. Caldwell et al.: Physics of Plasmas 18 (2011) 103101
55. Joshi, C., et al.: Nature 311 (1994) 525.
56. Amiranoff, F., Bernard, D., Cros, B., Jacquet, F., Matthieussent, G., Mine, P., Mora, P., Morillo, J., Moulin, F., Specka, A.E., Stenz, C.: Phys. Rev. Lett. 74 (1995) 5220.
57. Darrow, C., et al.: Phys. Rev. Lett. 56 (1986) 2629.
58. Everett, M., et al.: Nature 368 (1994) 527.
59. Tochitsky, S., et al.: Phys. Rev. Lett. 92 (2004) 095004.
60. Joshi, C., et al.: Phys. Rev. Lett. 47 (1981) 1285.
61. Mori, W.B., Decker, C.D., Hinkel, D.E., Katsouleas, T.: Phys. Rev. Lett. 72 (1994) 1482.
62. Coverdale, C.A., Darrow, C.B., Decker, C.D., Mori, W.B., Tzeng, K.-C., Marsh, K.A., Clayton, C.E., Joshi, C.: Phys. Rev. Lett. 74 (1995) 4659.
63. Modena, A., Najmudin, Z., Dangor, A.E., Clayton, C.E., Marsh, K.A., Joshi, C., Malka, V., Darrow, C.B., Danson, C., Neely, D., Walsh, F.N.: Nature 377 (1995) 606.
64. Faure, J., Glinec, Y., Pukhov, A., Kisetev, S., Gordienko, S., Lefebvre, E., Rousseau, J.-P., Burgy, F., Malka, V.: Nature 431 (2004) 541.
65. Geddes, C.G.R., Toths, C., van Tilborg, J., Esarey, E., Schroeder, C.B., Bruhwiler, D., Nieter, C., Cary, J., Leemans, W.P.: Nature 431 (2004) 538.
66. Mangles, S.P.D., Murphy, C.D., Najmudin, Z., Thomas, A.G.R., Collier, J.L., Dangor, A.E., Divall, E.J., Foster, P.S., Gallacher, J.G., Hooker, C.J., Jaroszynsk, D.A., Langley, A.J., Mori, W.B., Norreys, P.A., Tsung, F.S., Viskup, R., Walton, B.R., Krushelnick, K.: Nature 431 (2004) 535.
67. Tsung, F.S., Lu, W., Tzoufras, M., Mori, W.B., Joshi, C., Vieira, J.M., Silva, L.O., Fonseca, R.A.: Phys. Plasma 13 (2006) 56708.
68. Pukhov, A., Meyer-Ter-Vehn, J.: Appl. Phys. B 74 (2002) 355.
69. Gordienko, S., Pukhov, A.: Phys. Plasmas 12 (2005) 043109.

70. Lu, W., Tzoufras, M., Joshi, C., Tsung, F.S., Mori, W.B., Vieira, J., Fonseca, R.A., Silva, L.O. Phys. Rev. ST Accel. Beams 10 (2007) 061301.
71. Martins, S.F., Fonseca, R.A., Lu, W., Mori, W.B., Silva, L.O.: Nature Phys. 6 (2010) 311.
72. Froula, D.H., Clayton, C.E., Döppner, T., Marsh, K.A., Barty, C.P.J., Divol, L., Fonseca, R.A., Glenzer, S.H., Joshi, C., Lu, W., Martins, S.F., Michel, P., Mori, W.B., Palastro, J.P., Pollock, B.B., Pak, A., Ralph, J.E., Ross, J.S., Siders, C.W., Silva, L.O., Wang, T.: Phys. Rev. Lett. 103 (2009) 215006.
73. Osterhoff, J., Popp, A., Major, Zs., Marx, B., Rowlands-Rees, T.P., Fuchs, M., Geissler, M., Hörlein, R., Hidding, B., Becker, S., Peralta, E.A., Schramm, U., Grüner, F., Habs, D., Krausz, F., Hooker, S. M., Karsch, S.: Phys. Rev. Lett. 101 (2008) 085002.
74. Perry, M.D., Mourou, G.: Science 264 (1994) 917.
75. Umstadter, D., Chen, S.-Y., Maksimchuk, A., Mourou, G., Wagner, R.: Science 273 (1996) 472.
76. Clayton, C.E., Ralph, J.E., Albert, F., Fonseca, R.A., Glenzer, S.H., Joshi, C., Lu, W., Marsh, K.A., Martins, S.F., Mori, W.B., Pak, A., Tsung, F.S., Pollock, B.B., Rosse, J.S., Silva, L.O., Froula, D.H.: Phys. Rev. Lett. 105 (2010) 105003.
77. Rousse, A., et al.: Phys. Rev. Lett. 93 (2004) 135005.
78. Fuchs, M., et al.: Nature Phys. 5 (2009) 826.
79. Schenvoigz, H.P., et al.: Nature Phys. 4 (2008) 130.
80. Leemans, W.P., Esarey, E. Physics Today 62 (2009) 44.
81. Rosenzweig, J.B., et al.: Phys. Rev. Lett. 61 (1988) 98.
82. Barov, N., et al.: Phys. Rev. Lett. 80 (1998) 81.
83. Nakanishi, H., et al.: Phys. Rev. Lett. 66 (1991) 1870.
84. Hogan, M., et al.: Phys. Plasmas 7 (2000) 2241.
85. Joshi, C., et al.: Phys. Plasmas 14 (2007) 055501.
86. Muggli, P., et al.: IEEE Trans. Plasma Sci. 27 (1999) 791.
87. Blumenfeld, I., et al.: Nature 445 (2007) 741.
88. Hogan, M., et al.: New J. Phys. 12 (2010) 055030.
89. Muggli, P., et al.: New J. Phys. 12 (2010) 045022.
90. E. Gschwendtner et al.: *Nucl. Instr. and Meth. in Phys. Res. A***829** (2016) 76.
91. A. Caldwell et al.: *Nucl. Instr. and Meth. in Phys. Res. A***829** (2016) 3.
92. E. Öz et al.: *Nucl. Instr. and Meth. in Phys. Res. A***740** (2014) 197, E. Öz et al.: *Nucl. Instr. and Meth. in Phys. Res. A***829** (2016) 321.
93. M. Turner et al.: submitted to Nucl. Inst. Meth. Phys. Res. A, (2017), M. Turner et al.: *Nucl. Instr. and Meth. in Phys. Res. A***829** (2016) 314, M. Turner et al.: *Nucl. Instr. and Meth. in Phys. Res. A***854** (2017) 100.
94. K. Rieger et al.: *Review of Scientific Instruments***88** (2017) 025110.
95. M. Martyanov et al.: in preparation, F. Braunmueller et al., Nucl. Instr. and Meth. in Phys. Res. A, 909, 76 (2018).
96. Y. Fang et al.: *Phys. Rev. Lett.***112** (2014) 045001.
97. V. K. Berglyd Olsen et al.: accepted for publication in Phys. Rev. Accelerators and Beams (2017).
98. A. Caldwell et al.: *Eur. Phys. J. C***76** (2016) 463.
99. A. Martinez de la Ossa et al.: AIP Conference Proceedings 1507 (2012) 588, O.Lishilin et al.: *Nucl. Instr. and Meth. in Phys. Res. A***829** (2016) 37.
100. G. Budker: Proc. 7th Intern. Conf. High Energy Accel., Yerevan, (1969) 33.
101. S. Geer: Annu. Rev. Nucl. Part. Sci. 59 (2009) 347–365.
102. S.D. Holmes, in: Proc. 2010 Intern. Part. Accel. Conf., Kyoto, Japan, (2010) 1299.
103. V. Shiltsev: Mod. Phys. Lett. A 25 (2010) 567-577.
104. S. Geer: Phys. Rev. D 57 (1998) 6989.
105. *The Neutrino Factory Intern. Scoping Study Accelerator Working Group Report*, J. Instrum. 4 (2009)P07001.
106. Yu. Ado, V. Balbekov: Atomnaya Energiya, 31 (1971) 40; transl. in: Sov. Atomic Energy 31 (1971) 731.

107. A. Skrinsky, V. Parkhomchuk: Sov. J. Nucl. Phys. 12 (1981) 3.
108. D. Neuffer: Part. Accel. 14 (1983) 75.
109. R. Sandstrom (MICE Collab.): AIP Conf. Proc. 981 (2008) 107.
110. M. Bogomilov et al. (The MICE collaboration), Phys. Rev. Accel. Beams 20 (2017) 063501
111. Rogers, C. T. et al, in Proc. 2017 IPAC (Copenhagen, Denmark), (2017) 2874.
112. H. Kirk, et al. (MERIT Collab.), in: Proc. 2007 IEEE Part. Accel. Conf. (Albuquerque, NM, USA), (2007) 646.
113. H.J. Cai, et al.: Phys. Rev. Accel. Beams 20 (2017) 023401.
114. M. Chung, et al, Phys. Rev. Lett. 111 (2013) 184802.
115. D. Bowring, et al, in: Proc. 2016 IPAC (Busan, Korea), (2016) 444.
116. R. Palmer, et al: Phys. Rev. ST Accel. Beams 8 (2005) 061003.
117. Ya. Derbenev, R. Johnson: Phys. Rev. ST Accel. Beams 8 (2005) 041002.
118. Y. Alexahin, AIP Conference Proceedings 1222, no.1 (2010) 313.
119. D. Neuffer, et al, JINST 12 T07003 (2017)
120. G. Apollinari, S. Prestemon, A. Zlobin, Annu. Rev. Nucl. Part. Sci. 2015.65:355-377
121. V. Kashikhin, et al, IEEE Trans. Appl. Superconductivity, 18, no. 2 (2008) 938
122. Y. Alexahin, E. Gianfelice-Wendt, in: Proc. 2009 IEEE Part. Accel. Conf. (Vancouver, Canada), (2009) 3817.
123. M.H. Wang, et al, JINST 11 P09003 (2016).
124. E. Eichten, A. Martin, Physics Letters B 728 (2014) 125
125. R. Brock, et al, arXiv:1401.6081
126. A. Conway, et al, arXiv:1405.5910.
127. N. Chakrabarty, et al, Physical Review D, 91(1), 015008 (2015).
128. M. Antonelli, et al, NIM-A 807 (2016) 101.
129. T. Tajima, M. Cavenago Phys. Rev. Lett. 59 (1987) 1440.
130. V. Shiltsev, Physics-Uspekhi 55.10 (2012) 965.
131. X. Zhang, et al, Phys. Rev. Accel. Beams 19, 101004 (2016)

13

The Properties and Mechanisms of Cosmic Particle Accelerators

W. Hofmann and J. A. Hinton

13.1 Introduction

In the century since the measurements of Victor Hess [1]—considered as the discovery of cosmic rays—the properties of cosmic rays, as they arrive on Earth, have been studied in remarkable detail; we know their energy spectrum, extending to 10^{20} eV, their elemental composition, their angular distribution, and we understand the basic energetic requirements of cosmic ray production in the Galaxy. The energy density of cosmic rays in the Galaxy is known to be comparable to the energy density in Galactic magnetic fields and in the thermal energy of interstellar gas, hence cosmic rays play a non-negligible role in shaping the evolution of galaxies. Charged cosmic-ray particles are strongly deflected in the few μG interstellar magnetic fields—the radius of curvature of a $Z = 1$ particle in astronomical distance units of parsec is given by $R_{\mathrm{gyro,pc}} \sim E_{\mathrm{PeV}}/B_{\mu\mathrm{G}}$ (1 parsec (pc) = 3.1×10^{16} m = 3.26 light years). Therefore, except for energies in the 10^{20} eV range, cosmic rays cannot be traced back to their sources, the cosmic particle accelerators. Much of the current effort goes into identifying and quantitatively describing cosmic accelerators, with supernova remnant shocks as the likely dominant source of Galactic cosmic rays. Properties of cosmic particle accelerators can be studied and inferred in two ways: on the one hand based on the characteristics of local cosmic rays, as they arrive—more or less isotropically—on Earth; this is subject of Sect. 13.2, with a discussion of acceleration mechanisms given in Sect. 13.3. On the other hand, cosmic rays propagating through the Galaxy and in particular cosmic particle accelerators can be imaged using neutral particles and radiation created during the acceleration process and during propagation through the interstellar medium (ISM),

W. Hofmann (✉) · J. A. Hinton (✉)
Max-Planck-Institut für Kernphysik, Heidelberg, Germany
e-mail: wh@mpi-hd.mpg.de; Jim.Hinton@mpi-hd.mpg.de

with synchrotron radiation in the radio and X-ray regimes, high-energy gamma rays, high-energy neutrinos, and neutrons. So far, most information is obtained from electromagnetic probes, as discussed in Sect. 13.4. The small interaction cross section makes detection of high-energy cosmic neutrino sources challenging; only recently, the first detection was reported. Neutron decay limits their range. The flux on Earth of various high-energy cosmic messengers used to explore cosmic particle accelerators is illustrated in Fig. 13.1. The following discussion will mostly concentrate on Galactic particle populations and those particle accelerators which plausibly contribute to cosmic rays observed on Earth. For further details, we refer to reviews such as [2] on cosmic rays in the Galaxy, [3, 4] on supernovae as Galactic cosmic ray sources, [5] on cosmic rays above the knee, [6, 7] on ultra-high-energy cosmic rays, and [8, 9] on very high energy gamma ray astronomy to explore cosmic particle accelerators.

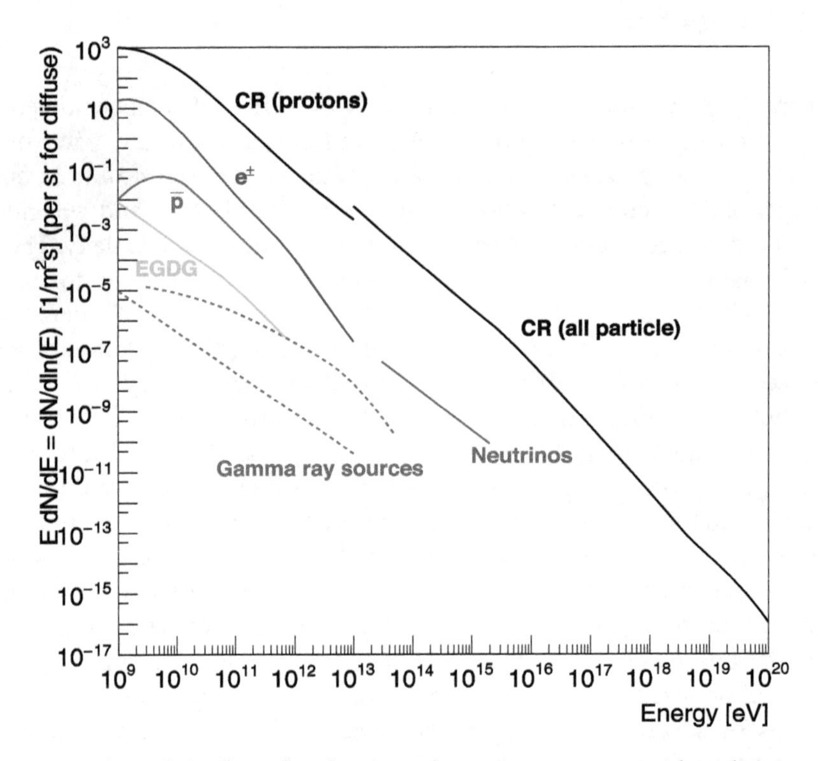

Fig. 13.1 Illustration of the flux of various cosmic messengers expressed as the rate per logarithmic energy interval $EdN/dE = dN/d\ln(E)$. For diffuse and close to isotropic fluxes, a solid angle of 1 sr is used. Indicated are the diffuse all-particle cosmic-ray flux resp. the proton flux (black), the antiproton flux (red), the flux of cosmic $e^{\pm}$ (blue), the flux of extragalactic diffuse gamma rays (EGDG, green), and the high-energy neutrino flux (magenta). Also shown are the gamma-ray fluxes from a localized strong source (RX J1713.7−3946) and from a faint source (NGC 253) (red dashed). Event numbers per spectral interval $\Delta\ln(E) = 1$ are obtained by multiplying the flux with the exposure (effective detection area × time × solid angle of the detector (for diffuse fluxes))

13.2 Cosmic Ray Properties and Implications for Cosmic Ray Sources

Certain properties of cosmic particle accelerators can be inferred from the cosmic ray spectrum measured on Earth, from the elemental composition of cosmic rays, from any anisotropies in their arrival directions, and also from the yield of electrons and antiparticles—positrons and antiprotons–among cosmic rays. The interpretation of data, however, assumes that cosmic rays measured on Earth are reasonably representative for cosmic rays in the Galaxy, and that one is able to disentangle effects reflecting properties of cosmic ray sources from those arising from cosmic ray propagation from their sources the Earth, over kpc distances.[1]

13.2.1 Cosmic Ray Spectrum

At lower energies, up to at most to 10^{14} eV, cosmic ray properties have been studied by space-based instruments, or instruments carried by giant balloons into the upper atmosphere, equipped with magnetic spectrometers or calorimeters for momentum/energy determination, with time-of-flight systems, transition radiation detectors and/or Cherenkov detectors for velocity determination, and with ionisation measurements for determination of the charge state. Examples of such instruments, illustrating the different techniques employed, include PAMELA [10], AMS [11], TRACER [12], CALET [13] and DAMPE [14]. While direct detection in space frequently provides superior performance, in particular regarding the measurement of elemental and isotopic composition, the steeply-falling energy spectrum of cosmic rays, coupled with at most m^2-scale detection areas and—in case of magnetic spectrometers—limited bending power, restricts the energy range of these detection systems. Starting at 10^{12} to 10^{13} eV, ground-based instruments take over, using the Earth's atmosphere as an absorbing medium and detecting the particle cascade created when a high-energy particle interacts in the atmosphere. Instruments detect either the shower particles reaching the ground—with an effective detection area determined by the size of the detector array, which can be as large as 3000 km^2 for the Pierre Auger Observatory [15]—or image the cascade by focusing onto a photosensor array the light emitted by shower particles in the atmosphere. The forward-beamed Cherenkov light illuminates areas of $\sim 10^5$ m^2 on the ground and is readily detectable for showers beyond 10^{11} eV; the isotropically emitted air fluorescence light—imaged e.g. by the Auger fluorescence telescopes [16] and the Telescope Array instrument [17]—allows detection over multi-km distances and provides detection areas in excess of 10^7 m^2, but only in the energy range above

[1] 1 kpc = 1000 pc $\approx$ 3300 light years; the distance from the Sun to the centre of the Galaxy is about 8 kpc.

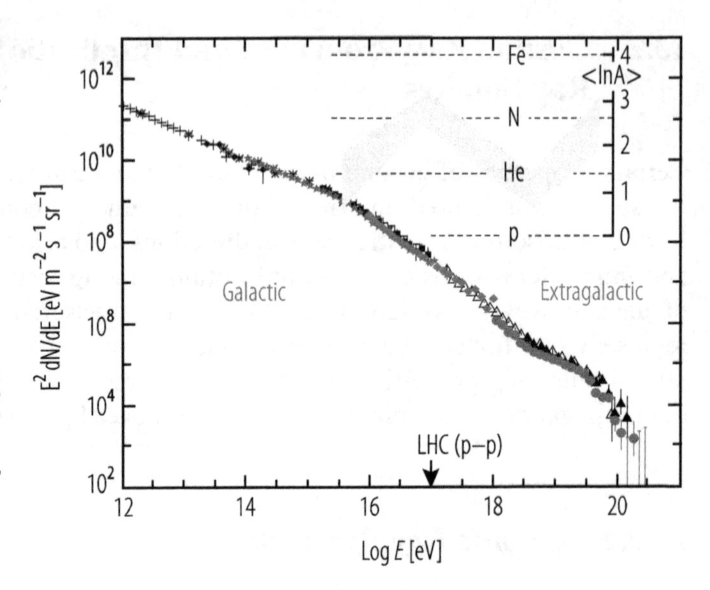

Fig. 13.2 Cosmic ray energy spectrum above 10^{12} eV, presented as a spectral energy distribution $E^2\,dN/dE = E\,dN/d\log(E)$ representing the energy contained per logarithmic energy interval (from [6]). The inset shows the variation of elemental composition—$\langle \ln A \rangle$—with energy, from [31]. While the absolute value of $\langle \ln A \rangle$ is depends on the model used to interpret the air shower data, the pattern of change—from 'light composition' to 'heavy' then back to 'light' and again 'heavy' persists

10^{17} to 10^{18} eV. Identification of primary particles is achieved via the longitudinal and transverse development of the air shower, the radial distribution of shower particles on the ground, and the ratio of nucleons, electrons and muons among shower particles. Compared to detection in space, cosmic-ray identification power is dramatically reduced, allowing essentially to distinguish light from heavy elements, and nuclei from electrons. Examples of air-shower arrays include KASKADE [18], optimised for measurements of cosmic ray composition, the Tibet array [19], optimised for low energy threshold, and the large LHAASO array under construction in China [20].

The cosmic ray energy spectrum shown in Fig. 13.2, compiled from numerous measurements using different techniques, is a mostly featureless power-law spectrum, $dN/dE \sim E^{-\Gamma}$, ranging from GeV energies to 10^{20} eV, covering over 30 orders of magnitude in flux, with the power-law index Γ varying at the "knee" from about 2.7 below 10^{15} eV to about 3 for energies from 10^{16} eV to 10^{18} eV, to flatten again at the "ankle" between 10^{18} eV to 10^{19} eV, and cutting off around 10^{20} eV. Below about 10^{10} eV, the spectrum is modulated by the solar wind; at these energies particles do not penetrate into the inner Solar system. Newer measurements indicate another spectral break in the 10^{11} eV to 10^{12} eV range, with a change in proton and Helium spectral indices [21–23]. The power-law shape, with its lack of features and characteristic energies in the spectrum, indicates that cosmic rays are of non-thermal origin. The local energy density in cosmic rays is dominated by lower-energy particles and amounts to about 1 eV/cm^3. Different mechanisms are discussed for the origin of the slight changes in spectral index: a harding of the spectrum can indicate the emergence of a new component/a new source of cosmic rays, a steepening the peak energy of accelerators, in particular if accompanied by a shift towards heavier composition, since the peak energy of accelerators will usually scale with the charge Z of the accelerated nuclei. Changes in spectral slope can, however, also indicate a change of cosmic ray propagation, i.e. of the energy-dependent diffusion coefficient. The spectral hardening observed at the ankle by 10^{19} eV is most easily explained as the emergence of a new component of cosmic rays, with a harder spectrum. Most models aiming at explaining the overall cosmic ray spectrum assume such a transition from Galactic sources to extragalactic sources in the 10^{17} to 10^{19} eV range (e.g. [24]). The steepening of spectra at the "knee" could be caused by a cutoff in the acceleration mechanism, characterising the upper end of the energy range of Galactic cosmic particle accelerators. A cutoff in the cosmic ray spectrum around 10^{20} eV—the Greisen-Zatsepin-Kuzmin (GZK) cutoff [25, 26]—has been predicted as a result of pion production by ultra-high-energy protons interacting with the cosmic microwave background radiation, limiting the range of protons beyond 10^{20} eV to about 100 Mpc or less (e.g. [27]). While early results questioned the existence of the GZK cutoff, both Auger [28, 29] and Telescope Array (TA) [17] have with much improved statistics established a cutoff in the spectrum of ultra-high-energy cosmic rays (UHECR). The energy of the cutoff is consistent with the GZK cutoff, indicating that most sources of UHECR are located at distances beyond 100 Mpc, but data are also well reproduced in terms of more nearby sources with a rigidity-dependent cutoff in the acceleration mechanism [30].

The cosmic-ray spectrum measured on Earth reflects the source spectrum (averaged over the lifetime of each source and over the population of sources), modified by factors arising from the energy-dependent propagation from sources throughout the Galaxy to Earth. Cosmic ray composition measurements discussed below have proven particularly useful in disentangling the two contributions.

13.2.2 Cosmic Ray Composition, Cosmic Ray Propagation, and Cosmic Ray Energetics

Lacking directional information, clues regarding the origin of cosmic rays (beyond those from shape of the energy spectrum) have been drawn mainly from the elemental composition (see e.g., reviews [2, 32]). In the GeV energy range, cosmic-ray chemical composition resembles the composition of solar-system material, with hydrogen and helium nuclei as dominant components (see Fig. 13.3). For some elements and isotopes, there are, however, marked differences between cosmic-ray composition and solar-system composition. This concerns e.g. boron (Z=5), which is suppressed by more than five orders of magnitude relative to carbon (Z=6) in the solar system, but is (within a factor of a few) equally represented among cosmic rays. The reason for this is that boron is not produced directly in stellar nucleosynthesis, but can easily be produced by spallation of heavier cosmic-ray nuclei interacting with interstellar gas. From the boron to carbon ratio, it is inferred that GeV cosmic rays arriving at the Earth have traversed about 10 g/cm^2 of interstellar medium (ISM) (e.g. [33]), with the amount of target material traversed

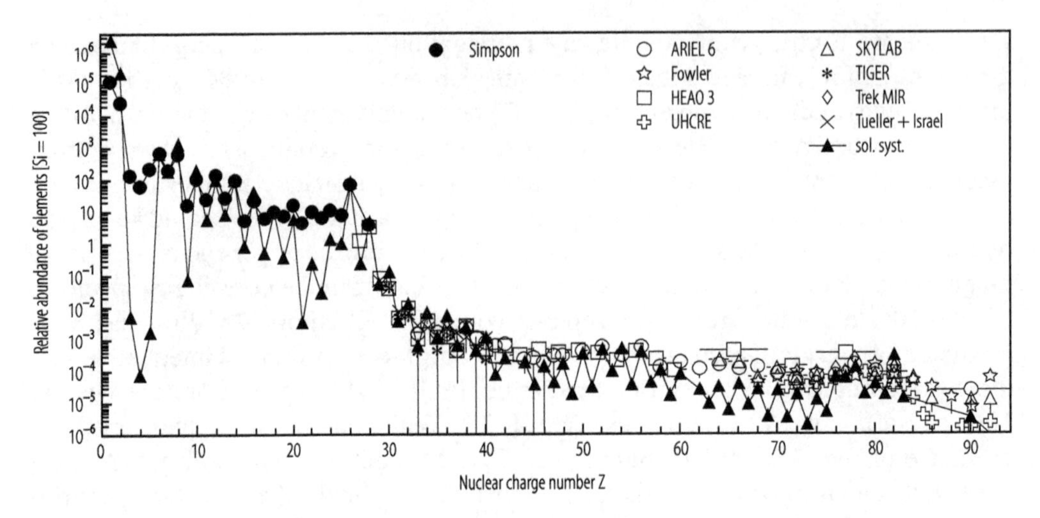

Fig. 13.3 Abundance of elements in cosmic rays as function of their nuclear charge Z at momenta around 1 GeV/c. Solar-system abundances are shown in grey for comparison. From [5]

decreasing roughly as $R^{-0.5}$ for higher rigidities R (defined as momentum/unit charge). Radioactive secondary nuclei, such as ^{10}Be with its half-life of about a million years, are also detected among cosmic rays and can be used to estimate the residence time of cosmic rays in the Galaxy: the observed ^{10}Be/^{9}Be ratio implies a typical age of GeV cosmic rays reaching the Earth in the 10^7 year range, with a trend towards lower ages at higher energy. Both the grammage traversed by cosmic rays and in particular the residence time, which is much larger than the time of about 10^5 years required to cross the entire Galaxy on a straight path, indicate that cosmic rays are trapped inside the Galaxy and propagate diffusively away from their sources, due to frequent deflection in interstellar magnetic fields. However, the combination of the age and the grammage traversed implies that cosmic rays spend a significant fraction of their lifetime outside the Galactic disc with its typical gas density of 1 atom/cm^3. This has lead to the development of "leaky box" models or diffusion models (see Fig. 13.4), where cosmic rays do not escape the Galaxy when leaving the disc with its scale height of $\mathcal{O}(100\text{ pc})$, but continue diffusive motion out to distances of several kpc from the disc, with the possibility of returning into the disc (for overviews see e.g. [2, 34]). Measurements of Galactic magnetic fields, via synchrotron emission or rotation measures (e.g. [35]), indeed indicate that particle-confining fields extend well beyond the height of the Galactic disc. For diffusive propagation, $<r^2> = 2Dt$. The diffusion coefficient D depends on the scale and degree of turbulence of Galactic magnetic fields, both poorly known. Assuming particles escape the system when reaching a halo height h above the disc, a particle will escape after a time of order $T \sim h^2/2D$. With h much larger than the disc scale height b, and a gas density in the halo which is negligible compared to the density ρ in the disc, the grammage traversed is $X \sim cT\rho(b/h) \sim h/D$. From the residence time and the grammage, or by directly comparing element ratios with model calculations, the halo height h and the diffusion coefficient D can hence be inferred;

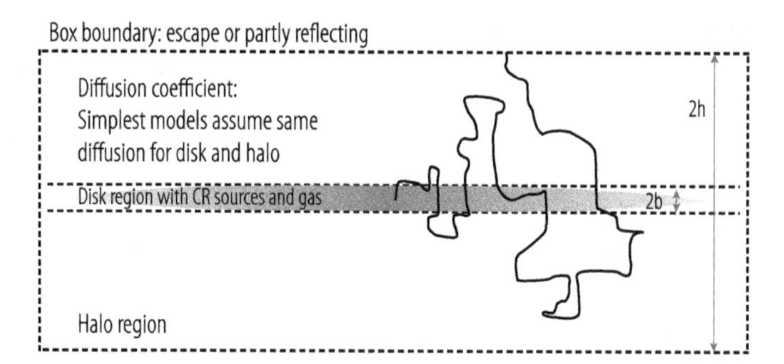

Fig. 13.4 "Leaky Box" cosmic-ray propagation models [34]. Cosmic-ray sources are located in the gaseous disc of the Galaxy, but diffuse propagation continues into the kpc-sized halo. Depending on the specific model, particles are assumed lost when reaching the halo boundary, or are partially reflected back

depending on model details, halo heights of a few kpc, up to 10 kpc, and energy-dependent diffusion coefficients of $D(E)$ of order $10^{28} E^{\alpha}_{GeV}$ cm^2/s are obtained, with α in the range from 0.3 to 0.6; $\alpha = 0.5$ is often used as a representative value. This value of D corresponds to an rms propagation distance of cosmic rays of $d_{pc} \sim 0.3 t_{\text{year}}^{1/2}$ at GeV energies; the diffusive approximation, however, holds only for reasonable large time and spatial scales, corresponding to many gyro radii and distances much larger than the coherence length of magnetic fields. Since the average spectrum of cosmic rays in the Galaxy is given by the source spectrum multiplied by the average residence time $T(E) \sim 1/D(E)$, the average source spectrum has to be harder than the observed spectrum, $\Gamma_{\text{source}} \approx \Gamma_{\text{CR}} + 0.5$. For the interpretation of cosmic rays spectra and composition, initial analytical models (e.g. [34]) have increasingly been replaced by numerical simulations such as GALPROP, allowing the inclusion of specific assumptions regarding the distribution of cosmic ray sources, of material in the Milky Way, and of energy loss and re-acceleration processes, see e.g. [2] for an overview and references. Beyond the effects of cosmic ray spallation products and radioactive decay, differences between solar system composition and cosmic ray composition are observed which seem to depend on the ionization potential or the volatility of elements (e.g. [36]), and which are attributed to the efficiency with which elements are injected into the acceleration process.

Assuming that cosmic rays more or less uniformly permeate the Galaxy, as confirmed by gamma-ray observations (see Sect. 13.4) and that they are not a temporary or local phenomenon, leaky box-models allow constraints to be placed on the energy requirements of Galactic cosmic-rays: to sustain the flux of cosmic rays in the Galaxy and its halo with a volume V of a few 10^{67} cm^3, a typical energy density $\rho \approx 1$ eV/cm^3 and a typical escape time $T \approx 10^7$ year, an energy input of $\rho V/T \approx 10^{41}$ erg/s needs to be provided by Galactic cosmic particle accelerators [37]. Only a few percent of this energy is dissipated in form of ionization and radiative or adiabatic losses; over 95% of the energy leaves the Galaxy into intergalactic space [38].

At higher energies, cosmic-ray composition is determined from ground-based measurements of the electron and muon content of air showers—heavy primaries have a higher fraction of muons—or from the depth of the shower maximum in the atmosphere, which is sensitive to the interaction cross section of the primary, which scales with mass number A roughly as the $A^{2/3}$, see e.g. [31]. As they are not sensitive to individual species, results are often presented in terms of the mean of $\log(A)$ and tend to be rather sensitive to the algorithms used to model the air shower (see e.g. [18, 31]), and inconsistencies between models and data are seen [29, 39]. Nearly all measurements, however, show a change from a dominantly light (H, He) composition up to the knee, to a heavy composition above the knee (see Fig. 13.2, and also Fig. 13.11 below), and while the knee is clearly seen in showers initiated by light elements, no change in the slope of spectra of heavy primaries is evident up to 10^{17} eV. Well beyond the knee, at energies of few 10^{18} eV, studies of the height of shower maximum again suggest a light composition, with a transition to a heavy composition at a few 10^{19} eV [40, 41], although details remain under discussion [42]. A common interpretation is in terms of two components, a Galactic component and an extragalactic component, the latter dominating above 10^{17} to 10^{18} eV, each component with a cutoff in the acceleration mechanism at fixed particle gyro radius (i.e. rigidity), corresponding to a peak energy which scales with the nuclear charge Z. In this scenario typical Galactic accelerators are required to reach an energy in the range of a few $Z \times 10^{15}$ eV (e.g. [43]).

13.2.3 Cosmic Ray Anisotropy

The diffusive propagation of all but the very highest energy cosmic rays, with a gyro-radius much smaller than the scale of the Galaxy, destroys almost all directional information in cosmic rays; nevertheless, cosmic rays will on average flow away from their sources, resulting in small anisotropies.

The arrival directions of cosmic rays are indeed almost uniform on the sky, with (dipole) anisotropies at the 10^{-3} level at energies below 0.1 PeV, an indication of a minimum in anisotropy in the 0.1–1 PeV range, and an increase up to 10^{-2} at higher energies (see e.g. [2, 44]). The phase of the anisotropy—i.e. the direction of maximum intensity—varies with energy. Effects which might cause anisotropies include [34]: (a) a diffusive flow of cosmic rays governed by gradients in cosmic ray density ρ, with a resulting anisotropy of order $3D\nabla\rho/c\rho$; since the cosmic ray density likely decreases with galactocentric radius, an outward flow results at the location of the Solar System. A density gradient and hence flow might also be caused by single nearby cosmic-ray sources, in which case the magnitude and direction cannot be predicted a priori. Given that the diffusion coefficient $D(E)$ increases with energy, density-gradient related anisotropies are expected to increase with energy. (b) Anisotropies of order $(\Gamma + 2)(v/c)$ caused by the motion of the Earth relative to the cosmic-ray rest frame (the Compton-Getting effect [45]). Cosmic-ray energies are also Doppler-shifted, for power-law spectra with index Γ the anisotropy hence depends on the spectral index, but not on the energy. Compton-Getting anisotropies could be caused by the 220 km/s motion of the Sun around the centre of the Galaxy, and/or—an order of magnitude smaller—by the orbital motion of the Earth around the sun. (c) Anisotropies arising from special magnetic field configurations near the Earth or sun, or from modulation by the solar wind. Such anisotropies should decrease with increasing rigidity (i.e. energy) of particles, but large-scale structures such as the heliotail caused by the motion of the sun relative to the local interstellar medium, possible extending beyond 1000s of AU,[2] could influence particles beyond multi-TeV energies. Decomposition of anisotropies into the different components is difficult also because most ground-based detectors measure the variation of rates along right ascension, for a fixed viewing direction, i.e., whereas the declination dependence is usually not measured directly. The observed energy dependence is non-trivial to explain and may include partial compensation between different contributions. In a study of a sample of simulated spiral galaxies and their cosmic-ray sources, most realisations show larger anisotropies than measured, and a uniform increase with energy [46]. The minimal anisotropy in the sub-PeV range can also be interpreted as evidence that the cosmic-ray "gas" co-rotates with the Galaxy [47]—quite plausible given that cosmic rays couple via magnetic fields to the interstellar plasma.

At TeV energies, anisotropies of a few 10^{-4} on smaller angular scales (few $10°$) are observed [48, 49]. The origin of these small-scale anisotropies is not well understood; explanations e.g. assume field lines connecting a cosmic ray source and the solar system [50], but the effect may also simply reflect the local concrete realisation of the turbulent magnetic field within the cosmic ray scattering length [51].

13.2.4 Electrons and Antiparticles Among Cosmic Rays

Electrons and antiparticles among cosmic rays play a special role, since their "natural" yields are quite low and hence their fluxes are most sensitive to contributions from "exotic" sources such as annihilation of Dark Matter particles. Contrary to cosmic ray nuclei, electrons and positrons suffer significant energy losses, mostly from synchrotron radiation, with an energy loss timescale of $\mathcal{O}(10^5\,\text{year}/E_{\text{TeV}})$. Combined with typical diffusion coefficients $D(E)$, this implies that electron and positron sources need to be within a distance of $d_{pc} \approx 500/E_{\text{TeV}}^{1/4}$ from Earth. Unlike cosmic ray nuclei, electrons and positrons therefore act as probes of local sources. While electrons are assumed to be accelerated together with nuclei in cosmic ray sources, antiparticle yields—antiprotons and positrons—were traditionally modeled as arising exclusively from nuclear interactions during cosmic ray propagation, resulting in highly suppressed yields.

[2] AU = Astronomical Unit, the mean Sun-Earth distance of $\approx 1.5 \times 10^{11}$ m.

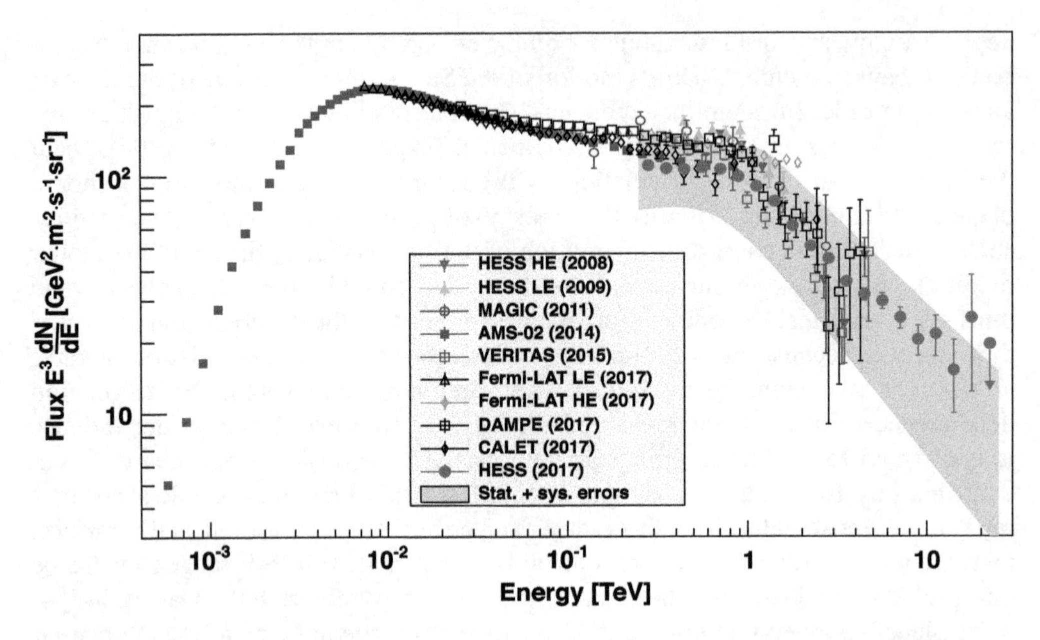

Fig. 13.5 Cosmic-ray electron spectrum, multiplied by E^3, as measured by H.E.S.S., MAGIC, VERITAS, AMS-02, Fermi, DAMPE and CALET ([13, 14, 53–55] and refs. given there)

Antiproton yields have been measured up to energies of 10^{11} eV. The antiproton to proton ratio rises with energy up to about 2×10^{-4} around 10 GeV, and then levels off, in good agreement with expectations for secondary antiprotons (e.g. [52]).

The cosmic-ray electron spectrum is illustrated in Fig. 13.5. The spectrum falls steeper than the spectrum of nuclei, with an index $\Gamma \approx 3$, and steepens further at around 10^{12} eV. The flux of cosmic-ray electrons at 10^{12} eV is about 0.1% of that of cosmic ray nuclei. The spectrum can be modeled by assuming that cosmic-ray sources accelerate electrons and nuclei in a ratio of about 1:100 at a given energy; with increasing energy, and hence shorter electron range, fewer and fewer sources contribute, resulting in both a steeper spectrum and increasing uncertainty in the predicted flux due to the stochastic distribution of sources (see e.g. [57]).

Recent measurements, however, reveal deviations from this picture for the electron (plus positron) flux (see Fig. 13.5), and in particular for the positron/electron ratio (Fig. 13.6). Beyond the range of solar modulation, the electron flux is predicted to decrease slightly faster than E^{-3}; the data suggest an additional component appears in the energy range between about 100 GeV and 1 TeV, before the flux cuts off. A similar, but much more dramatic effect is seen in the electron/positron ratio, which increases beyond 10 GeV, rather than continue to drop as predicted for secondary production of positrons. This feature, first detected by PAMELA [58] and Fermi [59], and studied with high statistics using AMS-02 [56], gives rise to considerable speculation regarding its origin (e.g. [60]). Both the electron yield and the positron/electron ratio can be described by assuming an additional, charge-symmetric source of electrons and positrons, with a (propagation-modified)

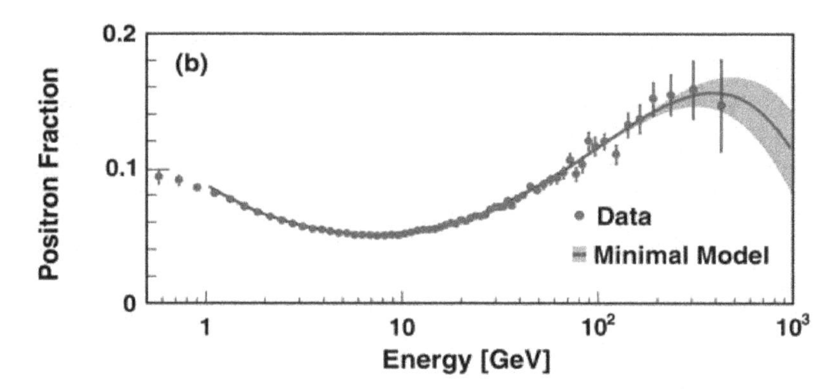

Fig. 13.6 Positron fraction, $N_{e^+}/(N_{e^-} + N_{e^+})$, as a function of energy, as measured by AMS-02 (from [56])

spectrum which rises faster than E^3 up to a few 100 GeV, and cuts off at about one TeV [60]. Dark matter annihilation of particles in the TeV mass range can account for the shape of the spectra, but would require enhanced annihilation rates compared to typical models, and annihilation modes which produce only leptons but no baryons; most conventional dark matter annihilation models predict the positron excess to be associated with an excess in antiprotons, which is not seen. An alternative explanation is provided by electrons escaping from nearby pulsars, or, more specifically, pulsar wind nebulae [61] (see also Sect. 13.4), although results regarding very low diffusion coefficients for cosmic-ray electrons disfavor this interpretation somewhat, making it difficult for electrons from known pulsars to reach the Earth [62].

13.2.5 Astronomy with Ultra High Energy Cosmic Rays

Due to the strong deflection of cosmic rays in Galactic and extragalactic magnetic fields, astronomical imaging of their sources is impossible over most of their energy range. Only at energies of several 10^{19} eV deflections for $Z = 1$ ultra high energy cosmic ray (UHECR) particles are predicted to become small enough—a few degrees—that particles can be traced back to their sources. The arrival directions of the highest-energy cosmic rays are essentially isotropic (see Fig. 13.7), but indications of correlations with astrophysical objects start to emerge. In the Auger data, the most significant over-density of UHECR lies roughly in the direction of Centaurus A, the closest AGN at 3.8 Mpc distance; the statistical significance of this fact is about 3σ after accounting for the number of trials [29]. Arrival directions also correlate with the distributions of starbust galaxies, and of gamma-ray AGN, at the $3-4\sigma$ level [29]. The 7-year Telescope Array data [17] exhibit at energies above $10^{19.2}$ eV a 3.7σ post-trial enhancement around RA = 9 h 16 min, Dec = 45°. The interpretation of these data is not obvious. One possibility is that few sources are

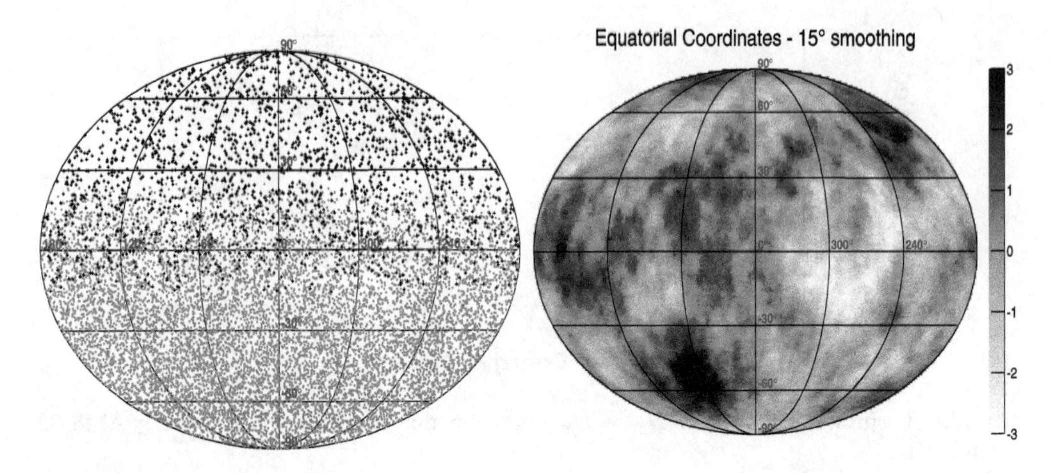

Fig. 13.7 Left: Arrival directions of Auger events (red points in the South hemisphere) and Telescope Array ones (black crosses in the Northern hemisphere) above 10^{19} eV in equatorial coordinates. Right: Excess significance sky map smoothed out at a 15° angular scale. From [63]

responsible for UHECR, emitting a mixture of light and heavy nuclei; the protons among those particles create the detected directional enhancements whereas the heavier nuclei are strongly deflected and are responsible for a uniform background, with similar spectra in both hemispheres. Verifying or disproving such a scenario requires particle-by-particle mass identification and increased statistics, both goals of next-generation UHECR experiments.

13.3 Particle Acceleration Mechanisms and Supernova Shocks as Cosmic Accelerators

A few general conditions can be imposed regarding sources of Galactic cosmic rays. To sustain the flux of cosmic rays in the Galaxy, an energy input of $\approx 10^{41}$ erg/s by Galactic cosmic ray sources is required (see Sect. 13.2), and a spectrum extending at least up to $\approx Z \times 10^{15}$ eV with an average source spectral index in the range $\Gamma \approx 2$ to 2.4. In addition, if cosmic accelerators use regular or turbulent magnetic fields to confine particles, the acceleration region has to have a size at least equal to the gyro radius of particles, $R_{gyro,pc} \sim E_{\mathrm{PeV}}/B_{\mu G}$ [7]. Figure 13.8 illustrates that there are astrophysical objects which fulfil this condition up to energies of 10^{20} eV, but not much beyond.

In Sect. 13.3.1 below we focus on the most established mechanism, diffusive shock acceleration, and in particular the well-studied case of supernova remnants. Note however, that acceleration associated with magnetic reconnection is now increasingly discussed, in particular in the cases of objects with relativistic bulk motions such as in the jets of active galaxies.

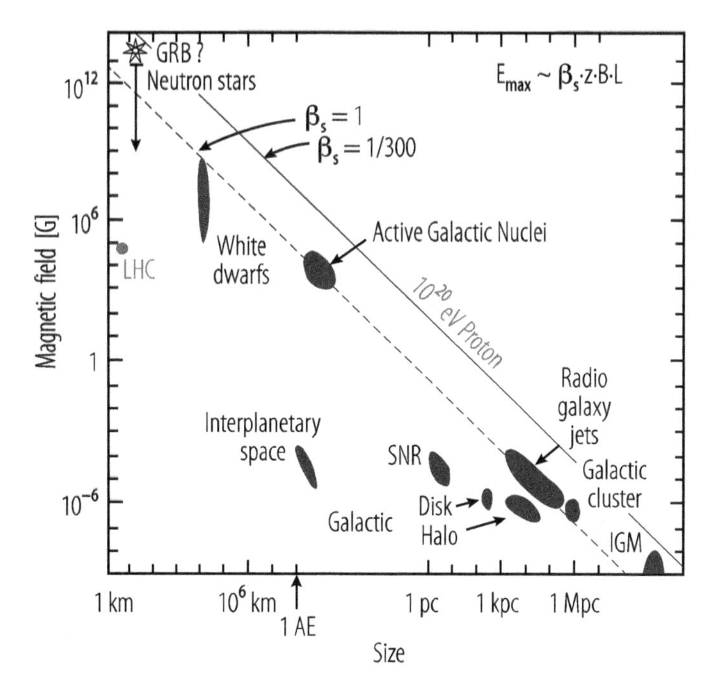

Fig. 13.8 The maximum energy attainable in a cosmic accelerator depends on the product of magnetic field strength B and size L [7]. The condition that the gyro radius of a particle of charge z is contained in the acceleration region yields $E_{max} \sim zBL$. For shock acceleration, see Sect. 13.3.1, one finds that the actually reachable E_{max} is lower by a factor of $\mathcal{O}(\beta_s)$, the shock speed in units of the speed of light. The lines corresponds to a maximum energy of 10^{20} eV, for $z = 1$ and $\beta_s = 1$ (dashed) or $\beta_s = 1/300$ (full). Potential sources range from very compact, high-field objects such as gamma-ray bursts (GRB) or neutron stars via the parsec-size supernova remnants (SNR) to the extended low-field intergalactic medium (IGM). From [5] and [7]

13.3.1 Shock Acceleration in Supernova Remnants

As sources of Galactic cosmic rays, supernova explosions were suggested very early as a suitable source [64], providing both sufficient energy—10^{51} erg kinetic energy per explosion, or 10^{42} erg/s for a supernova rate in the Galaxy of 1/30 year—as well as a plausible acceleration mechanism—first order Fermi acceleration (see e.g. [65, 66] and further references given in [3, 4]), the appropriate spectral index $\Gamma \approx 2$ and a peak energy around 10^{15} eV (see Fig. 13.8).

In a supernova explosion, stellar material of up to a few solar masses is ejected with initial speeds of up to 10^4 km/s or $\beta_{sh} = v_{sh}/c \sim 0.03$, and creates a shock where the ambient interstellar medium is compressed and piled up. While the piled-up ambient material slows down the ejecta, speeds remain supersonic for time scales of order 10^4 years (e.g. [67]). Shocks with high Mach number are characterized by a compression ratio $r = (\gamma + 1)/(\gamma - 1) = 4$, governed by the adiabatic index $\gamma = 5/3$ of the compressed monatomic gas. Charged particles which cross the shock in either direction find themselves in a medium moving with velocity $\approx v_{sh}$ relative

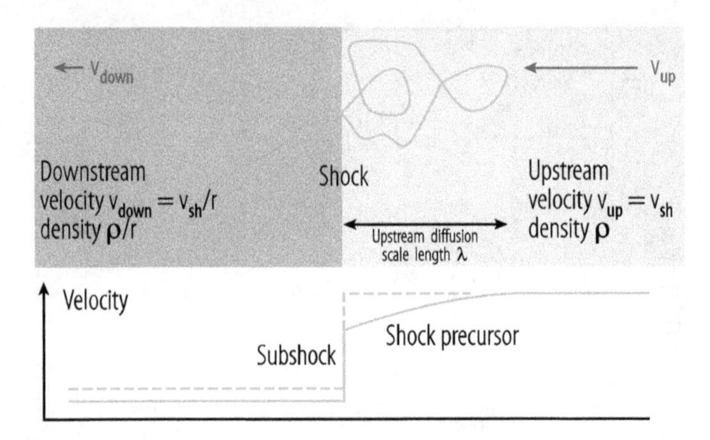

Fig. 13.9 Cartoon of cosmic-ray acceleration at shock fronts, viewed in the rest frame of the shock. In this frame, interstellar medium material streams into the shock with the shock propagation speed $v_{up} = v_{sh}$, is compressed by a factor r and streams out with $v_{down} = v_{sh}/r$. Charged particles can be scattered across the shock front and back. Particles will on average propagate a distance $\lambda \approx D/v_{sh}$ upstream before they are swept back through the shock. The lower part of the diagram illustrates how flow velocity varies across the shock for the case that the energy in accelerated particles is negligible ("test particle case", dashed line) and for the case where the shock is modified by the cosmic-ray pressure, see text for details

to the medium on the other side of the shock (Fig. 13.9). Particles are scattered off anisotropies in the magnetic field and isotropised in the new medium, with an average increase in energy $\Delta E/E$ of $\mathcal{O}(\beta_{sh})$. In their diffusive motion, particles can cross the shock front multiple times, their energy growing as $E \sim (1+k)^n$, where n is the number of crossing cycles and $k = (4/3)(1 - 1/r)(v_{sh}/c)$; here $(1 - 1/r)v_{sh}$ is the difference in flow speed before and after the shock and the factor 4/3 arises from averaging over shock crossing angles. Since, viewed in the rest frame of the shock, material is inflowing ("upstream") with speed v_{sh} and outflowing ("downstream") after shock compression with speed v_{sh}/r, in the long run particles tend to be carried into the downstream region, with a probability of order $4\beta_{sh}/r$ not to return to the shock at each cycle ($c\beta_{sh}/r$ is the downstream flow speed and hence loss rate, $c/4$ the angle-averaged shock crossing speed and hence rate of initiating another cycle). Given the gain per cycle and the loss probability per cycle, the energy spectrum of accelerated particles can be calculated as $dN/dE \sim E^{-\Gamma}$ with $\Gamma = (r+2)/(r-1)$. The time per acceleration cycle is governed by the diffusion coefficient D. Maximum magnetic field turbulence $\Delta B/B \sim 1$ implies that the mean free path of relativistic particles, between scattering off field inhomogeneities, is of order of the gyro radius. In this so-called 'Bohm diffusion' regime, a more detailed calculation gives $D = R_{gyro}c/3$. The cycle time ΔT can be estimated to be $4D(1+r)/(v_{sh}c) \approx (20/3)(R_{gyro}/v_{sh})$ (for $r = 4$). With a gain per cycle of order $\Delta E \approx \beta_{sh}E$, the resulting acceleration rate is hence $dE/dt \approx \Delta E/\Delta T \sim E\beta_{sh}^2/R_{gyro} \sim \beta_{sh}^2 B$, of order $0.05\beta^2 B_{\mu G}$ PeV/year. The maximum achievable energy is governed by the age of the system, by energy losses (in particular radiation losses of electrons), and by the scale λ of the upstream diffusion length of particles, $\lambda \approx R_{gyro}/3\beta_{sh}$ (determined from the balance between the upstream medium flow speed v_{sh} and the diffusion speed $v_{Diff} = D\nabla\rho/\rho \approx D/\lambda$), which has to be small compared to the size of the supernova remnant to give particles a chance to return to the shock. The age-limited peak energy is given by [4] as $E_{max}(age) \approx 0.5T_3 v_{sh,8}^2 B_{\mu G} f^{-1}$ TeV where T_3 is the remnant age in kyr, $v_{sh,3}$ is the shock speed in units of 10^3 km/s and f parametrizes diffusion effects, with $f \approx 1$ for Bohm diffusion. Injection efficiency, peak energy and particle acceleration rate also depend on the angle of the average magnetic field relative to the shock front. For heavier nuclei, the acceleration rate and peak energy scale with their charge Z, reflecting the reduced gyro radius for a given energy.

The acceleration process is predicted to be highly efficient, converting as much as 50% of the kinetic energy of the ejecta into non-thermal particles. This high efficiency makes the process non-linear (e.g. [68, 69] and further references in [4]); the accelerated cosmic ray currents induce turbulent magnetic fields—determined under certain assumptions as high as 300 μG (e.g. [70])—which in turn reduce the gyro radius and the diffusion coefficient D and increase the speed of particle acceleration. A shock precursor of scale λ develops since the in-streaming gas reacts to the upstream cosmic-ray pressure, reducing the compression ratio at the subshock, see Fig. 13.9. On the other hand, since the shock now compresses a mixture of normal gas with $\gamma = 5/3$ and relativistic cosmic ray gas with $\gamma = 4/3$ upstream of the shock, the total compression ratio increases up to $r = 7$ and the spectrum of accelerated particles becomes harder, up to $dN/dE \sim E^{-3/2}$. This hardening of spectra affects mainly the highest-energy particles, with gyro radii of order λ, which probe both the shock precursor and the subshock (see Fig. 13.10).

Strictly speaking, the calculated spectral index $\Gamma = (r + 2)/(r - 1)$ applies to the particles swept downstream and confined inside the remnant. The exact mechanism of cosmic-ray escape from supernova shocks into the upstream region is not well understood; it is usually assumed that the time-integrated spectrum of particles released from the remnant reflects the spectrum of accelerated particles, but that particles of highest energy are released early, and those of low energy late in the lifecycle of the remnant [71]. The escaping cosmic rays represent a current that generates turbulent magnetic fields outside the remnant, reducing the diffusion coefficient in the upstream region, thereby enhancing the rate of particle crossing of the shock front, and speeding up the acceleration process [72].

Cosmic ray spectra measured at the Earth result from the superposition of many sources, up to distances comparable to the scale height of the halo, and are essentially stationary, despite the stochastic nature of the sources. At any given time thousands of supernovae will contribute to the cosmic rays on Earth; an individual supernova will cause a change in cosmic ray intensity over a volume of only about 100 pc in radius; beyond that the energy density of it's cosmic rays falls below the 1 eV/cm^3 level of the cosmic ray sea. The spectrum of cosmic rays on Earth is given by the source spectra multiplied by the average residence time $T(E) \sim 1/D(E)$ in the Galaxy, qualitatively explaining the difference between the source spectral index $\Gamma \approx 2$ and the observed cosmic ray spectral index $\Gamma \approx 2.7$. The knee

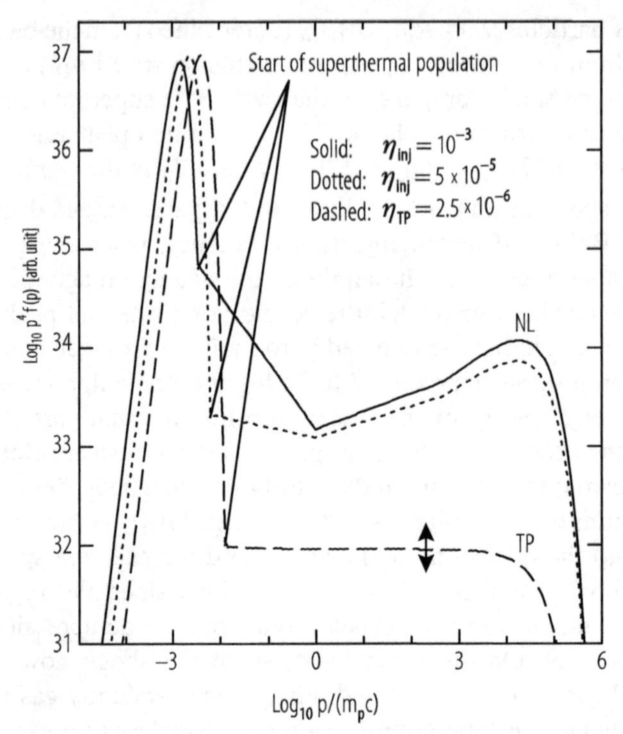

Fig. 13.10 Spectra of particles accelerated in a simulated supernova remnant shock, at a given time during the evolution of the remnant, as a function of momentum in units of the proton mass × c. The density in momentum space, $p^4 f(p)$, is shown, for relativistic particles equivalent to $E^2 dN/dE$. The 'TP' curve refers to the test particle case, where the energy carried by cosmic rays is modest compared to the kinetic energy of ejecta; in this case, the thermal distribution of unaccelerated protons extends into a power law with index $\Gamma = 2$. For efficient acceleration ('NL'), the shock is modified due to cosmic-ray pressure and the local spectral index of accelerated particles varies from $\Gamma > 2$ at low energy to $\Gamma < 2$ at high energy, with details depending on the efficiency η governing the rate of particle injection into the acceleration process. From [69]

in the cosmic ray spectrum is then associated with the peak energy of particles in the acceleration process; heavier nuclei of charge Z can be accelerated to Z-times higher energies than protons and dominate beyond the knee. With plausible parameters, supernova remnant-based models can reproduce the observed spectrum and variation of composition across the knee, see Fig. 13.11.

Shocks, and particle acceleration in shocks, can occur in all situations where non-relativistic or relativistic outflows exist with Mach numbers greater than unity; examples include stellar winds, Galactic outflows, winds driven by pulsars, or jets emerging from the vicinity of black holes driven by matter accretion. Shocks can also arise in collisions or from the infall of matter, e.g. during structure formation in galaxies and galaxy clusters. Less well understood than supernova shocks are relativistic shocks with $\beta_{sh} \approx 1$, here particles can gain significant energy in one or few shock crossings, but crossing the shock becomes increasingly difficult (e.g. [74]).

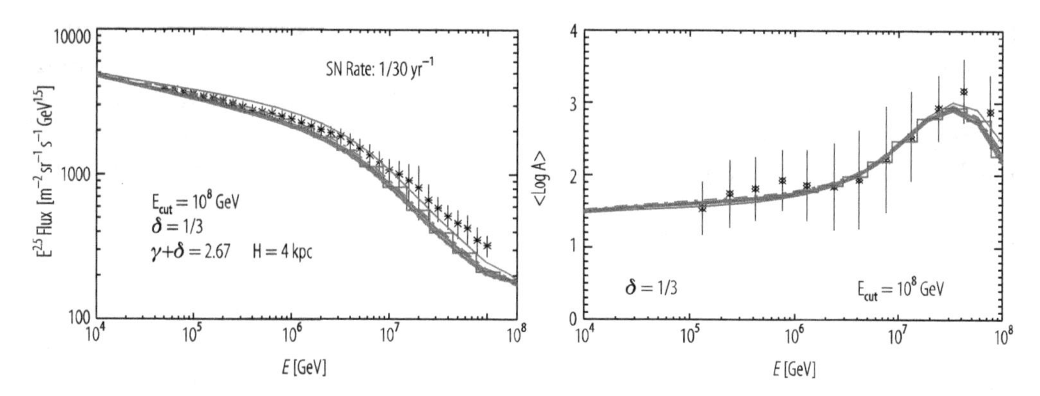

Fig. 13.11 Predicted cosmic ray spectra and change of composition—$\langle\log(A)\rangle$—with energy for ten simulated spiral galaxies populated with supernova remnants, with plausible acceleration parameters. The model includes an assumed extragalactic component which takes over at highest energies. From [73]

13.3.2 Pulsars as Particle Sources

An alternative mechanism for particle acceleration are neutron stars (or generally, rotating compact objects) with large magnetic fields, acting as unipolar inductors building up large electric fields and radiating Poynting flux.

A pulsar (a rotating magnetised neutron star) created in a supernova explosion radiates energy because its magnetic moment is misaligned by an angle α with respect to its rotation axis. The pulsar rotational energy $E = I\Omega^2/2 \approx 2 \times 10^{52}/P_{ms}^2$ erg serves as an energy source; here $I \approx 10^{45}$ cm^{2}g is the pulsar moment of inertia and P_{ms} is the rotation period in ms, with typical pulsar birth periods of order tens of ms. Assuming energy loss through dipole radiation, energy is radiated at a rate $\dot{E} = I\dot{\Omega}\Omega \approx (8\pi/3\mu_0 c^3)B_s^2 R^6 \Omega^4 \sin^2\alpha \approx 4 \times 10^{43} B_{12}^2 P_{ms}^{-4} \sin^2\alpha$ erg/s, causing the pulsar to spin down, with $\dot{\Omega} \sim \Omega^3$. Here, $R \approx 10$ km is the pulsar radius and B_{12} is the surface field in units of 10^{12} G. In case other dissipation mechanisms contribute, such as outflowing winds of particles, $\dot{\Omega}$ is parametrized as $\dot{\Omega} = -K\Omega^n$, where n is the braking index, $n = 3$ for dipole radiation. Measured values for n range from 2 to 3. Integrating the $\dot{\Omega}(\Omega)$ relation, one obtains $\dot{E}(t) = \dot{E}_0/(1+t/\tau)^{-(n+1)/(n-1)}$ for the general case, or $\dot{E}(t) = \dot{E}_0/(1+t/\tau)^{-2}$ for dipole radiation, with τ as the characteristic spin-down time. For dipole radiation, $\tau \approx (3\mu_0 c^3 I)/(16\pi R^6 B_s^2 \sin^2\alpha\ \Omega_0^2) \approx 15 P_{0,ms}^2/(B_{12}^2 \sin^2\alpha)$ years; in the time up to $t = \tau$, the pulsar loses half of its rotational energy, with the energy output diminishing like $1/t^2$ for $t >> \tau$.

In the near field, the rotating pulsar magnetic field with a surface strength in the 10^{12} G range creates electric fields with voltage drops of order $10^{17} B_{12}/P_{ms}^2$ volt. Electrons and positrons are generated by pair cascades near the pulsar surface and are accelerated until currents short-circuit the fields. Inside the light cylinder, $r < \Omega/c$, where the magnetic field and the currents co-rotate with the neutron star, particles either near the polar cap of the pulsar or in the 'outer gap' close to the light cylinder create beams of radiation swept across the sky by the pulsar rotation, see [75, 76]. In the far field, magnetic fields spiral up and cause a Poynting flux of electromagnetic energy. By a mechanism, which is yet to be understood in detail (e.g. [76]), this Poynting flux drives a particle wind, effectively converting much of the radiated energy into particle kinetic energy. It is usually assumed that the particle wind is dominated by electrons and positrons, but a component of nuclei extracted from the pulsar surface cannot be excluded. In the pulsar wind, reconnection of opposite magnetic field lines can provide a mechanism for energy release and particle acceleration [77]. The particle wind, initially assumed to be spherical [78, 79] but in more recent models concentrated in the equatorial plane, ends in a standing wind termination shock where the pressure of the wind is balanced by the ambient pressure; the termination shock is visible in high-resolution X-ray images of pulsars (see e.g. [80]). In the termination shock, particle velocities are randomised, particles are accelerated and emerge in a subsonic flow, creating a large and expanding magnetised bubble filled with high-energy electrons and positrons, see e.g. [78, 79, 81]. Pulsars are hence cosmic sources of high-energy electrons and positrons, possibly of nuclei, and of a complex mix of pulsed and beamed as well as of more or less steady and isotropic radiation spanning the range from radio to gamma rays, as elaborated in Sect. 13.4.3. However, since not all supernova explosions result in pulsars and since their initial rotational energy is usually much smaller than the kinetic energy released in a supernova explosion, the contribution of pulsars to overall cosmic-ray energetics should be modest.

13.4 Probing Cosmic-Ray Sources and Propagation Using Gamma-Rays and Neutrinos

For energies below $E \sim hZeB_{\mathrm{ISM}} \sim 10^{18}$ eV to 10^{19} eV (see e.g. [82]) cosmic ray protons are strongly deflected when propagating in the Galaxy on scales of the Galactic halo height h in typical interstellar fields B_{ISM}; their arrival directions therefore carry almost no information on their source locations. Directional messengers are therefore required to study Galactic cosmic-ray sources. Strong interactions of cosmic ray protons and nuclei with target protons and nuclei in the interstellar medium (ISM) lead to pion production and hence gamma-ray, neutrino and secondary electron and positron signatures (for a detailed description, see e.g. [83]). The following discussion will focus largely on the well-explored gamma ray signatures. Protons or nuclei with power-law spectra of index Γ_p generate gamma-ray spectra with $\Gamma_\gamma \approx \Gamma_p$, and a cutoff in proton spectra translates into a (smoother) cutoff in gamma-ray spectra about a decade in energy below the cutoff in primary spectra. For a typical ISM density, n, of 1 hydrogen atom per cm^3 the energy loss timescale for relativistic protons is $(f\,\sigma_{pp}\,n\,c)^{-1}$ or a few 10^7 years, comparable to or—in particular at high energy—longer than their residence time in the Galaxy. The fraction f of the primary energy lost in a typical collision ("inelasticity") is

$\sim$0.5 of which $\approx 1/3$ goes into the gamma-ray channel. Overall, about 1% of the energy injected into relativistic hadrons in the Galaxy in the end emerges in photons [38]. Inside, or in the vicinity of cosmic accelerators, particle density is strongly enhanced and the objects are visible as gamma-ray sources, assuming that sufficient amounts of target material (interstellar gas) are present. The intensity and extent of the gamma-ray or neutrino emission depends the distribution of target material, on whether accelerated particles are efficiently confined within the accelerator, and on how quickly particles diffuse away after escaping from the acceleration region (see Sect. 13.2.2) and merge into the cosmic-ray "sea".

For cosmic electrons and positrons, ionisation, bremsstrahlung, synchrotron radiation and Inverse Compton (IC) scattering of ambient radiation fields compete as energy-loss processes [84, 85]. For the highest energy electrons, at TeV energies and above, synchrotron and IC emission dominate and synchrotron X-rays and IC gamma-rays can be used as effective tracers of electron acceleration and propagation. The targets for IC scattering are typically the cosmic microwave background radiation (CMBR), starlight, and reprocessed starlight remitted in the far infrared, with typical energy densities of order 1 eV/cm^3. The typical lifetime of a high energy electron in the ISM is $5 \times 10^5 (B/5\,\mu\text{G} + U_{\text{rad}}/(\text{eVcm}^{-3}))^{-1}(E/\text{TeV})^{-1}$ years, much shorter than propagation time scales in the Galaxy. Radiative losses modify the energy spectra of very-high-energy electrons, introducing—for burst-like injection—an age-dependent cutoff in the electron spectra at the energy where the lifetime corresponds to the source age, at $E_{\text{cut,TeV}} \approx 3 \times 10^5/T_{\text{year}}$, or—for continuous injection—increasing the spectral index Γ_e by one unit, since at high energy only electrons injected within a period corresponding to the electron lifetime survive [86]. Electron spectra with power-law index Γ_e generate power-law IC and synchrotron spectra with index $(\Gamma_e + 1)/2$, and a cutoff energy E_c in electron spectra translates into cutoffs $E_{\gamma,\text{TeV}} \approx 10 E_{c,\text{TeV}}^2 E_{ph,\text{eV}}$ for IC gamma rays (in the Thomson regime where $E_{c,\text{TeV}} E_{ph,\text{eV}} \sim< 1$), where E_{ph} is the typical energy of the target photons, and $E_{X,\text{eV}} \approx 0.01 E_{c,\text{TeV}} B_{\mu G}$ for X-rays. With the rapid energy loss, emission by electrons is usually concentrated relatively close to the sites of acceleration. Figure 13.12 gives an example spectral energy distribution for emission dominated by energetic electrons.

Gamma-ray detection at high energies is based on pair-production and subsequent electromagnetic cascading. The most sensitive satellite-based gamma-ray detector currently operating is the Fermi Large Area Telescope (LAT), which has $\approx$1 m^2 detection area and $\approx$2.5 steradian field-of-view (FoV). The LAT combines a silicon-strip tracker for directional reconstruction and a 8.6 radiation-length thick calorimeter for energy determination [90]. The angular resolution achievable is strongly energy dependent: improving from 5° at 100 MeV to 0.25° at 10 GeV, where photon statistics become very limited for most sources. The most sensitive ground-based approach (see e.g. [9] for a review) is the Imaging Atmospheric Cherenkov Technique (IACT), which uses the Cherenkov light produced by electromagnetic cascade electrons and positrons in the atmosphere to establish the properties of the primary gamma-ray; the gamma-ray direction is determined by imaging the cascade, the gamma-ray energy is derived from the Cherenkov

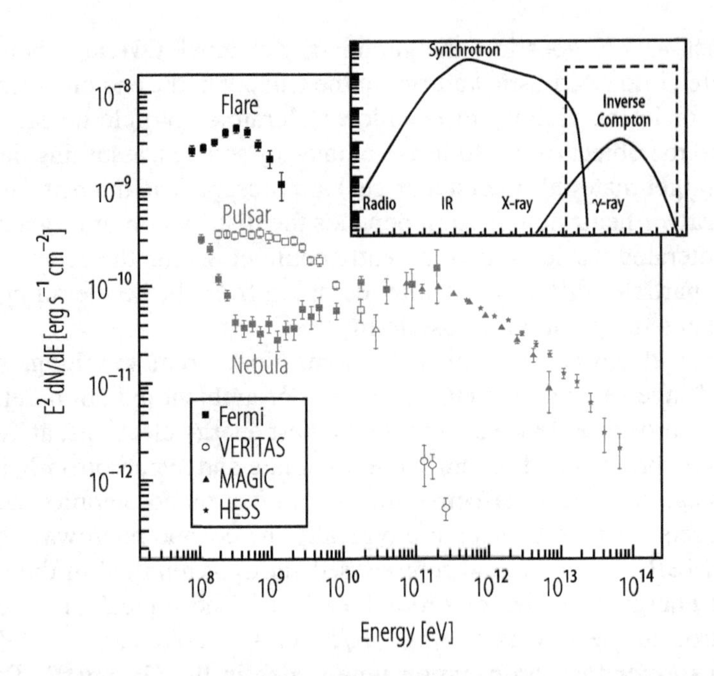

Fig. 13.12 The spectral energy distribution E^2dN/dE of the Crab Nebula, with its synchrotron and Inverse Compton (IC) components. The inset illustrates the full spectral range from radio to gamma rays [87], the main figure shows the steady-state gamma-ray flux from the nebula (blue solid symbols), the pulsed emission from the vicinity of the pulsar (red open symbols), and the brightest flare observed from the nebula so far (black solid symbols). Data from Fermi (squares), MAGIC (triangles), VERITAS (circles) and HESS (stars) are shown. Adapted from [88, 89]

light yield. The technique is in principle applicable for photon energies above $\sim$5 GeV, where the Cherenkov yield becomes significant. Current instruments HESS, MAGIC and VERITAS are sensitive from $\sim$20–50 GeV to $\sim$50 TeV, have $\sim 4°$ field-of-view and collection areas at TeV energies of $\sim 10^5\,\mathrm{m}^2$. The directional precision achievable from the ground is limited by shower fluctuations to $\approx 0.01°/(E/1\,\mathrm{TeV})^{-0.6}$ [91], with about 0.1° (and $\approx$15% energy resolution) achieved for current instruments [8]. Compared to Cherenkov telescopes, ground-level detection of shower particles allows large field-of-view and duty cycle, at the expense of higher energy threshold and reduced sensitivity and energy resolution [9]. The HAWC instrument, combining a high-altitude location at 4200 m asl. with the calorimetric detection of shower particle energy flow using large water Cherenkov detectors, instrumenting 60% of its 22,000 m^2 array area, has for the first achieved gamma-ray detection performance competitive with current IACTs [92].

At this time, over 3000 sources of GeV gamma rays have been detected using the space-based instruments Fermi [93] and AGILE, and well over 200 sources of TeV gamma rays [94] are seen with ground-based instruments, showing the abundance and indeed ubiquity of cosmic particle accelerators. Around one half of the TeV sources are of extragalactic nature, half or slightly more are associated with our

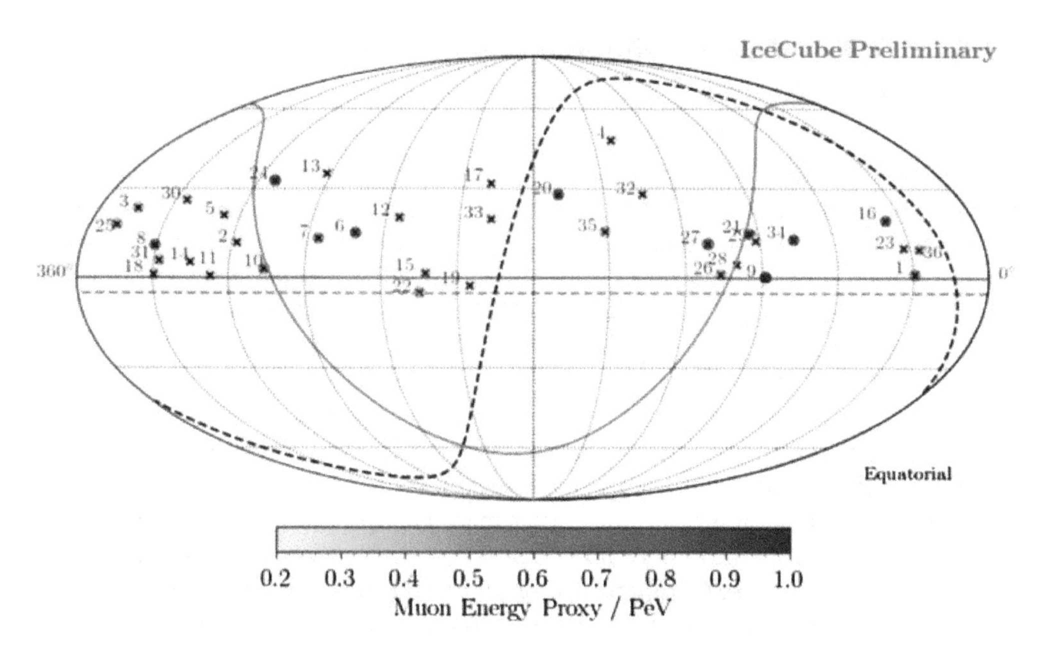

Fig. 13.13 Reconstructed arrival directions of observed IceCube neutrino events with estimated muon energies above 200 TeV, in equatorial coordinates. The marker color indicates the energy. The solid gray line indicates the galactic plane and the dashed black line the supergalactic plane. Reproduced from [96]

Galaxy. Of the Galactic TeV gamma ray sources, a handful are SNR where shells are resolved, a similar number are clearly associated with SNR but not resolved, about two dozen are associated with pulsars and their pulsar wind nebulae, and for the remaining 15–20 sources the identification is unclear, either because of a lack of counterparts or because multiple potential counterparts exist. Despite the much larger number of GeV sources, the number of well-identified Galactic objects is similar, with pulsars, identified by their pulsed emission, as the dominant class of Galactic GeV emitters.

For hadronic sources similar fluxes are generated in gamma-rays and neutrinos (see e.g. [83]). Neutrino telescopes primarily sensitive above $\sim$1 TeV now exist, with the best sensitivity reached by the IceCube detector beneath the South Pole. IceCube has for the first time detected astrophysical neutrinos [95, 99], emerging at energies beyond 100 TeV, from the strong background of atmospheric neutrinos. The neutrino flux appears to be of diffuse nature (see Fig. 13.13), and no consensus exists regarding its exact origin. No localized very-high-energy cosmic neutrino source has yet been detected [97]; a detection would provide completely unambiguous identification of hadronic accelerators and allow high-density environments, from which TeV photons may not emerge, to be probed. The KM3Net collaboration [98] is building a larger detector in the Mediterranean sea with greater sensitivity and the potential for detection of neutrino sources in the inner parts of the Galaxy, and IceCube is studying options for a tenfold increase of detection volume [96].

13.4.1 Diffuse Gamma Ray Emission: Tracing Cosmic Rays in the Galaxy

The Galactic cosmic rays interact with the material, radiation fields and magnetic fields in and around the Galaxy to produce broad-band diffuse emission. This diffuse emission peaks in the gamma-ray band due to strong contributions from π^0 decay, inverse Compton scattering and bremsstrahlung. Diffuse emission dominates the GeV sky (see Fig. 13.14) and provides a means to test the distribution and energy spectrum of cosmic rays in the Galaxy. Probing cosmic rays in this way requires an understanding of the distribution of gas and radiation fields in the Galaxy; historically the opposite has often been the case, with the gas distribution in the Milky Way being estimated from gamma-ray observations. The main components of the Galactic diffuse emission are, however, clear: π^0 decay and bremsstrahlung emission from interactions with molecular material with a scale-height in the disc of $\sim$50 parsecs, and with a more diffuse atomic component, plus a very extensive halo generated by inverse Compton scattering on Galactic radiation fields.

The diffuse GeV γ-ray emission suggests that the ISM is permeated with cosmic ray electrons, protons and nuclei, occupying a much greater volume than the Galactic disc (consistent with the picture presented in Sect. 13.2.2). In the outer parts of the Galaxy, where it is easiest to measure, this "sea" of cosmic rays has spectral properties similar to those measured at the Earth. The emissivity (gamma-ray flux per hydrogen atom, proportional to the cosmic-ray density) is seen to be roughly constant from the location of the sun up to $\sim$14 kpc from the Galactic Centre [101]. This relative uniformity suggests that the radial distribution of acceleration sites in the Galaxy is flatter than that of identified SNRs or pulsars.

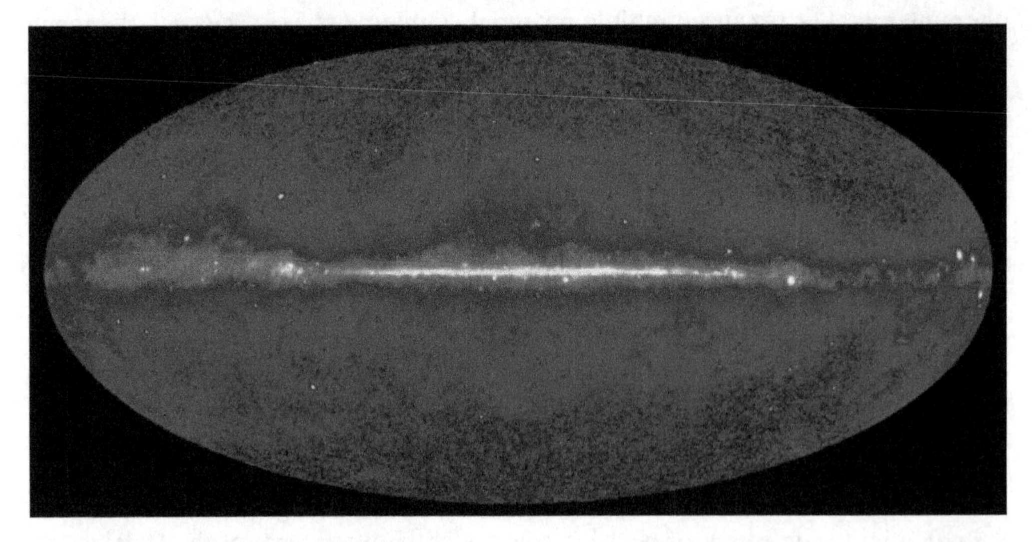

Fig. 13.14 The gamma-ray sky above 1 GeV energies as measured using the Fermi-LAT, plotted in Galactic coordinates [100]

Diffuse Galactic gamma-ray emission has also been detected at TeV energies: from the outer parts of the Galaxy, with the wide field-of-view shower-particle detector Milagro [102] and in the Galactic Centre region with HESS [103]. The level of emission seen is well above the extrapolation of the cosmic ray "sea" emission at lower energies and suggests the existence of recently injected cosmic rays and/or unresolved cosmic ray sources. The diffuse emission seen with HESS from the $\sim$100 pc radius Central Molecular Zone is correlated with the distribution of target material, illuminated by cosmic rays diffusing away from the Galactic Centre and hence suggesting a π^0-decay origin of the emission, with the central supermassive $(3 \times 10^6 M_{\odot})$ black hole Sgr A* as likely candidate for the origin of these cosmic rays. The local density of TeV cosmic rays at the centre of the Galaxy is enhanced by an order of magnitude compared to local cosmic rays.

Diffuse emission is seen with Fermi from beyond the Milky way, in satellite galaxies such as the Large and Small Magellanic Clouds and Local Group galaxies such as Andromeda. The gamma-ray fluxes seen from these objects support the idea that the energy input into cosmic ray acceleration is proportional to the star formation rate, with this trend continuing to more distant starburst galaxies, which are undergoing phases of enhanced star-formation [104–106].

A completely unexpected discovery were the Fermi bubbles—kpc-size emission regions of GeV gamma rays above and below the Galactic Centre, with relatively sharp boundaries, indicating confined populations of high-energy particles well beyond the Galactic disc and near halo [107–109]. Similar-shaped radio features hint at the presence of (primary or secondary) electrons. The origin of the Fermi bubbles is under discussion; possibilities discussed include (a) a Gyr-old stellar wind driven by star formation in the Galactic Centre region, similar to the outflows seen in starburst galaxies such as M82 or NGC 253, carrying cosmic rays along which then interact in the thin medium above and below the disc, (b) earlier jet activity of the supermassive hole at the Galactic Centre, or (c) in-situ acceleration of electrons by plasma-wave turbulence [108, 110, 111].

13.4.2 Supernova Remnants Viewed in Gamma Rays

As described in Sect. 13.3.1, the idea that supernova remnants (SNR) accelerate the majority of the Galactic cosmic rays has been with us for half a century, until relatively recently, however, the level of experimental support for this idea was relatively modest. Radio emission from the shells of SNRs was attributed to GeV electrons, but evidence for the acceleration of very-high-energy particles, in particular protons and nuclei, was essentially absent. Observations in the X-ray band from the 1990s onwards have established the presence of synchrotron emission from $>$100 TeV electrons in these objects. These observations also indicate the presence of enhanced magnetic fields [70], and for some objects "missing energy" is evident when comparing expected shock heating of the gas to measured thermal X-ray emission [70, 112]. A likely form for this missing energy is a population of cosmic-ray protons and nuclei, which may also be responsible for the amplification of magnetic fields in the SNR shell, via cosmic ray driven instabilities—the positive feedback process that leads to an increase in the efficiency and maximum energy of cosmic ray acceleration, provided that conversion of shock kinetic energy to cosmic rays occurs with a significant efficiency [113].

Detection of very-high-energy gamma rays from SNR was proposed in [114] as a test of cosmic-ray origin; the gamma-ray flux from interacting protons was predicted to be $F(>E)_{\mathrm{cm}^{-2}\,\mathrm{s}^{-1}} \approx 9 \times 10^{-11} \theta E_{\mathrm{TeV}}^{-1.1} E_{SNR,51} d_{kpc}^{-2} n_{\mathrm{cm}^{-3}}$, where θ is the efficiency of energy conversion, E_{SNR} the kinetic energy in the explosion in units of 10^{51} ergs, d the distance in kpc, and n the ambient density in hydrogen atoms per cm^3. With θ assumed as 0.1 or larger and E_{SNR} and n typically of order unity, supernova remnants within a few kpc were predicted to be bright enough for detection with air-Cherenkov instruments.

The subsequent discovery of resolved TeV emission from the shells of SNRs with H.E.S.S.—RX J1713.7−3946, RX J0852.0−4622, RCW 86 and SN 1006—(e.g. [8]) can be seen as the direct and definitive proof that very-high-energy particles are accelerated in the shells of SNR.

Figure 13.15 shows keV X-ray and TeV gamma-ray images of the nearby Galactic SNR RX J1713.7−3946 [117], the best-studied gamma-ray remnant. However,

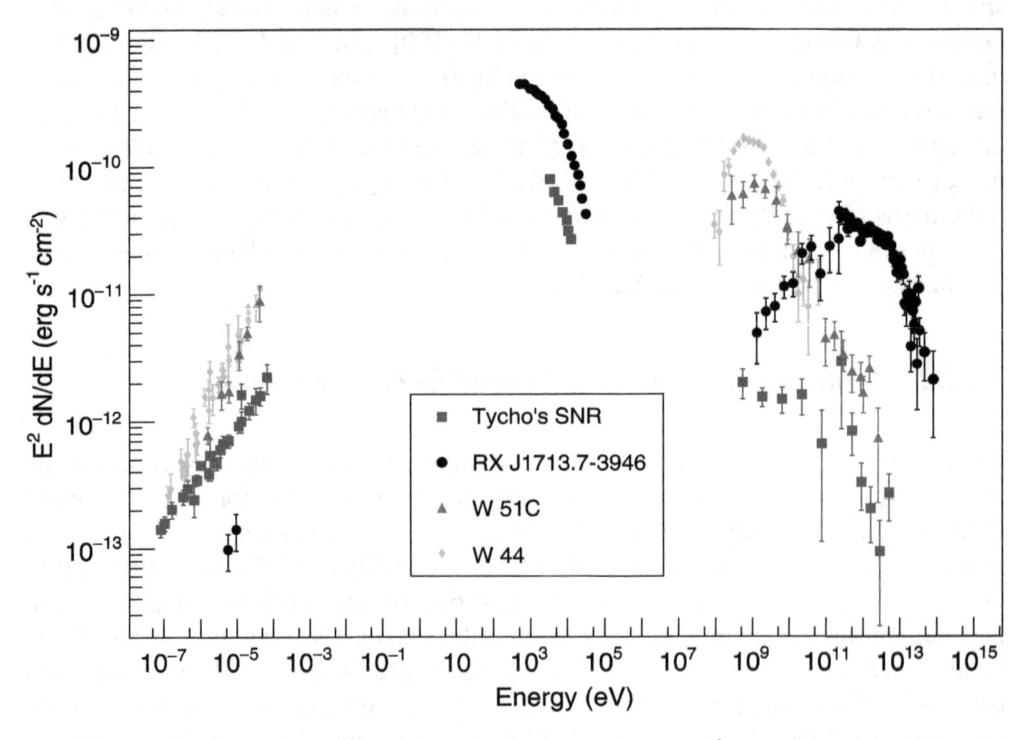

Fig. 13.15 The radio to gamma-ray spectral energy distributions E^2dN/dE of three archetypal Galactic supernova remnants of increasing age: Tycho's SNR [115, 116], RX J1713.7−3946 [117], W 51 [118, 119] and W 44 [120]

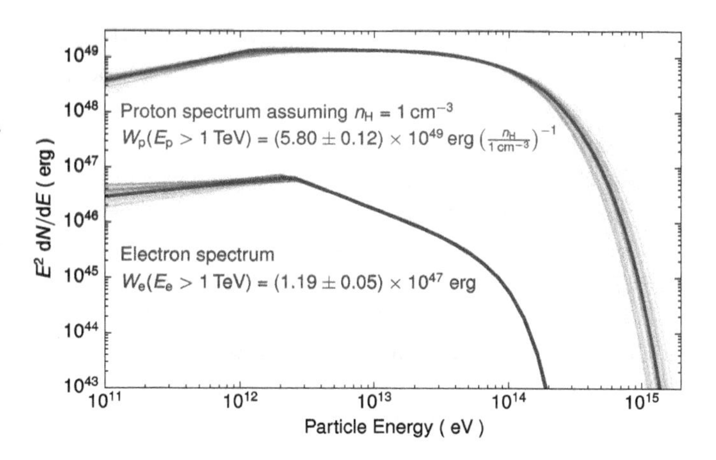

Fig. 13.16 Spectrum of primary particles required to reproduce the wide-band gamma ray and X-ray spectra of RX J1713.7−3946, assuming either dominant emission by a population of accelerated protons, or of electrons. From [117]

there is still significant debate regarding the nature of the parent particle populations. The close resemblance of the X-ray and gamma-ray images, taken a factor 10^9 apart in energy, has been taken as implying a common parent population, i.e. electrons with spectra extending to ~100 TeV, with the observed TeV emission arising from Inverse Compton scattering rather than from proton interactions. This picture is supported by the spectral shape of the gamma-ray emission, see Fig. 13.15; in the GeV energy range, the spectral index of gamma rays is $\Gamma \approx 1.5$, as expected from an electron population with the canonical E^{-2} energy spectrum resulting from shock acceleration, whereas proton interactions should result in a gamma-ray index $\Gamma \approx 2$. These data do not exclude efficient proton acceleration in this SNR; in the low-density environment, protons may simply not find enough targets to create a gamma-ray flux which is competitive with the flux produced by an energetically sub-dominant, but much more efficiently-radiating population of accelerated electrons. This is illustrated in Fig. 13.16, which shows the spectrum of primary particles that reproduce the measured wide-band gamma ray spectrum: either about 6×10^{49} ergs are required in protons above 1 TeV (assuming a target density of 1/cm^3), or only about 10^{47} ergs in electrons.

Other gamma-ray detected SNRs have dramatically different spectra; this may be due to a combination of the environment and evolutionary stage of the system. TeV emission is for example seen also from extremely young Galactic SNRs such as Tycho's SNR and Cassiopeia A, although the angular resolution of TeV instruments is insufficient to resolve their shells. These objects are firmly associated with historically observed supernova and are bright X-ray and radio sources. Figure 13.15 shows the spectral energy distribution of Tycho's SNR [116], which exhibits a spectral softening at a few hundred GeV. The maximum energy to which a particle can be accelerated inside an SNR shell is very likely time-dependent, and determined by the shock velocity and magnetic field strength. Very young SNRs with fast shocks and strong magnetic fields may accelerate extremely high energy particles (up to PeV) which can escape ahead of the shock at later times [71], leading to a decrease in the maximum or break energy with time. In addition, the presence or absence of dense target material (in the form of molecular gas) in the neighbourhood of the SNR will affect the balance between inverse Compton emission and Bremsstrahlung and π^0-decay emission. The nature of the environment depends strongly on the nature of the explosion. Type Ia supernova occur in evolved systems, with the explosion occurring far from the birth place of the star and typically far from molecular gas clouds. The majority of core-collapse supernova explosions occur inside massive stellar clusters, born from molecular clouds, but where powerful stellar winds have already redistributed the molecular material to produce a highly non-uniform density environment.

Emission correlated with the distribution of target material, rather than following the emission seen in radio or X-rays, would be a clear sign for a hadronic origin of gamma rays, and is seen for several Galactic SNRs (see e.g. [121]), all of which are significantly older than the ~1000 year age of RX J1713.7−3946. W 51 is an example of this class of object: as for RX J1713.7−3946 it is thought to be the remnant of the core-collapse of a very massive ($>8\,M_{\odot}$) star but its age is estimated to about 10^4 y and it is clearly interacting with dense molecular material. The spectrum of W 51 is rather steep at TeV energies (Fig. 13.15), with the peak energy output in the GeV domain [118], in marked contrast to RX J1713.7−3946. The centroid of the GeV and TeV emission is also consistent with the point of interaction of the SNR with gas clouds rather than the centre of the remnant. Both these facts point to a hadronic origin of the gamma-ray emission, with 5×10^{50} ergs of cosmic rays present in the SNR [118], consistent with the picture that a significant fraction of the energy of a typical SNR goes into the acceleration of cosmic ray protons and nuclei. The steep spectrum of the emission suggests that particles with TeV energies may already have escaped the SNR. Hadronic origin is also demonstrated for remnants such as W 44 and IC 443, where spectra show a clear pion-decay feature—a break in spectral index at about the pion mass [120], which is not expected in spectra from the inverse Compton process.

There is now compelling evidence that SNRs are effective accelerators of both electrons and nuclei, but due to uncertainties on the population and evolution of particle-accelerating SNRs one cannot yet confidently conclude that they are the dominant sources of the Galactic cosmic rays.

13.4.3 Pulsars and Pulsar Wind Nebulae

The newly-formed neutron stars left behind in some types of supernova explosion are generally rapidly rotating and highly magnetised. The rotational energy of such pulsars is converted into pulsed emission (primarily and gamma-ray energies) and into an ultra-relativistic outflow of electron-positron pairs, see Sect. 13.3.2. The termination shock where this wind is halted by external pressure is thought be a site of acceleration up to very high energies. After the shock the direction of particle motions become randomised and synchrotron and Inverse Compton emission is produced in a pulsar wind nebula (PWN). There is as of yet no clear physical picture of how particle acceleration operates in these systems. None the less, there is a huge body of evidence that acceleration to PeV energies occurs in these systems. The Crab Nebula, the most prominent example of a PWN, is a unique object which is bright and well-studied in every waveband of the electromagnetic spectrum, with synchrotron emission dominating from the radio up to 1 GeV and inverse Compton emission seen above, up to almost 100 TeV (see Fig. 13.12). The Crab pulsar was born in a historically observed supernova explosion in 1054 and has the most extreme rate of conversion of rotational energy ($\dot{E} \approx 5 \times 10^{38}$ erg/s) of any Galactic pulsar. In 1989 it became the first source to be detected at TeV energies [122]. Until very recently the gamma-ray emission from the Nebula was thought to be steady in time, but recent dramatic flaring activity has been seen at GeV energies, apparently corresponding to a rapid increase in the number of synchrotron-emitting >PeV electrons (Fig. 13.12) [89, 123, 124]. It seems very difficult to explain these flares in terms of diffusive shock acceleration, as this process seems to be too slow to offset catastrophic synchrotron energy losses. A single shot acceleration mechanism such as magnetic reconnection at the termination shock is an attractive option.

The gamma-ray emission from the Crab Nebula appears almost point-like with the $\sim 0.1^\circ$ resolution of current instruments, but this situation is atypical. In the last decade, about two dozen TeV gamma-ray sources were discovered and firmly or tentatively identified as PWN, including objects such as Vela X, MSH 15–52, G 21.5−0.9, G 0.9+0.1, N 175B, and the nebula surrounding PSR B1706–44 (for summaries, see e.g. [8, 125–127]). Most of these PWN are extended TeV gamma-ray sources, with size of a fraction of a degree, often accompanied by a significantly smaller-scale X-ray nebula surrounding the pulsar. While X-ray luminosities tend to correlate quite well with the instantaneous spin-down energy loss $\dot{E}$ of the pulsar, no clear correlation is observed between the gamma-ray luminosity and $\dot{E}$ [126]. Gamma-ray luminosities range from a fraction of a percent of the pulsar spin-down energy loss $\dot{E}$ to tens of percent, indicating a relatively efficient conversion of (rotational) kinetic energy into high-energy particles. A likely explanation for the difference in size between X-ray and gamma-ray PWN and for the different $\dot{E}$-dependence of luminosities is that in the typical μG fields derived for extended PWN, keV X-ray emitting electrons have energies of 100s of TeV and cooling times of order 1000 years, whereas TeV gamma ray emitting electrons have energies in the 10 TeV range and cooling times beyond a few 10,000 years. For pulsars with ages between a 1000 and a few 10,000 years (most of the gamma-ray PWN population), only "recently" accelerated electrons with number $\sim \dot{E}$ therefore contribute to the X-ray emission, whereas all electrons ever accelerated contribute to the gamma-ray emission, reflecting essentially the initial rotational energy E_0 of the pulsar (half of which is lost during the initial few 100 years of spin-down history, see Sect. 13.3.2), rather than $\dot{E}$. The sizes of gamma-ray PWN tend to increase with the age of the pulsar, saturating at sizes of a few tens of pc. The gamma-ray PWN are frequently displaced from the pulsar, locating the pulsar (and the associated X-ray PWN) at the edge of the gamma-ray PWN (e.g. Fig. 13.15). One explanation is that PWN are often crushed and/or displaced by the supernova reverse shock [128], another that—as discussed above—the gamma-ray PWN reflects relic electrons abundantly created in the early history of the pulsar, whereas now the pulsar may have moved away from its birth place due to a kick from the explosion. However, in the few cases where pulsar motion is known, it does not line-up well with the vector connecting the pulsar and the centroid of the VHE PWN. In a few cases, such as for the source HESS J1825−137 [129] shown in Fig. 13.15, gamma-ray PWN (as well as X-ray PWN) show energy-dependent morphology, with the nebula shrinking towards the pulsar with increasing gamma-ray energy, presumably due to radiative cooling of electrons as the propagate away from the pulsar. GeV-TeV gamma-ray emission can therefore be used to measure the time-integrated particle injection of the pulsar and to understand the propagation of relativistic particles away from their sources.

PWN represent the bulk of Galactic TeV gamma-ray sources, by far outnumbering emission traced to SNR shock-accelerated protons. At first, this seems surprising. However, contrary to supernova remnant shocks, where acceleration of very-high-energy particles stalls after a few 10^3 to at most 10^4 years, a pulsar can supply the nebula with energy for many 10^4 years. In addition, under typical conditions, electrons and positrons are more efficient TeV gamma-ray emitters than are SNR-accelerated protons—radiative energy loss timescales are smaller by one to two orders of magnitude. Therefore, PWN dominate the Galactic population of TeV gamma-ray sources even though their energy reservoir—the rotational energy of the pulsar—is typically an order of magnitude smaller than the $\approx 10^{51}$ ergs released in a supernova explosion. In fact, it seems likely that a sizeable fraction of the currently unidentified TeV gamma-ray sources—where no counterpart is seen in other wavebands—are PWN where the pulsar is not detected (due to beaming effects or simply a lack of sensitive observations) and where radiation-cooled electrons no longer have sufficient energy to produce keV X-ray synchrotron photons.

PWN also occur inside binary systems, where the huge radiation fields and stellar wind/outflow of the companion star dramatically modify the PWN properties. PSR B1259−63 is young and powerful pulsar in an eccentric 3.4 year orbit around a $\approx$10 solar mass companion. Variable and point-like TeV and GeV emission is seen around the periastron passage of the neutron star when radiation densities are highest but the time-profile of the emission and the spectral energy distribution are complex and very poorly understood (e.g. [130]). Extended radio emission is seen on milliarcsecond scales supporting the idea that this system is a PWN "compactified" by the high pressure environment and rapid radiative losses. In other well-established TeV binaries systems, LS 5039, LS I +61 303 and HESS J0632+057, the nature of the compact object is not certain and the systems may be accretion-, rather than rotation-powered.

PWN naturally accelerate positrons and electrons in equal number. An increase in the proportion of cosmic ray electrons and positrons contributed by PWN (and/or the presence of a small number of dominant local/recent accelerators) may explain the increase in the fraction of positrons seen in the locally measured cosmic rays at high energies, as discussed in Sect. 13.2.4. Given a suitable pulsar age T, energy-dependent diffusion $D(E)$ of electrons causes the detected spectrum to harden compared to the source spectrum—high energy particles reach Earth faster—and radiative energy losses during propagation over a time T cause the spectrum to cut off at a certain energy. The two effects combine to produce an excess in an E^3

weighted spectrum, the peak position and peak level being adjustable via pulsar age, distance, and energy output, and matching the detected excess contribution for plausible parameters. For example, positrons released $T \approx 10^5$ years ago exhibit a cutoff at $E_{\mathrm{cut,TeV}} \approx 3 \times 10^5/T_{\mathrm{year}} \approx 3\,\mathrm{TeV}$ and at TeV energies travel over a distance $d \approx (2DT)^{1/2} \approx 500\,\mathrm{pc}$ (for $D \approx 10^{28} E_{\mathrm{GeV}}^{0.5}$ cm^2/s, see Sect. 13.2.2). Assuming that about 10% of the rotational kinetic energy of $\approx 10^{50}$ erg of a 10 ms pulsar is released in electrons and positrons, the resulting average electron energy density in the 500 pc volume is of order 10^{-3} eV/cm^3, comparable to the density of secondary electrons from nuclear interactions of cosmic rays. In an energy range where such a source dominates the flux of electrons and positrons, the positron fraction is 1/2. While many details of this scheme remain to be clarified, such an explanation of the effect in terms of conventional astrophysics would need to be ruled out first, before more exotic schemes such as Dark Matter annihilation are invoked. Local PWN, for which the signature of electron escape and subsequent propagation to the Earth may be apparent in the cosmic ray electron spectrum at very high energies, include Vela-X [131] and Geminga. A recent measurement of the cosmic-ray diffusion around from Geminga [62], however, revealed unusually small diffusion coefficients, which make it difficult for those electrons to reach Earth during the relevant time scale.

In addition to the unpulsed emission from nebulae, pulsed GeV emission from many pulsars is observed, with cutoffs at a few GeV e.g. [88], as expected since higher-energy gamma rays have difficulty escaping from the pulsar magnetosphere with its huge magnetic fields. In this context the recent detection of pulsed emission up to energies of a few 100 GeV, following a steep power law rather than the expected super-exponential cutoff [132, 133] was very surprising. Cascade processes may be responsible for transporting the pulsed signal away from the pulsar, reducing the suppression at higher energies.

13.4.4 Other Galactic Systems as Sources of High-Energy Radiation

Several additional classes of cosmic particle accelerators have recently been identified in our galaxy. These objects are generally stellar binary systems of various types, or related to the collective effects of clusters of stars. Stellar binaries containing a normal star and a black-hole are known to (episodically) host accretion-powered jets which can be relativistic. These systems are the Galactic analogs of the active galactic nuclei described in Sect. 13.4.5, and have been dubbed "micro-quasars". Cygnus X-3 appears to be black hole with a massive stellar companion, and periodic emission has been detected using Fermi. The emission is correlated with the appearance of radio features and seems to be associated with the formation of a jet in the system. The only TeV detection of emission from a well-established black-hole binary is that of a single flare from Cygnus X-1 with the MAGIC telescope [134]. Whilst intriguing, further TeV detections will be required to confirm Cygnus X-1 as a TeV source.

The well-studied stellar binary Eta Carina contains two very massive ($M > 30M_{\odot}$) stars which both produce powerful (radiatively driven) winds. The collision of these stellar winds results in strong shocks and a situation akin to a supernova explosion, except in a much denser (in terms of both matter and radiation) and higher magnetic field environment (e.g. [135]). Gamma-ray emission is seen from this system up to ~100 GeV [136], with two distinct components, and variability observed in the higher energy component which emerges above 20 GeV [137]. Whilst many questions remain, it now seems clear that Eta Carinae is a cosmic particle accelerator and a hadronic origin of one of the components seems plausible.

Evidence for acceleration associated with the collective effects of stellar winds is provided by the detection of very extended emission from the massive stellar cluster Westerlund 1 [138]. Degree scale emission is seen stretching well beyond the stellar cluster, which is one of the most massive in the Galaxy. However, a supernova remnant, unseen at other wavelengths due to the unusual environment, is not excluded as the origin of the TeV emission. Systems such as Westerlund 1 can be seen as the Galactic analogues of the starburst galaxies described in Sect. 13.4.1.

A single Nova (powered by a thermonuclear explosion on the surface of a degenerate white dwarf star) has been detected in high energy gamma-rays [139]. In an ~10 day flare from the white dwarf V407 Cyg in 2010 emission was seen up to ~5 GeV.

13.4.5 Particle Acceleration Driven by Supermassive Black Holes

The accretion of material onto supermassive ($M \sim 10^4 M_{sol} - 10^{10} M_{sol}$) black holes leads to the formation of oppositely-directed and highly-collimated jets of material. The mechanism for launching of these jets is still hotly debated, but magnetic fields in the accretion disc around the black hole look to be playing a crucial role [140, 141]. Radio emission associated with these jets has been know about for a long time, and is most dramatic in the case of Radio Galaxies where the jets are seen at a large angle to the observer and their true scale of (often) 100s of kiloparsecs can be seen, often dwarfing the host galaxy. Particle acceleration is clearly taking place in these objects, at several different locations, for example inside the inner (relativistic) jets, at the termination shocks of powerful jets and at the edges of the so-called radio "lobes" which result when jets are decelerated and spread out. Inverse Compton X-ray emission has been detected from several such systems, tracing the same population of electrons as seen in radio and allowing magnetic field strength and total energy densities to be estimated. X-ray Synchrotron emission, indicating the presence of very high energy electrons, is also seen in several places within systems powered by an active galactic nucleus (AGN). Despite the well established relativistic particle populations in these objects the nature of the acceleration mechanism is still not known. Diffusive shock acceleration does not seem to be the whole story in these objects, with a more distributed acceleration mechanism apparently needed to explain the X-ray synchrotron emission from the inner jets of nearby AGN [142].

Active galaxies are prime candidates for the acceleration of UHECRs, as can be seen from Fig. 13.8, with both the nucleus and the extended jets as possible acceleration sites for 10^{20} eV particles. As discussed in Sect. 13.2.5 there are hints of UHECR anisotropy correlated with the distribution of matter in the nearby universe (within the GZK horizon) and in particular an excess in the direction of the very nearby active galaxy Cen A (and also the direction of the more distant Centaurus cluster of galaxies). Gamma-ray emission his been detected from Cen A on a wide range of spatial scales. Figure 13.17 shows emission from the giant (10°) lobes of Cen A in radio synchrotron emission and (very likely) inverse Compton gamma-ray emission in the Fermi band [143]. Non-thermal emission is also seen in Cen A from the nucleus, inner jets and inner lobes. The synchrotron X-ray emission seen from the termination shock of the inner lobe closely resembles the situation for supernova remnants, but on a very much larger scale [144]. TeV emission is seen from the inner parts of Cen A (see the right-hand panel of Fig. 13.17) with a position consistent with the inner parts of the jet or the nucleus itself [145]. The giant lobes, jets and nucleus are all candidate acceleration sites for UHECR [146–148].

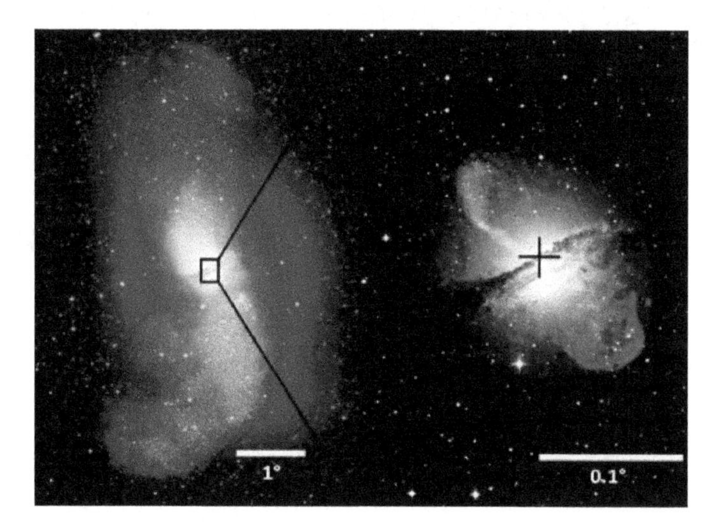

Fig. 13.17 The nearby active galaxy Centaurus A on scales of many degrees (left) and fractions of a degree (right). The left-hand plot is a multi-wavelength composite showing visible light overlaid with radio data in orange and gamma-ray data from Fermi in purple. The right-hand plot indicates the centroid of the TeV emission (cross) on a MWL composite of optical, X-ray (Chandra: blue) and microwave (orange) data. (Credit—left hand—NASA/DOE/Fermi LAT Collaboration, Capella Observatory, and Ilana Feain, Tim Cornwell, and Ron Ekers (CSIRO/ATNF), R. Morganti (ASTRON), and N. Junkes (MPIfR); right hand—ESO/WFI (visible); MPIfR/ESO/APEX/A.Weiss et al. (microwave); NASA/CXC/CfA/R.Kraft et al. (X-ray))

The bulk of the ~60 know extragalactic TeV gamma ray sources is of blazer type, as is the bulk of the extragalactic GeV sources. In blazers, jets are oriented close (within ~10°) to the line-of-sight to the observer, presenting a dramatically different perspective compared to radio galaxies. Apparently super-luminal motions observed in very-long-baseline-interferometer (VLBI) images of the central regions of these objects indicate bulk relativistic motion. Particle acceleration is presumably powered by this bulk motion, related to inhomogenities and shocks arising in the flow. Blazers are characterised by rapidly variable emission and complete non-thermal dominance of their spectral energy distributions, with the peak energy output usually in the gamma-ray domain. Markarian 421 was the first extragalactic object to be discovered in TeV gamma-rays [150] and is a dramatic example of a high-energy-peaked blazar. The measured energy spectra of these objects are significantly influenced by gamma-ray absorption by pair production with infrared or optical extragalactic background light (EBL) (e.g. [151]); vice versa, with assumptions regarding the intrinsic blazar spectra the absorption features can be used to constrain the level of EBL (e.g. [152]), which traces the history of star formation in the Universe and which, in certain spectral ranges, is difficult to measure directly due to overwhelming foregrounds [153]. TeV Blazars exhibit extreme variability, down to minute time scale [154]. This extreme variability places tight constraints on the size r of the emission region, from causality arguments: $r < \delta c t_{var}$, where δ is a Doppler boost factor reflecting the bulk motion of the particle acceleration region in the blazer jet. Blazars usually exhibit double-humped spectral energy distributions (Fig. 13.18), most likely reflecting a synchrotron component at

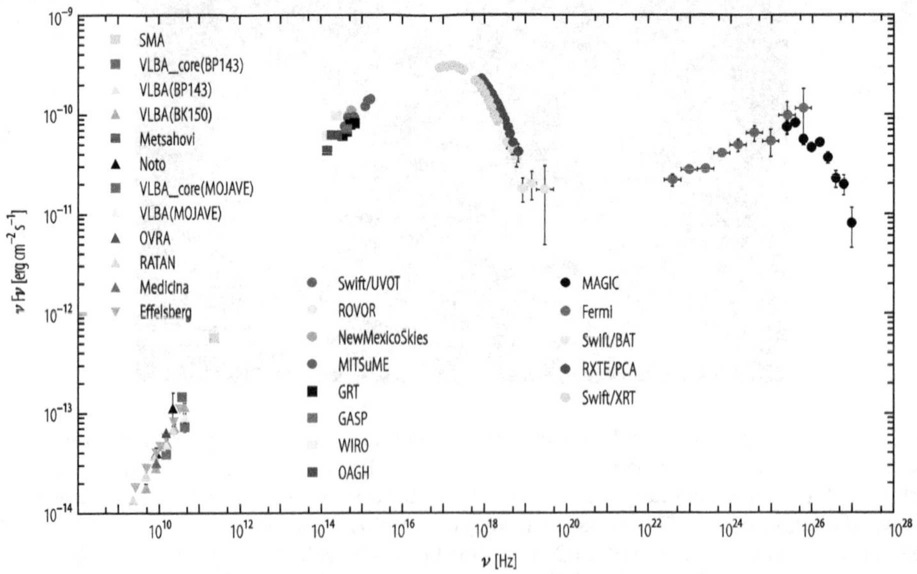

Fig. 13.18 The spectral energy distribution $\nu F_\nu \sim E^2 dN/dE$ of Mrk 421 as a function of frequency ν (where 1 TeV $\approx 2 \times 10^{26}$ Hz), measured simultaneously with a variety of instruments. F_ν is the energy flux per unit frequency. Reproduced from [149]

radio and X-ray energies, and an IC component at GeV/TeV energies, hinting at electrons as the parent particles. Often, the synchrotron X-ray intensity is such that X-rays form the dominant target for IC scattering. The parameters of the system—the electron spectrum, size, boost factor and magnetic field—can be determined from the location of the peaks in the radiation spectra, the exact spectral shape, the variability timescale and the additional condition that the photon density in the source region must be low enough to allow gamma rays to escape without pair-producing. Combined with short variability timescales this approach usually enforces large Doppler factors well beyond $\delta = 10$; how such bulk motions are achieved is not fully understood. While blazers obviously act as cosmic particle accelerators, it is unclear if they accelerate significant amounts of protons in addition to the electrons evident from their spectra, and most likely they have little impact on the cosmic rays detected on Earth.

13.5 Outlook

Significant progress has been achieved in the study of cosmic particle accelerators, in particular through the improved detection and imaging of high energy (GeV) and very high energy (TeV) gamma rays emitted in interactions of accelerated charged particles; also the first detection of very high energy cosmic neutrinos was reported. Particle acceleration is a ubiquitous feature in the Universe, associated especially with the evolution of massive stars and with black holes. Over 3000 multi-GeV accelerators are detected, and over 200 multi-TeV accelerators, with maximum energies well beyond 100 TeV. A variety of acceleration mechanisms seems to be realised in nature, including acceleration in non-relativistic SNR shocks and colliding-wind shocks, in relativistic pulsar wind shocks, in the unipolar inductors of pulsars, and in the compact and extended jets of AGN. In many systems, the acceleration process is highly efficient in converting bulk kinetic energy. The observed rapid variability in systems such as AGN or in the Crab Nebula represent a challenge to models and may require additional acceleration schemes. Despite this significant progress, a quantitative confirmation that SNR represent the sources of the bulk of nucleonic cosmic rays in our own galaxy and a first-principle calculation of their yield and spectrum is still lacking; other open questions concern the details of diffusive propagation of cosmic rays and of the resulting cosmic-ray anisotropies, the origin of the high-energy cosmic-ray positrons, and the origin of the highest-energy cosmic rays. A variety of new instruments aim to address these issues, including next-generation air-Cherenkov instruments such as CTA [155], ground-based cosmic-ray detectors like LHAASO [20], ultra-high-energy cosmic ray detectors such as JEM-EUSO [156] and a larger version of the Pierre Auger detector, or the very-high-energy neutrino detectors KM3Net [98] and the planned IceCube Gen2 instrument [157].

References

1. V.F. Hess: Phys. Z. 21 (1912) 1084.
2. A.W. Strong, I.V. Moskalenko, V.S. Ptuskin: Annu. Rev. Nucl. Part. Sci. 57 (2007) 285.
3. A.M. Hillas: J. Phys. G 31 (2005) 95.
4. S.P. Reynolds: Annu. Rev. Astron. Astrophys. 46 (2008) 89.
5. J. Blümer, R. Engel, J.R. Hörandel: Prog. Part. Nucl. Phys. 63 (2009) 293.
6. K. Kotera, A.V. Olinto: Annu. Rev. Astron. Astrophys. 49 (2011) 119.
7. A.M. Hillas: Annu. Rev. Astron. Astrophys. 22 (1984) 425.
8. J.A. Hinton, W. Hofmann: Annu. Rev. Astron. Astrophys. 47 (2009) 523.
9. F.A. Aharonian, et al.: Rep. Prog. Phys. 71 (2008) 096901.
10. P. Picozza, et al.: Astropart. Phys. 27 (2007) 296.
11. M. Aguilar, et al. (AMS Coll.): Phys. Rep. 366 (2002) 331.
12. S. Wissel, et al.: arXiv:1107.3272 (2011).
13. O. Adriani et al.: Phys. Rev. Lett. 119 (2017) 181101.
14. G. Ambrosi et al.: Nature 552 (2017) 63.
15. I. Allekotte, et al. (Pierre Auger Coll.): Nucl. Instrum. Meth. A 586 (2008) 409.
16. J. Abraham, et al. (Pierre Auger Coll.): Nucl. Instrum. Meth. A 620 (2010) 227.
17. U. Abbasi et al.: Astropart. Phys. 80 (2016) 131.
18. K.H. Kampert, et al.: Nucl. Phys. B Proc. Suppl. 136 (2004) 273.
19. M. Amenomori, et al.: Astrophys. J. 678 (2008) 1165.
20. G. Di Sciascio: Nucl. Part. Phys. Proc. 279–281 (2016) 166.
21. O. Adriani, et al.: Science 332 (2011) 69.
22. M. Aguilar et al.: Phys. Rev. Lett. 114 (2015) 171103.
23. M. Aguilar et al.: Phys. Rev. Lett. 115 (2015) 211101.
24. D. Allard, E. Parizot, A. V. Olinto: Astropart. Phys. 27 (2007) 61.
25. K. Greisen: Phys. Rev. Lett. 16 (1066) 747.
26. G.T. Zatsepin, V.A. Kuz'min: JETP Lett. 4 (1966) 78.
27. V. Berezinsky, A.Z. Gazizov, S.I. Grigorieva: Phys. Rev. D 74 (2006) 043005.
28. J. Abraham et al.: Phys. Lett. B685 (2010) 239.
29. A. Aab et al.: arXiv:1708.06592 and Proceedings, 35th International Cosmic Ray Conference (ICRC 2017): Bexco, Busan, Korea, July 12–20, 2017.
30. A. Aab et al.: JCAP 1704 (2017) 038.
31. K.H. Kampert, M. Unger: Astropart.Phys. 35 (2012) 660.
32. J.A. Simpson: Annu. Rev. Nucl. Part. Sci. 33 (1983) 323.
33. M. Simon, A. Molnar, S. Roesler: Astrophys. J. 499 (1998) 250.
34. V.L. Ginzburg, V.S. Ptuskin: Rev. Mod. Phys. 48 (1976) 161, (Erratum) Rev. Mod. Phys. 48 (1976) 675.
35. R. Jansson, G.R. Farrar: Astrophys. J. 761 (2012) L11.
36. L.O'C. Drury, D.C. Ellisson, J.-P. Meyer: Nucl. Phys. A 663 (2000) 843–843.
37. B.D. Fields, et al.: Astron. Astrophys. 370 (2001) 623.
38. A.W. Strong, et al.: Astrophys. J. Lett. 722 (2010) L58.
39. A. Aab et al.: Phys. Rev. Lett. 117 (2016) 192001
40. J. Abraham, et al.: Phys. Rev. Lett. 104 (2010) 091101.
41. A.Aab et al: Phys. Rev. D90 (2014) 122006.
42. R.U. Abbasi, et al.: Astropart. Phys. 64 (2015) 49.
43. J.R. Hörandel: Adv. Space Res. 41 (2008) 442.
44. M. Amenomori, et al.: Astrophys. J. 626 (2005) L29.
45. A.H. Compton, I.A. Getting: Phys. Rev. 47 (1935) 817.
46. P. Blasi, E. Amato: arXiv:1105.4529 (2011).
47. M. Amenomori, et al.: Science 314 (2006) 439.
48. A.A. Abdo, et al.: Phys. Rev. Lett. 101 (2008) 221101.
49. R.U. Abbasi, et al.: Astrophys. J. 740 (2011) 16.

50. L.O'C. Drury, F.A. Aharonian: Astropart. Phys. 29 (2008) 420.
51. G. Ciacinti, G. Sigl: arXiv:1111.2536 (2011).
52. O. Adriani, et al.: Phys. Rev. Lett. 102 (2009) 051101.
53. M. Aguilar et al.: Phys. Rev. Lett. 113 (2014) 121102.
54. S. Abdollahi et al.: Phys. Rev. D95 (2017) 082007.
55. D. Kerszberg: 5th International Cosmic Ray Conference (ICRC 2017): Bexco, Busan, Korea, July 12–20, 2017.
56. L. Accardo et al.: Phys. Rev. Lett. 113 (2014) 121101.
57. M. Pohl, et al.: Astron. Astrophys. 409 (2003) 581.
58. O. Adriani, et al.: Nature 458 (2009) 607.
59. M. Ackermann, et al.: arXiv:1109.0521.
60. D. Grasso, et al.: Astroparticle Phys. 32 (2009) 140.
61. F.A. Aharonian, A. Atoyan, H.J. Völk: Astron. Astrophys. 294 (1995) L41.
62. A.U. Abeysekara et al.: Science 358 (2017) 911.
63. A. Aab et al.: Astrophys. J. 794 (2014) 172.
64. F. Zwicky: Phys. Rev. 55 (1939) 986.
65. R.D. Blandford, J.P. Ostriker: Astrophys. J. Lett. 221 (1978) L29; Astrophys. J. 237 (1980) 793.
66. A.R. Bell: Mon. Not. R. Astron. Soc. 182 (178) 147; Mon. Not. R. Astron. Soc. 182 (1978) 443.
67. J.K. Truelove, C.F. McKee: Astrophys. J. Suppl. 120 (1999) 299.
68. L.O'C. Drury, H.J. Völk: Astrophys. J. 248 (1981) 344.
69. D.C. Ellison, A. Decourchelle, J. Ballet: Astron. Astrophys. 413 (2004) 189.
70. J. Vink: arXiv:1112.0576 (2011).
71. V.S. Ptuskin, V.N. Zirakashvili: Astron. Astrophys. 429 (2005) 755.
72. M.A. Malkov et al.: Astrophys. J. 768 (2013) 73.
73. P. Blasi, E. Amato: arXiv:1105.4521 (2011).
74. G. Pelletier, M. Lemoine, A. Marcowith: arXiv:0811.1506 (2008).
75. A.K. Harding, arXiv:0710.3517 (2007).
76. J. Arons: arXiv:0708.1050 (2007).
77. Y. Lyubarsky, J.G. Kirk: Astrophys. J. 547 (2001) 437.
78. C.F. Kennel, F.V. Coroniti: Astrophys. J. 283 (1984) 694.
79. C.F. Kennel, F.V. Coroniti: Astrophys. J. 283 (1984) 710.
80. C.-Y. Ng, R.W. Romani: Astrophys. J. 601 (2004) 479; Astrophys. J. 673 (2008) 411.
81. M.J. Rees, J.E. Gunn: Mon. Not. R. Astron. Soc. 167 (1974)1.
82. M. Nagano, A.A. Watson: Rev. Mod. Phys. 72 (2000) 689.
83. S.R. Kelner, F.A. Aharonian: Phys. Rev. D 78 (2008) 034013.
84. G.R. Blumenthal, R.J. Gould: Rev. Mod. Phys. 42 (1970) 237.
85. M.S. Longair: High Energy Astrophysics, Cambridge University Press.
86. N.S. Kardashev: Sov. Astron. 6 (1962) 317.
87. A.M. Atoyan, F.A. Aharonian: Mon. Not. R. Astron. Soc. 287 (1996) 525.
88. A.A. Abdo, et al.: Astrophys. J. 708 (2010) 1254.
89. R. Bühler, et al.: arXiv:1112.1979 (2011).
90. W.B. Atwood, et al.: Astrophys. J. 697 (2009) 1071.
91. W. Hofmann: arXiv:astro-ph/0603076 (2006).
92. U. Abeysekara et al.: Astrophys. J. 843 (2017) 40.
93. F. Acero et al.: Astrophys. J. Suppl. 218 (2015) 23.
94. TeVCat, http://tevcat.uchicago.edu/
95. M.G. Aartsen et al.: Phys. Rev. Lett. 115 (2015) 081102
96. M.G. Aartsen et al., arXiv:1710.01191.
97. M.G. Aartsen et al.: Astrophys. J. 835 (2017) 151
98. P. Bagley, et al.: KM3NeT Technical Design Report (ISBN 978-90-6488-033-9).
99. M. Ackermann et al.: arXiv:1710.01207 and Proceedings, 35th International Cosmic Ray Conference (ICRC 2017): Bexco, Busan, Korea, July 12–20, 2017

100. https://fermi.gsfc.nasa.gov/ssc/observations/types/allsky/
101. M. Ackermann, et al.: Astrophys. J. 726 (2011) 81.
102. A.A. Abdo, et al.: Astrophys. J. 688 (2008) 1078.
103. A. Abramowski et al.: Nature 531 (2016) 476.
104. A.A. Abdo, et al.: Astrophys. J. Lett. 709 (2010) L152.
105. F. Acero, et al.: Science 326 (2009) 1080.
106. V.A. Acciari, et al.: Nature 462 (2009) 7274.
107. G. Dobler, et al.: Astrophys. J. 717 (2010) 825.
108. M. Su, T.R. Slatyer, D.P. Finkbeiner: Astrophys. J. 724 (2010) 1044.
109. M. Ackermann et al.: Astrophys.J. 793 (2014) 64
110. R.M. Crocker, F.A. Aharonian: Phys. Rev. Lett. 106 (2011) 101102.
111. P. Mertsch, S. Sarkar: Phys. Rev. Lett. 107 (2011) 091101.
112. E.A. Helder, et al.: Science 325 (2009) 719.
113. S.G. Lucek, A.R. Bell: Mon. Not. R. Astron. Soc. 314 (2000) 65.
114. L.O'C. Drury, F.A. Aharonian, H.J. Völk: Astron. Astrophys. 287 (1994) 959.
115. H.J. Völk, E.G. Berezhko, L.T. Ksenofontov: Astron. Astrophys. 483 (2008) 529.
116. S. Archambault et al.: Astrophys. J. 836 (2017) 23.
117. H. Abdalla et al.: Astron. Astrophys. 612 (2018)
118. A.A. Abdo, et al.: Astrophys. J. 706 (2009) L1.
119. J. Aleksić, et al.: Astron. Astrophys. 541 (2012) 11.
120. M. Ackermann et al.: Science 339 (2013) 807.
121. Y. Uchiyama, et al.: Astrophys. J. Lett. 723 (2010) L122.
122. T.C. Weekes, et al.: Astrophys. J. 342 (1989) 379.
123. A.A. Abdo, et al.: Science 331 (2011) 739.
124. M. Tavani, et al.: Science 331 (2011) 736.
125. P.M. Gaensler, P.O. Slane: Annu. Rev. Astron. Astrophys. 44 (2006) 17.
126. F. Mattana, et al.: Astrophys. J. 694 (2009) 12.
127. O. Kargaltsev, G. Pavlov: arXiv:1002.0885 (2010).
128. J.M. Blondin, R.A. Chevalier, D.M. Frierson: Astrophys. J. 563 (2001) 806.
129. F.A. Aharonian, et al.: Astron. Astrophys. 460 (2006) 365.
130. S.W. Kong, et al.: Mon. Not. R. Astron. Soc. 416 (2011) 1067.
131. J.A. Hinton, et al.: Astrophys. J. Lett. 743 (2011) L7.
132. E. Aliu, et al.: Science 334 (2011) 69.
133. J. Aleksić, et al.: arXiv:1109.6124 (2011).
134. J. Albert, et al.: Astrophys. J. 665 (2007) L51.
135. E.R. Parkin, et al.: Astrophys. J. 726 (2011) 105.
136. A.A. Abdo, et al.: Astrophys. J. 723 (2010) 649.
137. C. Farnier, R. Walter: Mem. Soc. Astron. Italiana 82 (2011) 796.
138. A. Abramowski, et al.: arXiv:1111.2043 (2011).
139. A.A. Abdo, et al.: Science 329 (2010) 817.
140. R.D. Blandford, R.L. Znajek: Mon. Not. R. Astron. Soc. 179 (1977) 433.
141. R.D. Blandford, D.G. Payne: Mon. Not. R. Astron. Soc. 199 (1982) 883.
142. M.J. Hardcastle, et al.: Astrophys. J. 670 (2007) L81.
143. A.A. Abdo, et al.: Science 328 (2010) 725.
144. J.H. Croston, et al.: Mon. Not. R. Astron. Soc. 395 (2009) 1999.
145. F.A. Aharonian, et al.: Astrophys. J. Lett. 695 (2009) L44.
146. M.J. Hardcastle, et al.: Mon. Not. R. Astron. Soc. 393 (2009) 1041.
147. M. Honda, Astrophys. J. 706 (2009) 1517.
148. F.M. Rieger, F.A. Aharonian: Astron. Astrophys. 506 (2009) L41.
149. A.A. Abdo, et al.: Astrophys. J. 736 (2011) 131.
150. M. Punch, et al.: Nature 358 (1992) 477.
151. A. Franceschini, G. Rodighiero, M. Vaccari: Astron. Astrophys. 487 (2008) 837.
152. D. Mazin, M. Raue: Astron. Astrophys. 471 (2007) 439.
153. M.G. Hauser, E. Dwek: Annu. Rev. Astron. Astrophys. 39 (2001) 249.

154. F.A. Aharonian, et al.: Astrophys. J. 664 (2007) L71.
155. M. Actis, et al.: Exp. Astron. 32 (2011) 193.
156. T. Ebisuzaki, et al.: AIP Conf. Proc. 1367 (2011) 120–125.
157. M. Ackermann, et al.: Intern. Cosmic Ray Conf. 2017, arXiv:1710.01207

Permissions

All chapters in this book were first published in PPRLV3AC, by Springer; hereby published with permission under the Creative Commons Attribution License or equivalent. Every chapter published in this book has been scrutinized by our experts. Their significance has been extensively debated. The topics covered herein carry significant information for a comprehensive understanding. They may even be implemented as practical applications or may be referred to as a beginning point for further studies.

The contributors of this book come from diverse backgrounds, making this book a truly international effort. We would like to thank all the contributing authors for lending their expertise to make the book truly unique. They have played a crucial role in the development of this book. Without their invaluable contributions this book wouldn't have been possible. They have made vital efforts to compile up to date information on the varied aspects of this subject to make this book a valuable addition to the collection of many professionals and students.

This book was conceptualized with the vision of imparting up-to-date and integrated information in this field. To ensure the same, a matchless editorial board was set up. Every individual on the board went through rigorous rounds of assessment to prove their worth. After which they invested a large part of their time researching and compiling the most relevant data for our readers.

The editorial board has been involved in producing this book since its inception. They have spent rigorous hours researching and exploring the diverse topics which have resulted in the successful publishing of this book. They have passed on their knowledge of decades through this book. To expedite this challenging task, the publisher supported the team at every step. A small team of assistant editors was also appointed to further simplify the editing procedure and attain best results for the readers.

Apart from the editorial board, the designing team has also invested a significant amount of their time in understanding the subject and creating the most relevant covers. They scrutinized every image to scout for the most suitable representation of the subject and create an appropriate cover for the book.

The publishing team has been an ardent support to the editorial, designing and production team. Their endless efforts to recruit the best for this project, has resulted in the accomplishment of this book. They are a veteran in the field of academics and their pool of knowledge is as vast as their experience in printing. Their expertise and guidance has proved useful at every step. Their uncompromising quality standards have made this book an exceptional effort. Their encouragement from time to time has been an inspiration for everyone.

The publisher and the editorial board hope that this book will prove to be a valuable piece of knowledge for students, practitioners and scholars across the globe.

Index

R

S

T

U